THEORETICAL ASPECTS OF PHYSICAL ORGANIC CHEMISTRY:
The S_N2 Mechanism

THEORETICAL ASPECTS OF PHYSICAL ORGANIC CHEMISTRY

The S_N2 Mechanism

Sason S. Shaik

Ben Gurion University of the Negev
Beer Sheva, Israel

H. Bernhard Schlegel

Wayne State University
Detroit, Michigan, USA

Saul Wolfe

Simon Fraser University
Burnaby, British Columbia, Canada

A WILEY-INTERSCIENCE PUBLICATION

JOHN WILEY & SONS, INC.

New York / Chichester / Brisbane / Toronto / Singapore

In recognition of the importance of preserving what has been
written, it is a policy of John Wiley & Sons, Inc., to have books
of enduring value published in the United States printed on
acid-free paper, and we exert our best efforts to that end.

Library of Congress Cataloging in Publication Data:

Shaik, Sason S., 1948–
    Theoretical aspects of physical organic chemistry : the S_N2
mechanism / Sason S. Shaik, H. Bernhard Schlegel, Saul Wolfe.
        p. cm.
    On t.p. "N" is subscript.
    Includes index.
    ISBN 0-471-84041-6
    1. Physical organic chemistry. I. Schlegel, H. Bernhard, 1951–
II. Wolfe, Saul, 1933– . III. Title. IV. Title: S_N2
mechanism.
    QD476.S4 1992
    547.1'3—dc20   91-16855
                                                     CIP

Printed in the United States of America

10 9 8 7 6 5 4 3 2 1

To Our Wives

PREFACE

The first experimental studies of gas-phase S_N2 displacement reactions were performed in the early 1970s. Using the flowing afterglow or ion–cyclotron resonance techniques, which are forms of mass spectrometry, the reactions of various anions with methyl chloride were found to proceed at nearly the collision rate, with rate enhancements encompassing 17–20 orders of magnitude compared to aqueous solution.

Since the postulates of physical organic chemistry had evolved mainly as the result of studies of reactions in solution, the implication of the gas-phase kinetic experiments was that solvent effects and intrinsic reactivity effects would now have to be distinguished.

Concurrently, the first ab initio molecular orbital calculations of the energies and structures of gas-phase S_N2 transition states began to appear in the literature. However, the results of such calculations initially encountered some skepticism, because transition states, in which the central carbon atoms are pentacoordinate, were often found to have lower energies than the reactants themselves.

Several discoveries and developments then ensued, which led to an understanding and reconciliation of these puzzling experimental and theoretical results. On the experimental side it was established that, in the gas phase, anions can form stable ion–molecule complexes or clusters with substituted methanes. Although the geometries of such complexes could not be determined rigorously, the complex of methyl chloride with chloride ion could be seen to have nonequivalent halogens; since bridgehead halides did not form stable complexes with anions, it seemed probable that the species formed in the reaction of Cl^- with CH_3Cl had C_{3v} symmetry, with the anion on the backside of the methyl group. Depending on the stabilities of such complexes, and provided that there was a central barrier separating reactant and product ion–molecule complexes, it followed that the pentacoordinate structure could indeed have a lower energy than the separated reactants and still comprise a genuine S_N2 transition state.

On the theoretical side, the ability to compute and to characterize ion–molecule complexes and transition structures as minima or saddle points on energy surfaces became feasible following the development of efficient gradient methods for geometry optimization. Computations which employed these procedures afforded results consistent with the view that the gas-phase S_N2 reaction coordinate is double-welled.

The seminal experimental work of this period was published by Olmstead and Brauman in 1977, and by Pellerite and Brauman in 1980. These papers suggested, firstly, that experimental reaction efficiencies and ion–molecule complexation energies could be employed, in conjunction with RRKM theory, to deduce the magnitudes of the central barriers of the gas-phase reactions; and, secondly, that these central barriers varied in the manner predicted by Marcus theory, a powerful rate–equilibrium treatment originally developed for the analysis of electron transfer reactions.

These findings prompted a comprehensive theoretical study by David Mitchell, in Kingston, of more than two dozen gas-phase S_N2 reactions. This study provided a uniform data set of geometries and energies of reactants, ion–molecule complexes, and transition states. The data were found to be consistent with the predictions of the Marcus treatment, and also with certain aspects of the Leffler–Hammond postulate.

The Marcus treatment was not at that time as popular nor as well understood as it has since become. The essence of this treatment is the assertion that the rate of a reaction is determined by an intrinsic barrier, modified by a thermodynamic driving force. Since the reasons for the exo- or endoergicities of reactions could, in general, be understood, it appeared that the rate of a general S_N2 reaction X^- + R–Y → X–R + Y$^-$ could now be predicted and explained, if the trends in intrinsic barriers could be understood and explained. However, although trends could be calculated, their origins could not be rationalized in terms of any known qualitative molecular orbital arguments. In addition, as was subsequently demonstrated by the outstanding work of W. L. Jorgensen, extension of the theoretical work to include solvent effects is a formidable computational problem.

Nonetheless, Mitchell's PhD thesis aroused much interest. Over 100 copies were eventually distributed worldwide, and, during the summer of 1981, the possibility of an article in *Accounts of Chemical Research* was discussed with Professor J. F. Bunnett. The Kingston group learned, from Bunnett, of the "ideas about S_N2 transition states recently developed by Addy Pross and Sason Shaik of Ben Gurion University of the Negev, Beer Sheva, Israel," which were about to appear in the *Journal of the American Chemical Society*. By chance, Professor Pross was in Detroit, and was eager to visit Kingston to explain the basis of the state correlation diagrams of what was then termed the *valence-bond configuration mixing model*. This was the description employed by Pross and Shaik in their seminal 1983 *Accounts* article. Later the terms VBCM (valence bond configuration mixing) and SCD (state correlation diagram) evolved separately to describe two complementary aspects of a general curve-crossing model of chemical reactivity.

Although couched in a relatively unfamiliar valence-bond terminology, the model appeared to make sense. More importantly, the model led directly to the notion of an intrinsic barrier, and allowed trends in such barriers to be predicted in terms of well-defined thermochemical properties. All of these predictions were consistent with the gas-phase experimental data, and with Mitchell's results. Most importantly, the continuing work of Shaik suggested that solvent effects could be incorporated into the model, using the Marcus theory of nonequilibrium polarization.

Accordingly, when one of us (Wolfe) was invited in 1983 to edit a monograph in the general area of transition state calculations, it seemed that a more challenging project might involve an attempt to unify the quantitative molecular orbital data of Mitchell with the qualitative and semiquantitative valence-bond insights of the Pross–Shaik model.

This book is the result of that decision. The work began with discussions in Beer Sheva in December 1983, by which time Shaik had completed a comprehensive review of the S_N2 reaction, including the treatment of this reaction in terms of the model. This review, subsequently published in 1985 in *Progress in Physical Organic Chemistry*, appeared to complement nicely the PhD research of Mitchell, and the original conception was that the two works could be combined. It remained to be determined whether the seemingly different molecular orbital and valence-bond languages could be treated together in a pedagogically pleasing manner.

Discussions continued in the summer of 1984, at Jones Falls on the Rideau Canal near Kingston. These discussions, involving the three authors and also Mitchell, led to the preparation of a proposed Table of Contents. During this meeting, we found that the two groups had independently arrived at the conclusion that the energetics of the S_N2 process are dominated by the distortion of the substrate along the reaction coordinate. A joint manuscript, published in 1985, resulted, and persuaded us that the more ambitious work was indeed feasible.

With the financial support of an International Scientific Exchange Award from the Natural Sciences and Engineering Research Council of Canada, Shaik was able to spend six months in Kingston during 1985. This visit produced first drafts of four chapters written by Shaik, and one chapter written by Schlegel. After a number of revisions and updates, these became Chapters 2, 5, and 6.

However, it quickly became clear that a satisfactory whole could not be created from two disparate parts, and that a major effort would be needed to unify the material. A lengthy introductory chapter was needed, to place the work within the context of current thought, and it was realized that the valence-bond arguments of the Pross–Shaik model required, on the one hand, clearer understanding by a naive organic chemist (e.g., Wolfe), and on the other hand, presentation in a form likely to gain qualitative acceptance by a trained theoretician.

Extensive rewriting was not possible for Mitchell, who withdrew from the project at this point. Chapter 1 was finally produced by Wolfe, following lengthy conversations among the authors, which were taped and transcribed. Chapters 3 and 4 proved to be the most recalcitrant. In the end, versions prepared by Wolfe from many earlier drafts were accepted, with modifications, by the other authors. The logistical work, and the preparation of the penultimate and final drafts of the entire book were carried out by Wolfe.

The lengthy gestation, between mid-1985 and early 1990, has, we feel, been helpful. Numerous conceptual difficulties were clarified, and these, along with novel applications of the ideas, mainly by Shaik, have now appeared in the primary chemical literature. The model and its applications has also been presented to students in graduate and undergraduate physical organic chemistry courses at the Universities of Alabama, Goteborg, Lund, Padova, and Paul Sabatier (Toulouse), at Ben Gurion and Queen's Universities, at the University of Rochester, and at the NATO Summer School in Spain in 1988.

Our experience suggests that the Introduction and Chapter 1, supplemented by portions of Chapter 2, provide the concepts and historical evolution of physical organic chemistry in a manner suitable for presentation in the fourth semester of organic chemistry. The entire work can be offered as the basis of an advanced undergraduate or introductory graduate course in physical organic chemistry.

Numerous colleagues, co-workers, and students have provided us with comments and encouragement over the years. Of these, acknowledgment is made by Shaik to Addy Pross and Philippe Hiberty, for their friendship and intense collaborations, and to the late Professor David Ginsburg, whose involvement made possible the leave of absence at Queen's University. Wolfe thanks J. I. Brauman, V. Bierbaum, C. H. DePuy, and P. Kebarle for their insights and advice concerning the nature of gas-phase experimental work. All of the authors thank J. F. Bunnett for his foresight. Discussions with Shmaryahu Hoz regarding the role of electron transfer in polar reactions were most helpful, as were comments and participation in various published works by R. Bar, E. Buncel, E. Canadell, D. Cohen, L. Eberson, J. M. Lefour, P. Maitre, G. Ohanessian, G. Sini, I. H. Um, and F. Volatron.

Finally, we thank Sharon Lummis, who converted taped conversations and arguments, and multiply edited text and equations into a word-processed first draft, whose existence convinced us that the project might some day be completed.

SASON S. SHAIK
H. BERNHARD SCHLEGEL
SAUL WOLFE

Beer Sheva, Israel
Detroit, Michigan
Burnaby, British Columbia

CONTENTS

THEORETICAL ASPECTS OF PHYSICAL ORGANIC CHEMISTRY:

The S_N2 Mechanism

INTRODUCTION—A
HISTORICAL OVERVIEW

The transfer of a moiety M between two centers X and Y is a process of some importance in chemistry. When M is an electron, the reaction, termed electron transfer, comprises the simplest kind of redox behavior. If M is a proton, the process, termed proton transfer, comprises the simplest kind of acid–base behavior. When M is an alkyl group, and X and Y carry lone electron pairs, the process comprises the most thoroughly studied reaction of physical organic chemistry, the nucleophilic displacement reaction.

Physical organic chemistry is the branch of organic chemistry that seeks to provide a quantitative description of how reactions occur, from an analysis of the rate–equilibrium relationships, the structure–reactivity relationships, the solvent effects, and the isotope effects that are exhibited within reaction series. The subject had its genesis in early research on proton, alkyl, and electron transfer reactions, and the experimental and conceptual developments of this period were then found also to be applicable to the study of other reactions.

One of the earliest physical organic chemical investigations was carried out by N. Menschutkin in 1890.[1] He examined the alkyl transfer reactions of triethylamine with ethyl iodide in 23 solvents (equation 1).

$$(C_2H_5)_3N + C_2H_5I \leftrightarrow (C_2H_5)_4N^+ \ I^- \tag{1}$$

Subsequent workers suggested that the solvent dependence of the rate constants for the forward reaction of equation 1 was caused by a solvent dependence of the equilibrium constants. Although a number of attempts were made to dissect the thermodynamic from the kinetic factors for the process, the discovery of rate-equilibrium relationships is usually attributed to Brönsted and Pedersen. Their work, in 1924,[2] was concerned with proton transfer reactions (equation 2). Brönsted's findings greatly influenced L. P. Hammett and, through Hammett's efforts culminating in the 1940 textbook in which the term "physical organic chemistry" was introduced,[3] all subsequent research on the subject of linear free-energy relationships.

$$B + HA \rightarrow BH^+ + A^- \qquad (2)$$

The first successful physical model of a linear free-energy relationship employed valence bond (VB) theory,[4] and evolved independently in the mid-1930s through work by R. P. Bell[5] and by M. Polanyi and M. G. Evans[6] on the general process $A + BC \rightarrow AB + C$. What is now termed the Bell–Evans–Polanyi (BEP) principle relates changes in barrier heights to changes in the heats of reaction for a series of related reactions.

These pioneering developments overlapped with attempts by kineticists to give physical meaning to activation energies and preexponential factors determined experimentally through Arrhenius or van't Hoff relationships. Such efforts eventually led to a synthesis of the ideas developed in collision theory with the results of the newly formulated quantum mechanics, and culminated in the development of the transition-state theory of chemical reactions.[7]

Some 20 years later, the BEP principle was extended by J. E. Leffler and G. S. Hammond to a consideration of the structures of transition states.[8] Concurrently, the work of the chemical physicist R. A. Marcus led to the novel concept of the intrinsic barrier,[9] the barrier associated with a reaction whose free-energy change is zero. The Marcus approach, first developed to treat electron transfer reactions in solution, was later found also to be applicable to the treatment of proton and alkyl transfer reactions, both in solution and in the gas phase. Figures 1 and 2 exemplify the power of the Marcus treatment. In Figure 1 are plotted rate-equilibrium data for 27 gas-phase nucleophilic displacement reactions of the type $X^- + CH_3Y \rightarrow XCH_3 + Y^-$. Figure 2 refers to the same set of reactions, but with the intrinsic barriers now taken into account. These figures will be seen again in Chapters 1 and 6.

Figure 3a depicts an energy surface appropriate to the general process R(reactants) $\rightarrow I$(intermediates) $\rightarrow P$(products). The surface exhibits minima for R, I, and P, and it is equivalent to, but more informative than, the contour plot, Figure 3b, in which the viewer is positioned above the three-dimensional surface of Figure 3a. As can be seen, there are valleys on the surface of Figure 3a which meet at cols, or saddle points, so named because of their shapes. The minimum energy reaction path for the process $R \rightarrow I \rightarrow P$, also termed the reaction coordinate, is shown as the solid line in Figures 3a–c. There is a saddle point at the highest energy point on the path between R and I, and another saddle point at the highest energy point on the path between I and P. These saddle points are termed transition states or transition structures.

The characteristics of saddle points will be examined in Chapter 2, but we may note here that a saddle point has an energy maximum in only one direction, the

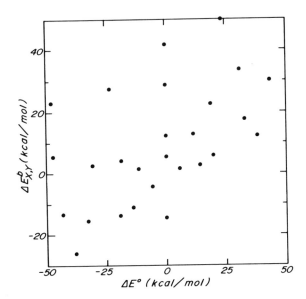

Figure 1   A plot of $\Delta E^{b}_{X,Y}$ versus ΔE^{0} for various combinations of X and Y in the gas-phase reaction $X^{-} + CH_3Y \rightarrow XCH_3 + Y^{-}$. The quantity ΔE^{b} refers to the energy difference between the separated reactants and the transition state. (Reprinted, by permission, from S. Wolfe, D. J. Mitchell, and H. B. Schlegel, *J. Am. Chem. Soc.* **103**, 7694 (1981). Copyright © 1981 by The American Chemical Society).

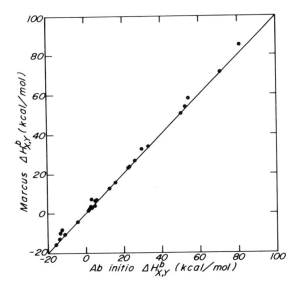

Figure 2   The data of Figure 1, replotted according to the Marcus equation (see Chapters 1 and 6 for details). (Reprinted, by permission, from S. Wolfe, D. J. Mitchell, and H. B. Schlegel, *J. Am. Chem. Soc.* **103**, 7694 (1981). Copyright © 1981 by The American Chemical Society).

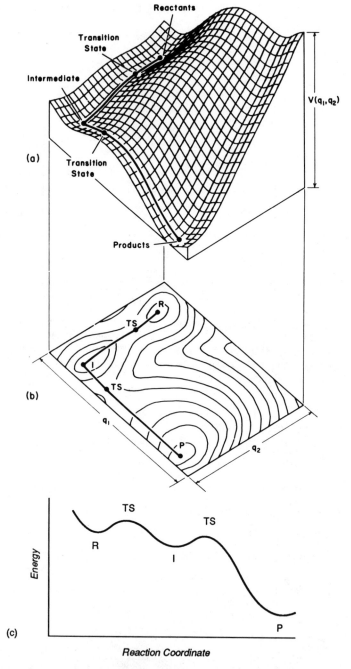

Figure 3 (a) Potential energy surface for a general reaction $R \rightarrow I \rightarrow P$. (b) Contour plot that corresponds to a. (c) Reaction coordinate for the process depicted in a and b, corresponding to the solid lines of a and b. (Figures 3a and 3b reprinted, by permission, from G. M. Maggiora and R. E. Christoffersen. In *Transition States of Biochemical Processes*. R. D. Gandour and R. L. Schowen, Eds., page 122. Copyright © 1978 by Plenum Press, New York).

4

direction of the reaction coordinate, and it is a minimum in all other directions. The BEP principle, the Leffler–Hammond hypothesis, and Marcus theory are concerned with parallel processes, that is, changes in the energy and position of the transition state that occur in the direction of the reaction coordinate that connects reactants and products. In 1967,[10] E. R. Thornton pointed out that substituent effects operate in chemical reactions to alter the position of the transition state not only in the parallel direction, in which the energy is a maximum, but also in the perpendicular direction, in which the enegy is a minimum. In the 1970s this idea was taken further by R. A. More–O'Ferrall[11] through the assignment of specific chemical structures to the termini of the perpendicular direction, and by W. P. Jencks,[12] who developed protocols to probe the main experimental features of these potential-energy surface diagrams.

The foregoing discussion has introduced a number of the components of a traditional (empirical) treatment of physical organic chemistry. These will be examined in Chapter 1. They have a common objective, namely, to try to achieve understanding of a dynamic process, the rate of a chemical reaction, in terms of static, thermodynamic, properties of reactants and products, intrinsic properties of reactions, and certain qualitative features of energy surfaces.

The central problems, in all cases, are the proper descriptions of multidimensional energy surfaces, and the concept of the barrier that must be surmounted when bonds are interchanged and electrons are redistributed as reactants are transformed into products. By 1990 it became clear that theoretical organic chemistry, in the form of ab initio[13] or semiempirical[14] molecular orbital (MO) calculations, can provide reliable information concerning the main features of energy surfaces, and that such information can gainfully be employed not only to complement and evaluate conclusions based on experiment, but also to develop trends and interpretations that are not directly accessible by experiment.

Independently, a qualitative MO treatment has emerged that rationalizes reactivity trends in terms of orbital symmetry, nodal character, and orbital interactions or orbital mixings. However, for the most part, these qualitative and quantitative MO methods have not been applied in a systematic manner to the treatment of reactivity in the traditional language of physical organic chemistry. With the noteworthy exception of the BEP principle, there has been little direct interaction between the empirical and theoretical methods of physical organic chemistry. However, the BEP principle is based on the VB method, and there is no obvious relationship between frontier molecular orbital theory,[15] or the principle of conservation of orbital symmetry,[16] and the BEP principle. Indeed, there seems to be an absence of real communication not only between the experimental and theoretical approaches to physical organic chemistry, but also between the seemingly different MO and VB theoretical approaches themselves.

A fusion of the methods of theoretical organic chemistry with each other and with the concepts of physical organic chemistry is one of the objectives of this book. The general aim is to develop a theoretical model of chemical reactivity that is compatible with the results of contemporary MO calculations, and at the same time consistent with chemical experience.

In the VB method, a molecule is thought to consist of distinct atoms or groups of atoms held together by valence bonds. This idea was not unfamiliar to organic chemists, because of the well-established concept of the functional group. This concept, which dates from the origins of the structural theory of organic chemistry, had evolved from the empirical observation that the characteristics of certain groups of atoms were more or less transferable from molecule to molecule. Indeed, the "chemical intuition" of an organic chemist could be described as an ability to predict many of the properties and reactions of a molecule once the nature and arrangement of its functional groups had been determined.

Nonetheless, despite these conceptual advantages of the VB method, it must be recognized that the theory has difficulties in providing insights that emerge more easily from MO considerations based on orbital symmetry, energy, and interaction, overlap, and nodal properties and coefficients.

The model of reactivity developed in this book is a hybrid of the VB and MO methods, which incorporates the qualitative insights of each. The model is based generally on the construction of avoided crossing diagrams and, in particular, on the construction of state correlation diagrams (SCD's). Chapter 3 is concerned with the development of a qualitative understanding of VB wave functions, and the construction from these wave functions of the SCD's that describe the mechanism of barrier formation in a chemical reaction and, therefore, provide a protocol for the discussion of reactivity trends. The required insights are developed in stages: the first comprises an examination of the relationship between VB and MO wave functions; in the second, it is shown how to construct potential energy profiles from VB structures of reactants, products, and intermediates; then, using the topological features of an SCD, equations are developed that relate the intrinsic barrier to the energy gaps of the SCD and the curvatures of its constituents. In the most general case, the final barrier is found also to depend on the thermodynamic driving force of the reaction.

Chapters 4–6 apply the ideas developed in Chapter 3 to the analysis of reactivity patterns in a single reaction, the S_N2 process. Chapter 4 assembles the experimental data concerning energy gaps, curvatures, and reaction energies, where these exist, and suggests ways to obtain the required information when the experimental data are not available. Chapters 5 and 6 use these reactivity factors in a comprehensive examination of the S_N2 reactivities of CH_3X molecules, both in the gas phase and in solution.

The decision to focus on a single reaction in the application section of the book represented the final choice of a number of possible versions of the work and many discussions. It was felt that such a focused approach ultimately conveys most convincingly the power of the model and its unifying potential. For example, it will be shown that the model leads not only to linear free-energy relationships, the Marcus equation, and the Leffler–Hammond postulate, but also accounts for the role of frontier MO's, orbital symmetry, and so forth.

The model has already been applied with success to the treatment of other classes of reactions and some structural problems.[17] It has also become possible to compute SCD's rigorously by ab-initio-based VB methods.[18] These continuing empirical and computational tests of the model reveal subtleties that are gradually becoming understood and controlled. We hope that interested readers will be encouraged by these results.

REFERENCES

1. N. Menschutkin. *Z. Phys. Chem.* **6,** 41 (1890).

2. J. N. Brönsted and K. J. Pedersen. *Z. Phys. Chem.* **108,** 185 (1924).

3. L. P. Hammett. *Physical Organic Chemistry.* McGraw-Hill, New York, 1940.

4. W. Heitler and F. London. *Z. Physik.* **44,** 455 (1927).

5. R. P. Bell. *Proc. Roy. Soc. London.* **154A,** 414 (1936).

6. M. G. Evans and M. Polanyi. *Trans. Faraday Soc.* **34,** 11 (1938).

7. H. Eyring. *J. Chem. Phys.* **3,** 107 (1935).

8. J. E. Leffler. *Science,* **117,** 340 (1953); G. S. Hammond. *J. Am. Chem. Soc.* **77,** 334 (1955).

9. R. A. Marcus. *Discuss. Faraday Soc.* **29,** 21 (1960).

10. E. R. Thornton. *J. Am. Chem. Soc.* **89,** 2915 (1967).

11. R. A. More-O'Ferrall. *J. Chem. Soc. B.* 274 (1970).

12. D. A. Jencks and W. P. Jencks. *J. Am. Chem. Soc.* **99,** 7948 (1977).

13. W. J. Hehre, L. Radom, P. v. R. Schleyer, and J. A. Pople. *Ab Initio Molecular Orbital Theory.* John Wiley, New York, 1986.

14. See, for example, M. J. S. Dewar, E. G. Zoebisch, E. F. Healey, and J. J. P. Stewart. *J. Am. Chem. Soc.* **107,** 3902 (1985).

15. K. Fukui, T. Yonezawa, and H. Shingu. *J. Chem. Phys.* **20,** 722 (1952); K. Fukui. In *Molecular Orbitals in Chemistry, Physics and Biology.* Edited by P.-O. Löwdin and B. Pullman, Academic Press, New York, 1964.

16. R. B. Woodward and R. Hoffmann. *The Conservation of Orbital Symmetry.* Verlag Chemie, Weinheim/Bergstrasse, 1970.

17. For a general review of applications of SCD's to S_N2 reactivity see S. S. Shaik. *Progr. Phys. Org. Chem.* **15,** 197 (1985). For other reactions see (a) D. Cohen, R. Bar, and S. S. Shaik. *J. Am. Chem. Soc.* **108,** 231 (1986); (b) S. S. Shaik, *J. Org. Chem.* **52,** 1563 (1987); (c) E. Buncel, S. S. Shaik, I.-H. Um, and S. Wolfe, *J. Am. Chem. Soc.* **110,** 1275 (1988); (d) S. S. Shaik and E. Canadell. *J. Am. Chem. Soc.* **112,** 1446 (1990); (e) S. S. Shaik and A. Pross. *J. Am. Chem. Soc.* **111,** 4306 (1989); (f) F. Volatron and O. Eisenstein. *J. Chem. Soc. Chem. Commun.* 301 (1986); (g) A. Demolliens, O. Eisenstein, P. C. Hiberty, J. M. Lefour, G. Ohanessian, S. S. Shaik, and F. Volatron. *J. Am. Chem. Soc.* **111,** 5623 (1989); (h) S. S. Shaik, P. C. Hiberty, G. Ohanessian, and J. M. Lefour. *J. Phys. Chem.* **92,** 5086 (1988); (i) A. J. Shusterman, I. Tamir, and A. Pross. *J. Organomet. Chem.* **340,** 203 (1988); (j) A. Pross. *Acc.*

Chem. Res. **18,** 212 (1985); (k) I. H. Williams. *Bull. Soc. Chim. No. 2.* 192 (1988); (l) S. S. Shaik and R. Bar. *New. J. Chim.* **8,** 411 (1984).

18. (a) A. Sevin, P. C. Hiberty, and J. M. Lefour. *J. Am. Chem. Soc.* **109,** 1845 (1987); (b) O. K. Kabbaj, F. Volatron, and J. P. Malrieu. *Chem. Phys. Lett.* **147,** 353 (1987); (c) G. Sini, P. C. Hiberty, S. S. Shaik, G. Ohanessian, and J. M. Lefour. *J. Phys. Chem.* **93,** 5661 (1989); (d) G. Sini, P. C. Hiberty, and S. S. Shaik. *J. Chem. Soc. Chem. Commun.* 772 (1989); (e) G. Sini, G. Ohanessian, P. C. Hiberty, and S. S. Shaik. *J. Am. Chem. Soc.* **112,** 1407 (1990).

SOME CONCEPTS IN PHYSICAL ORGANIC CHEMISTRY

1.1 INTRODUCTION

In the Introduction to this work, we commented briefly upon the evolution of some of the concepts that have become part of a traditional treatment of physical organic chemistry. These concepts are examined more closely in this chapter. They have a common objective, namely, to try to achieve understanding of a dynamic process, the rate of a chemical reaction, in terms of static (thermodynamic) properties of reactants and products, intrinsic properties of reactions, and certain qualitative features of energy surfaces.

1.2 THE BRÖNSTED RELATIONSHIP

The carboxylate- and phosphate-anion-catalyzed decomposition of nitramide proceeds according to equation 1.1, with 1.1a rate-determining, and 1.1c required to complete a catalytic cycle.[1-3]

$$H_2N_2O_2 + B \xrightarrow{k_B} HN_2O_2^- + BH^+ \qquad (1.1a)$$

$$HN_2O_2^- \rightarrow N_2O + OH^- \qquad (1.1b)$$

$$BH^+ + OH^- \rightarrow B + H_2O \qquad (1.1c)$$

$$H_2N_2O_2 \xrightarrow{k_{obs}} N_2O + H_2O \qquad (1.1d)$$

The reaction is catalyzed by the base, that is, $k_{obs} = k_o + k_B [B]$ and, in addition, the catalytic constant k_B is related to K_B, the base strength, according to equation 1.2. Equation 1.2a, with G_B a constant, is termed a Brönsted relationship,

$$k_b = G_B(K_B)^\beta; \quad G_B = 6.2 \times 10^{-5}, \quad \beta = 0.83 \quad (1.2a)$$

$$\log k_B = -\beta \, pK_B + C; \quad pK_B = -\log K_B \quad (1.2b)$$

and a graphical presentation of $\log k_B$ versus pK_B is termed a Brönsted plot. Brönsted and Pedersen, who made these observations,[1] predicted that an analogous relationship, equation 1.3, should exist for reactions that are subject to general

$$k_A = G_A(K_A)^\alpha; \quad \log k_A = -\alpha \, pK_A + C \quad (1.3)$$

acid catalysis. A typical such Brönsted plot is shown schematically in Figure 1.1, for a reaction catalyzed by acids, where K_A is the dissociation constant of the acid and α is the Brönsted parameter.

Equation 1.1a refers to a proton transfer reaction, and the existence of a Brönsted relationship for this process means that relative changes in rate constants are proportional to relative changes in the basicities of the series of bases B. Expressed in the logarithmic form, equations 1.2b and 1.3, and connecting $\log k_B$ ($\log k_A$) and pK_B (pK_A) to the appropriate Arrhenius equations, a Brönsted relationship states that changes in the free energy of activation of the kinetic process are directly proportional to the overall change in the free energy of the equilibrium process of acid or base dissociation. *A Brönsted relationship is, therefore, an example of a linear free-energy relationship.*

Brönsted's work led to numerous studies of acid- and base-catalyzed reactions, and these were often found to obey equations 1.2 or 1.3. Early examples included the decomposition of nitramide by aniline bases,[4] the mutarotation of glucose,[5] the iodination of aqueous acetone,[6] the decomposition of diazoacetate,[7] and the depolymerization of dihydroxyacetone.[8] An important advance occurred with the discovery[9] of a Brönsted relationship for a process that does not involve a rate-determining proton transfer. Thus, the quaternization of trimethylamine by a series of methyl esters, equation 1.4, is characterized by a linear relationship between

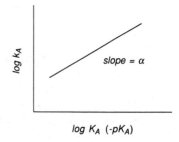

Figure 1.1 A typical example of a Brönsted plot, showing the relationship between K_A, the dissociation constant of an acid, and k_A, the rate constant of a reaction catalyzed by this acid.

$\log k_1$ and the pK_A of the carboxylic acid RCO_2H.

$$RCO_2CH_3 + N(CH_3)_3 \rightarrow RCO_2^- + {}^+N(CH_3)_4 \qquad (1.4)$$

In these latter reactions there is a formal transfer of a methyl cation from oxygen to nitrogen. Linear Brönsted plots were also found during this period for substitution reactions of acyl halides[10] and nitroaromatics.[11] Many such results are contained in a seminal 1935 review by Hammett.[12] This review, and Hammett's 1940 text,[13] greatly influenced the development of the subject of linear free-energy relationships (or rate-equilibrium relationships), *and the notion that substituent effects upon the energetics of one type of reaction could be related to substituent effects in a different reaction soon became central to the emerging discipline of physical organic chemistry.*

However, it was already evident in these early experimental studies that the Brönsted behavior that characterizes a particular chemical reaction does not extend to all possible acids or to all possible bases. Instead, it was more often observed that the individual acids or bases that form the series have to be closely related; but the meaning of the phrase "closely related" could be determined only by experiment. For example, carboxylate anions and aniline derivatives fall on different lines in nitramide decomposition;[1,4] the Brönsted behavior of pyridine and quinoline differs from that of the anilines; and primary, secondary, and tertiary amines exhibit different relationships.[14]

Nevertheless, an important step had been taken toward understanding one of the influences upon the energetics and rates of chemical reactions. Much of this work coincided with the beginnings of the transition state theory,[15] and it was natural to proceed from studies aimed at the characterization of transition structures to questions concerning the origins of the barriers associated with such structures, and to attempt to provide a physical model of a rate–equilibrium relationship that would account not only for the existence of Brönsted behavior, but also for deviations from such behavior, if such deviations were found.

1.3 THE BELL–EVANS–POLANYI PRINCIPLE

The Bell–Evans–Polanyi (BEP) model[16,17] emerged from early quantum mechanical studies by Evans and Polanyi of the energy surfaces that describe reactions of the type $A + BC \rightarrow AB + C$. These calculations were based on the empirical London–Eyring–Polanyi–Sato (LEPS) method (see Section 2.2.1), and the complete potential energy surface for such a process is shown in Figure 1.2a.

The energy of the reactants, $A + BC$, is calculated as the sum of the B–C bond energy and the repulsive interaction between A and B. The B–C bond energy can be expressed as a Morse function in terms of the B–C distance, $r(BC)$, equation

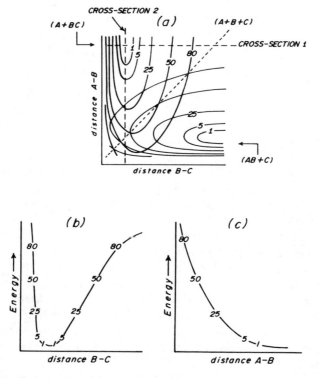

Figure 1.2   (a) Potential energy surface for the reaction A + BC → AB + C (A = C). The dashed line refers to all points at which the reactant and product curves cross. (b) A cross section of a (cross-section 1) which shows the variation in energy as a function of the B–C distance at a constant A · · · B distance. (c) A cross section of a (cross-section 2) which shows A · · · B–C repulsion at a constant B–C separation (Adapted, by permission, from reference 16a).

1.5,[18] where D_e is the bond dissociation energy, $r_e(BC)$ is the equilibrium bond

$$V(r(BC)) = D_e[1 - \exp(-a(r(BC) - r_e(BC))]^2 \qquad (1.5)$$

length of BC, and a is adjusted by fitting the potential V to the harmonic stretching force constant. This is illustrated by Figure 1.2b. The A · · · B term is shown in Figure 1.2c, and represents the repulsive interaction between nonbonded and closed-shell species.

In an alkyl transfer (S_N2) reaction, in which A = Y:, and B–C = R–X, A · · · B repulsion results from the approach of the closed-shell nucleophile Y: to the R group of the closed-shell molecule RX. This repulsion can be parameterized in a number of ways. Although, at the time, these could only be approximate, the requirement for such a term was clear. These methods provide the contours of the

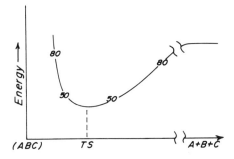

Figure 1.3 A plot of energy as a function of geometry along the dashed line of Figure 1.2a (Adapted, by permission, from reference 16a).

vertical valley in Figure 1.2a. Similar calculations for the products, AB + C lead to the horizontal valley of Figure 1.2a.

For most values of $r(AB)$ and $r(BC)$, the reactant and product surfaces differ in energy. Only along the diagonal dashed line in Figure 1.2a are the energies of the reactants and products equal. The transition state is the lowest energy point along this line of intersection, as is shown in Figure 1.3.

From the transition state one can draw a reaction path descending into the reactant valley and into the product valley. Such a path is indicated by the dotted line in Figure 1.4. Figure 1.5 shows the energies of the reactant and product surfaces along this path and illustrates clearly that the transition state results from the crossing of the two surfaces. The rise in energy of the reactant curve is mainly due to the breaking of the BC bond; the descent of the product curve along the reaction path is due mainly to the formation of the AB bond. Associated with the interchange of bonds is a corresponding interchange in the wave function of the system, from reactant-like to product-like, at the crossing point. Within the framework of

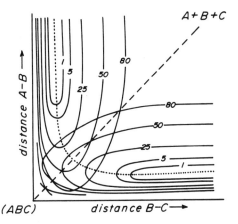

Figure 1.4 Potential energy surface for the reaction A + BC → AB + C. The dotted line is the minimum energy reaction path which connects reactants and products (Adapted, by permission, from reference 16a).

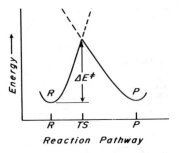

Figure 1.5 Formation of the *TS* for the reaction A + BC → AB + C via the crossing of reactant (*R*) and product (*P*) curves. The barrier is $\Delta E^{\ddagger}$ (Reprinted, by permission, from reference 16a).

this model, this concept of curve crossing is a quantum mechanical counterpart of Robert Robinson's curved arrow mnemonic **1.1**.

$$N: \quad R-X \longrightarrow N-R \quad :X^-$$

1.1

Two factors contribute to the height of the barrier (Figure 1.5). First, the breaking of B–C leads to an increase in the energy of the system (Figure 1.2*b*). The upper limit of the barrier will then correspond to the bond dissociation energy D_{BC}. In such a case, A, B, and C will be infinitely separated at the crossing point of the reactant (*R*) and product (*P*) curves, as in **1.2**. However, because this process is

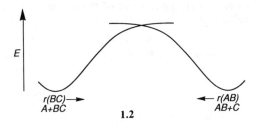

1.2

energetically costly, bond breaking will not be complete at the transition state, and a lower-energy crossing point should exist, in which B–C is partly broken and A–B is partly formed, as in **1.3**. The difference between **1.2** and **1.3** resides in

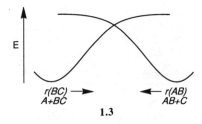

1.3

the horizontal separation of the BC and AB curves along the reaction coordinate. The second factor that contributes to the barrier is the nonbonded repulsion shown in Figure 1.2*c*. Since bond breaking is energetically costly, one might expect a lower energy transition state to be achieved when A is brought to an equilibrium bonding distance from B, without breaking of B–C. However, closed-shell repulsion between A and B does not permit this. The result is a structure that represents a compromise between bond breaking and the A $\cdots$ B and B $\cdots$ C nonbonded repulsions. This is the crossing point having the lowest energy in Figure 1.3. Strictly speaking, the curves in Figure 1.5 do not cross; instead, there is an avoided crossing, as will be discussed in Chapter 3.

This model was extended[16d] to the treatment of what are now termed "cycloaddition reactions" between ethylene and butadiene ($4\pi + 2\pi$), and between two ethylenes ($2\pi + 2\pi$). It was argued that the two reactions differ because of more extensive mixing of ionic structures into the crossing point in a ($4\pi + 2\pi$) cycloaddition.

1.3.1 The VBCM Model

The valence bond configuration mixing (VBCM) model of Pross and Shaik[19] is a related generalization which uses a many-curve modelling of the reaction profile. In the VBCM model, configurations additional to those of reactants and products are taken into account. These additional configurations are VB structures that are not important for the description of the reactant and the product, but can mix into the crossing point[19a,d] and influence the energy and character of the transition state. This is illustrated in Figure 1.6 in terms of a single intermediate configuration. An example is the carbanionic structure **1.4**, which mixes into the transition state of

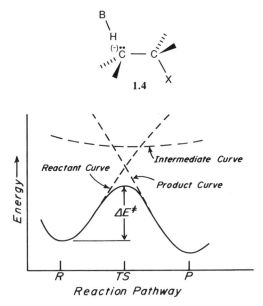

Figure 1.6 Potential energy profile (solid curve) for a reaction in which, in addition to *R* and *P*, there is a contribution to the *TS* from an intermediate *I*. The effect of the mixing of curve *I* into the point of intersection is to introduce a large avoided crossing (see Chapters 3 and 4).

the concerted E2 process **1.5**. In general, however, several such configurations

1.5

may become important.[19a]

The essence of the VBCM model has been reviewed,[19d, g] and the model can also be found in a current text on organic reaction mechanisms.[19h]

The foregoing remarks concerning the mechanism of barrier formation are summarized in the following three statements.[16, 19]

Statement 1.1   The reaction profile results from the crossing of reactant-like and product-like energy curves. This crossing reflects the electronic reorganization that accompanies the transformation of reactants and products.

Statement 1.2   Energy curves associated with VB structures that are not important for the description of the reactants or the products can mix into the crossing point, and thereby contribute to the description of the transition state.

Statement 1.3   The height of the barrier is augmented by bond breaking and by nonbonded repulsive interactions in the transition state, and it is diminished by bond making.

At this point we may inquire into the factors that cause barriers to vary within a reaction series. The barrier height will be determined not only by bonding and nonbonding interactions in the transition state (Statement 1.3), but also by the relative stabilities of the reactants and the products. While the contributions of bond breaking and nonbonded repulsions can be discussed only after the transition state has been located, a knowledge of the energy change for the reaction allows predictions to be made without detailed information concerning the reaction profile. As is seen in Figure 1.7, a vertical translation of the product curve relative to that of the reactant causes a change in the reaction energy (ΔE) and, concomitantly, a proportionate change in the energy of the crossing point (i.e., the energy of the transition state). The proportionality constant is given by $S_R/(|S_R| + |S_P|)$, where S_R and S_P are the instantaneous slopes of R and P at the crossing point. This leads to equation 1.6, which can be expressed as Statement 1.4, the BEP principle.

$$\Delta(\Delta E^{\ddagger}) = \frac{S_R}{|S_R| + |S_P|} \Delta(\Delta E) = \alpha \, \Delta(\Delta E) \qquad (1.6)$$

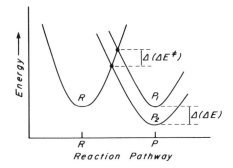

Figure 1.7 A rate–equilibrium relationship, illustrating the effect of product stabilization upon the barrier (from reference 20).

Statement 1.4 An increase of the reaction exoergicity in a series of related reactions will result in a proportional reduction of the barrier (the BEP principle).

Although Statement 1.4 refers to the effect of the reaction energy ΔE, with certain assumptions the statement can be extended to enthalpies and free energies,[16a, 17, 20] and, in this manner, the BEP principle comprises a physical model for a Brönsted relationship.

It is interesting that the BEP principle is the only feature of the model of Bell, Evans, and Polanyi that has survived to the present day. The reason for this may be that the model was designed to locate the transition state as the lowest energy crossing point of the R and P curves and, therefore, focused on these two curves, up to their crossing point, but not beyond this point (e.g., Figure 1.5).

As we shall demonstrate in Chapter 3, when the reactant and product curves are examined not only up to the crossing point but along the entire $R \to P$ reaction coordinate, it becomes possible to describe the process in terms of an SCD that allows, *in addition to the reaction ergicity*, a *qualitative* discussion of barrier height in terms of vertical gaps between the curves, their curvature along the reaction coordinate, and their extent of avoided crossing. This approach allows the conditions for existence and breakdown of the BEP principle to be defined, as will be seen in Chapters 3 and 6.

1.4 THE LEFFLER–HAMMOND POSTULATE AND THE REACTIVITY–SELECTIVITY PRINCIPLE

Evans and Polanyi did not at first ascribe physical significance to the α parameter of equation 1.6. The use of Morse potentials for R and P of Figure 1.5 for a proton transfer reaction, and a lateral separation of the curves appropriate for the heavy atom distances in a hydrogen bond led to unacceptably high activation energies and a too-narrow range of α values.[20] Alteration of the curves of Figure 1.5, to include mixing of ionic states, also failed to produce an acceptable quantitative model.[21] Consequently, for some time α remained an experimental parameter with no special physical meaning.

In his original exposition[1,22] Brönsted had predicted that (a) the linear relationships observed in his work should hold over only a limited range; (b) in the general case, log k would be found to be a curved rather than a linear function of log K; and (c) the proportionality constant α should vary between limiting values of zero and one for a complete set of acids or bases. Thus, the rate of proton transfer from an acid to a base could not continue to follow equation 1.3 indefinitely. If the acid were made stronger and stronger, the rate would become faster and faster until the reaction became encounter-controlled. Further increases in acid strength would have no effect on the rate constant, and α would be equal to zero. Conversely, for very weak acids, the rate should be directly proportional to the acid strength, and α should be unity. The linear plots observed could, therefore, be attributed to the narrow range of log K that had been employed.

The belief that limiting values exist for the slopes of Brönsted plots, and that there may be a continuous variation between these limits, eventually led to an *interpretation of the Brönsted slopes as a measure of the progress along the reaction coordinate and, therefore, as a measure of the structure of the transition state*. This principle is commonly termed the Leffler–Hammond postulate.[23,24]

In this approach the transition state is regarded as a configuration which lies on an energy profile between the reactants and the products, along a reaction coordinate (RC) (Figure 1.8) which is composed of successive structures that change progressively from reactant-like to product-like. The geometry and charge distribution of the transition state are assumed to be a weighted average of the geometry and charge distribution of the reactant (R) and the product (P). An exergic (exothermic) reaction will then possess an ''early'' transition state which is reactant-like, as shown in Figure 1.8a. Likewise, an endoergic (endothermic) reaction will possess a ''late'' transition state which is product-like, as shown in Figure 1.8b. The transition state of an ergoneutral reaction should possess ''intermediate'' character, and lie near the midpoint of the reaction coordinate, Figure 1.8c.

Such thinking leads to a simple rule for the prediction of substituent effects upon the transition structure: any substituent which lowers the energy of the product

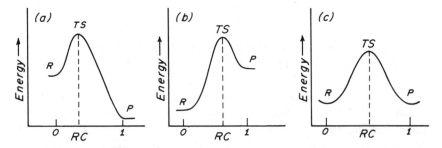

Figure 1.8 Examples of reactions that have reactant-like (''early'') (a), product-like (''late'') (b), and intermediate (''central'') (c) transition states.

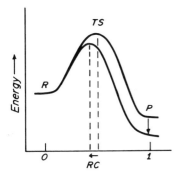

Figure 1.9 Effect of stabilization of the product upon the position of the TS along a reaction coordinate.

relative to that of the reactant is expected to shift the transition structure to an "earlier" position along the reaction coordinate, as depicted by the arrows of Figure 1.9.

Changes in the energy of a transition state resulting from a perturbation, for example, a substituent change in one of the reactants, can be represented as a linear combination of changes in the energies of the reactants and the products.[24] If the weighting coefficients for this mixing are given by α and $(1 - \alpha)$ (equation 1.7a),

$$\delta E^{\ddagger} = \alpha \, \delta E_P + (1 - \alpha)\delta E_R \qquad (1.7a)$$

α can be interpreted as a position along the reaction coordinate between the reactants and the products (equations 1.7a and 1.7b, where E is a generalized energy

$$\delta(\Delta E^{\ddagger}) = \alpha \, \delta(\Delta E); \qquad \Delta E = E_P - E_R \qquad (1.7b)$$

function). These ideas were expanded by Leffler and Grunwald in 1964[25] and, despite the original qualifications,[23,24] the postulate was adopted by many workers as a way to reach conclusions concerning the structures of transition states.

If we compare equations 1.6 and 1.7b, it is clear that these have the same mathematical form, so that there is a link between the Leffler–Hammond postulate and the BEP principle. For example, for any reaction series in which ΔE changes from positive to negative, the transition state will change from product-like (large α) to reactant-like (small α). In addition, a plot of $\Delta E^{\ddagger}$ versus ΔE should exhibit a variable slope α and be curved. This implies that all Brönsted plots should be curved, as originally stated by Brönsted.

It also follows that the reaction rate will be sensitive to changes in equilibria caused by changes in substituents. When α approaches zero, as in a highly exoergic reaction, the rate constant will be so large as to be insensitive to substituent effects. Such a reaction can be termed "unselective" because differently substituted reactants should exhibit no difference in reaction rates. Conversely, when α approaches unity, as in a highly endoergic reaction, changes in ΔE will affect the reaction rate significantly. Such a reaction can be termed "selective." *The notion that reactivity and selectivity are inversely related is termed the reactivity–selectivity principle (RSP).*[26]

Much effort has been expended to provide support for the existence of curved Brönsted plots and the RSP. Early support for the former was provided by R. P. Bell, from a study of the deprotonation of carbonyl compounds by carboxylate bases.[21] A more complete compilation of data from many other laboratories, with variation of both the carbonyl compound and the base, is shown as in Figure 1.10.[27]

Evidence for more pronounced nonlinear Brönsted behavior was provided by investigations of fast proton transfers.[28] This work by Eigen demonstrated that for proton transfer between, typically, oxygen, nitrogen, and sulfur acids and bases, the Brönsted parameter varies smoothly between one and zero (zero and one), often over a rather small range of log k. Figure 1.11 illustrates this for proton transfer to ammonia from a series of oxygen acids. In Figure 1.11 X represents ammonia and HY the oxygen acids. It can be seen that the logarithm of the protonation rate constant is almost directly proportional to pK_{HY} for the weakest acids ($\alpha = 1$), and is nearly independent of pK_{HY} for the strongest acids ($\alpha = 0$). The limiting rates are consistent with those expected for encounter-controlled processes. Examples of sharply curved Brönsted plots have also been found for certain carbon acids such as malononitrile, chloroform, and phenylacetylene.[29]

However, other interpretations can be offered for such results, which cast doubt on the predictions of the Leffler–Hammond–Brönsted hypothesis.[30] When curvature is observed in Brönsted plots for carbon acids, it is usually accompanied by changes in the nature of the donor atom of the base, and it can be argued that the curvature is the result of changes in solvation rather than changes in the extent of

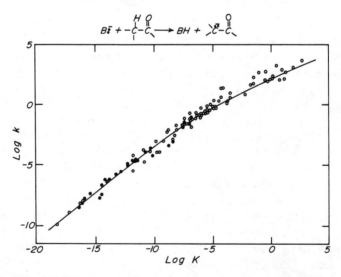

Figure 1.10 Curved Brönsted plot for the reaction of carbonyl compounds with bases (Adapted, by permission from reference 17b. Copyright © 1973 by Cornell University Press).

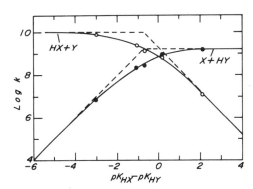

Figure 1.11 Eigen's demonstration of curvature in a Brönsted plot (Adapted from reference 28).

C–H bond breaking. For the work of Bell,[21] it has been noted that although the pK_a ranges for mono- and β-dicarbonyl substrates overlap, the Brönsted slopes are nearly constant within each structural class, so that the different slopes could correspond to different families, each of which maintains the integrity of its transition structure. Even in the work of Eigen it could be argued[28] that curvature was the result of a change in mechanism, from a rate-determining proton transfer at one extreme to a diffusion-controlled reaction independent of pK_a at the other.

Moreover, the number of curved Brönsted plots is actually quite small and there are many more examples of strictly linear Brönsted behavior, not only for proton transfer reactions,[31] but also for alkyl and other group transfer reactions[32] in which the reaction energy changes by more than 10 kcal/mol. In such cases, constancy of the Brönsted parameter would imply, in terms of the Leffler–Hammond hypothesis, that the structure of the transition state remains constant throughout the series. The implications of this assertion are examined later.

A useful experimental probe for variation of transition state structure within a series is the alpha deuterium isotope effect (e.g., equation 1.8). According to current interpretation,[33] the alpha deuterium isotope effect is a probe of the extent of crowding in the transition state relative to the reactant. A decrease in crowding is thought to result in an isotope effect k_H/k_D greater than unity, and an increase in crowding results in an isotope effect less than unity. A gradual change in this isotope effect within a series would, therefore, be interpreted in terms of a gradual change in the transition state structure, whereas a constant isotope effect would imply an invariant transition state structure. A series of particular interest is provided by the Menschutkin reaction of equation 1.8, because the Brönsted plot for this series[32a] suggests an invariant transition state. The alpha deuterium isotope effects do vary within this series,[34] so that the physical significance of the constant Brönsted parameter becomes uncertain.

This point has been reemphasized by Hoz et al.[32e] from a theoretical study of the addition of water to formaldehyde in the gas phase. As the intermolecular C–O distance is shortened, the thermodynamic driving force for the proton transfer reactions increases, and linear Brönsted plots encompassing a range of 80 kcal/mol in ΔE result. However, despite the linearity of the Brönsted plots, the transition state structures for the proton transfer reactions vary significantly. Analogous findings have been made by Williams[32b] and by Yamataka and Nagase,[32c] and a critical general analysis of the Brönsted parameter has been published.[35a]

A more serious problem for the Leffler–Hammond treatment is the existence of reaction series for which the Brönsted exponent is negative or greater than unity.[35] Such results are incompatible with the hypothesis that the Brönsted α is equal to the degree of transfer at the transition state, because negative transfer and more than complete transfer clearly have no meaning. In some cases,[35g,h] desolvation causes apparent Brönsted coefficients to depart from the normal range.

Among the systems that produce these anomalous results are proton transfers from C–H acids.[35b-f] For the series CH_3NO_2, $CH_3CH_2NO_2$, $(CH_3)_2CHNO_2$, equation 1.9, the effects of the progressive methyl substitution are to lower the specific

$$R_2CHNO_2 + HO^- \rightarrow R_2C{=}NO_2^- + H_2O \qquad (1.9)$$

rate of the reaction with HO^-, and to raise the equilibrium constant for the process. This leads to a Brönsted $\beta = -0.5$. Moreover, since the Brönsted exponents for the forward and reverse reactions must add to unity ($\alpha + \beta = 1$), the exponent for the reverse reaction must be $\alpha = 1.5$.

That $\alpha + \beta = 1$ may be seen from the following argument, based on equation 1.10. For this reaction, the equilibrium constant $K = (A^-)(HS^+)/(HA)(S) =$

$$HA + S \underset{k_B}{\overset{k_A}{\rightleftharpoons}} A^- + HS^+ \qquad (1.10)$$

K_{HA}/K_{HS^+}, and the ratio of the Brönsted relations for the forward and reverse reactions, $K = k_A/k_B = G_A(K_{HA})^{\alpha}/G_B(K_{HA})^{-\beta} = (G_A/G_B)(K_{HA})^{\alpha+\beta}$. Since K_{HA} is the only quantity varied, its exponent must be the same in both expressions for K, and $\alpha + \beta = 1$.

The existence of strictly linear Brönsted plots, and exponents outside the limits of zero to unity, also create problems for the reactivity-selectivity principle. A constant Brönsted exponent implies that the selectivity remains constant, and independent of the energetics of the reaction. An α value greater than unity implies that rates are more selective than are equilibria. These contradictions have led some workers to argue that the RSP be abandoned as a useful concept.[36]

The problems just discussed do not constitute failures of the Bell–Evans–Polanyi treatment, because those workers had already foreseen that ΔE could not represent the only rate-determining factor. At this point in our analysis, it seems clear that transition structures are determined by properties other than those of just the reactants and the products.[19,35a]

1.5 QUALITATIVE POTENTIAL ENERGY SURFACES AND TRANSITION STATE STRUCTURE

A transition state is a point on the energy surface that is a maximum in one and only one direction, and a minimum in all other directions. Such a point is also termed a saddle point or col. The region of the energy surface around the saddle point can be described in terms of the $3N - 6$ normal coordinates, where N is the number of atoms. Figure 1.12 shows a portion of the potential energy hypersurface for a reaction A + BC → AB + C, plotted as a function of the two distance coordinates r_{AB} and r_{BC}, for a constant ABC angle. Two of the normal modes of the transition state are the asymmetric stretch **1.6** and the symmetric stretch **1.7**.

A→ ←B   C→   ←A   B   C→

1.6   **1.7**

The asymmetric stretch has a negative force constant, and it describes the reaction coordinate or parallel mode that connects the transition state to the reactants and the products. The changes in energy along this parallel mode (Q_{RC}) comprise the reaction profile of Figure 1.13a, on which the transition state is an energy maximum.

The symmetric stretch has a positive force constant, and it describes the direc-

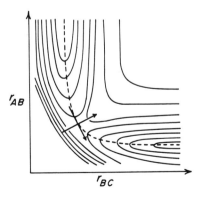

Figure 1.12 Contour surface for the reaction A + BC → AB + C. The reaction is assumed to proceed without any bending of the ABC angle, so that the potential energy depends only on r_{AB} and r_{BC}. The reaction coordinate is the dashed line (Adapted, by permission, from reference 37a).

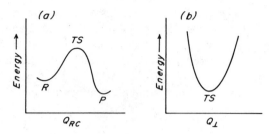

Figure 1.13 (a) Energy changes along the parallel direction (Q_{RC}), corresponding to the dotted line of Figure 1.12. (b) Energy changes along the perpendicular direction ($Q_{\perp}$) of Figure 1.12.

tion perpendicular to the reaction coordinate.[37] In this mode ($Q_{\perp}$) the transition state is an energy minimum, as shown in Figure 1.13b.

A transition state that has N atoms will have $3N - 7$ perpendicular modes, but only one parallel mode. A perturbation of a transition state, for example, a substituent change, will cause a change in the forces acting on the nuclei, and the transition state will shift to a new location on the hypersurface at which the forces acting on the nuclei are again zero. To treat this situation, it is necessary to determine the effect of the perturbation on the parallel mode and also on the perpendicular mode(s).

The new position of the transition state will be determined by the form of the potential energy curves that describe these different modes. If it is assumed that these curves are parabolic, and that the perturbation is linear, it is found that the transition state shifts in opposite directions along the parallel and perpendicular modes.[37] This result is shown in Figure 1.14.

For the parallel mode, we begin with an unperturbed reaction profile V_0, and a reaction coordinate Q_{RC} that ranges from zero to one. The transition state is located at $Q_{RC} = 0.5$. For the region near $Q_{RC} = 0.5$, we can write

$$V_0 = -\tfrac{1}{2} k(Q_{RC} - \tfrac{1}{2})^2 \qquad (k > 0) \qquad (1.11)$$

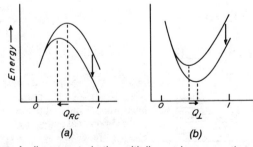

Figure 1.14 Effect of a linear perturbation with linear slope upon the parallel mode (a) and the perpendicular mode (b).

For the linear perturbation V', we can write

$$V' = m \left(Q_{RC} - \tfrac{1}{2} \right) + b \qquad (1.12)$$

The result of the perturbation is a shift of the transition state to

$$Q_{RC}^{\ddagger} = \frac{1}{2} + \frac{m}{k} \qquad (1.13)$$

When the substituent stabilizes the product relative to the reactant ($m < 0$), the transition state shifts to an earlier position along the reaction coordinate ($Q_{RC}^{\ddagger} < 0.5$). The parallel mode or parallel effect is thus equivalent to the Leffler–Hammond hypothesis: along the parallel mode, the transition state shifts away from the species (R or P) that is stabilized by the substituent.[37]
For the perpendicular effect, the equations are

$$V_0 = \tfrac{1}{2} k \left(Q_{\perp} - \tfrac{1}{2} \right)^2 \qquad (k > 0) \qquad (1.14)$$

$$V' = m \left(Q_{\perp} - \tfrac{1}{2} \right) + b \qquad (1.15)$$

$$Q_{\perp}^{\ddagger} = \frac{1}{2} - \frac{m}{k} \qquad (1.16)$$

In this case, when the substituent favors the looser species ($Q_{\perp} = 1$) relative to the tighter species ($Q_{\perp} = 0$), that is, $m < 0$, the transition state becomes looser $Q_{\perp}^{\ddagger} > 0.5$). Thus, along the perpendicular mode, the transition state shifts *toward* the species stabilized by the substituent, and the perpendicular and parallel effects are opposite in nature.[37]
A consideration of substituent effects upon any reaction must focus upon the saddle point of the energy surface and must, therefore, consider both the parallel and the perpendicular effects.[37] The new position of the transition state will be obtained by vector addition of the individual shifts along the parallel and the various perpendicular modes, as shown schematically in Figure 1.15. Such an analysis will obviously be complex for a polyatomic transition state, because of the large number of perpendicular modes. In one possible simplification, the problem is truncated to the parallel effect and a single perpendicular effect that directly involves the reaction centers.
For example, in the S_N2 reaction $PhO^- + PhCH_2Cl \rightarrow PhOCH_2Ph + Cl^-$, one might focus upon the reacting bonds (**1.8a** and **1.8b**).

O→ ←C   Cl→   ←O   C   Cl→

1.8a   **1.8b**

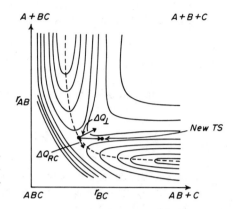

Figure 1.15 Substituent effect upon the position of the *TS* that results from the combination of the parallel and perpendicular effects.

Although problems exist, this approach has considerable conceptual merit, and its conclusions can be summarized by the following series of statements:

Statement 1.5 A substituent change may lead to both parallel and perpendicular changes of the position of the transition state on the hypersurface.

Statement 1.6 Along the parallel coordinate, the transition state moves away from the species that is stabilized, in harmony with the Leffler–Hammond postulate.

Statement 1.7 Along the perpendicular coordinate, the transition state moves toward the species that is stabilized.

Statement 1.8 Shifts in the position of the transition state are large when the unperturbed potential energy curves have low curvature (*k* in equations 1.13 and 1.16).

Statement 1.9 The new position of the transition state is obtained by vector addition of the shifts along the parallel and perpendicular coordinates (Figure 1.15).

The distinction between parallel and perpendicular modes can be incorporated into a three-dimensional potential energy surface diagram (PESD), by taking into account reactants, products, and two intermediates that may be present during the reaction.[38,39] An example of a possible PESD is shown in Figure 1.16 for the elimination reaction of equation 1.17. The parallel mode is the E2 process, and

$$B^- + \overset{\displaystyle H}{\underset{\displaystyle X}{\overset{\displaystyle |}{C}}}-\overset{\displaystyle |}{C} \longrightarrow \overset{\displaystyle }{C}=\overset{\displaystyle }{C} + X^- \qquad (1.17)$$

involves synchronous breaking of the C–H and C–X bonds (**1.9a**).

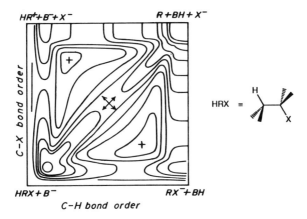

Figure 1.16 Potential energy surface diagram for a concerted elimination reaction, showing the parallel and perpendicular directions: R refers to the $R^1R^2C-CR^3R^4$ moiety (Reprinted, by permission, from reference 38a).

The perpendicular mode involves preferential breaking of either the C–H or the C–X bond (**1.9b**). The termini of the perpendicular modes are the reactive intermediates corresponding to ElCB (**1.10a**) and El (**1.10b**) elimination processes, respectively.

The corners of the PESD of Figure 1.16 now represent the reactants and the products, together with **1.10a** and **1.10b.** A substituent change may affect the energy at one or more of the four corners. If the energies of only the reactants and the products are affected, only the parallel effect need then be considered and, in accord with Statement 1.6, stabilization of the product would cause the transition state to shift to an earlier position along the reaction coordinate. On the other hand, if the energies of only the intermediates are affected, only the perpendicular effect need be considered, and the transition state will shift toward the intermediate that is stabilized by the substituent, as summarized in Statement 1.7.

The extension of this procedure to other reactions seems, in principle, straight-forward, and the general form of a PESD will be as in Figure 1.17. In this Figure, R and P are reactants and products, and I_1 and I_2 are viable reactive intermediates; energy contours have been deleted for simplicity.

For example, in the nucleophilic displacement reaction $N:^- + R–X \rightarrow N–R + X:^-$, the intermediate corners comprise the triple ion species $(N:^- \; R^+ \; :X^-)$ and the hypothetical hypervalent species $(NRX)^-$, as shown in Figure 1.18. The axes of the diagram are n_{RX} and n_{NR}, the R $\cdots$ X and N $\cdots$ R bond orders. The transition state will have some combination of bond orders, and will lie somewhere in the square. This structure can, therefore, have the character of all of these species, with a dominant effect from the closest corner.

Substituent effects upon the position of the transition state in Figure 1.17 follow from the arguments given previously:

1. A substituent that stabilizes either of the R or P corners of Figure 1.17 will induce a parallel effect, and the transition state will shift away from the corner that is stabilized (directions 1 or 2).

2. A substituent that stabilizes either of the intermediate corners of Figure 1.17 will cause a shift of the transition state toward the corner that is stabilized (directions 3 or 4). For example, in an S_N2 reaction upon CH_3OCH_2X, the triple ion $(N:^- \; R^+ \; :X^-)$ will be stabilized and the transition state will shift

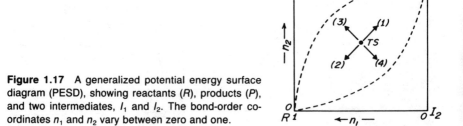

Figure 1.17 A generalized potential energy surface diagram (PESD), showing reactants (R), products (P), and two intermediates, I_1 and I_2. The bond-order coordinates n_1 and n_2 vary between zero and one.

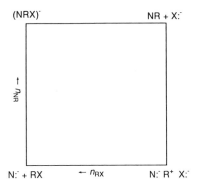

(NRX)⁻   NR + X:⁻

N:⁻ + RX   ← n_{RX} N:⁻ R⁺ X:⁻

Figure 1.18 PESD for a nucleophilic displacement reaction.

toward this corner of Figure 1.18. This would imply that the transition state has significant R^+ character, and a loose or "exploded" geometry.

3. The overall result of the substituent change is obtained by summation of the parallel and perpendicular effects, and the transition state would then be expressed as a resonance hybrid of four structures (equation 1.18).

$$TS = R \leftrightarrow P \leftrightarrow I_1 \leftrightarrow I_2 \qquad (1.18)$$

The potential energy surface diagrams of Figures 1.16–1.18 have become known as More O'Ferrall–Jencks diagrams.[38,39] Related models have been developed and applied independently by other workers, including Harris and Kurz,[40] Critchlow,[41] and Albery.[42] Alternative treatments that also take into account configurations besides those of just the reactants and the products have been given by many other workers.[43]

Attempts to quantitate the PESD were initiated by Jencks and Jencks,[44] who used the appropriate selectivity parameters to describe the coordinates of the diagram. For example, the general acid-catalyzed addition of a nucleophile to a carbonyl group, equation 1.19, can be described by the PESD of Figure 1.19

$$N^- + \quad \diagdown \!\!\! C\!\!=\!\!O + HA \longrightarrow N\!-\!\overset{\displaystyle |}{\underset{\displaystyle |}{C}}\!-\!OH + A^- \qquad (1.19)$$

(energy contours are deleted for simplicity), whose axes are the $O \cdots H$ and $N \cdots C$ bond orders. Each of these bond orders refers to a unique chemical event: a horizontal movement on Figure 1.19 describes a proton transfer; a vertical movement describes a nucleophilic addition.

It is now supposed that the extent of the proton transfer is given by, or can be related to the Brönsted α, and that the progress of the nucleophilic addition is given by, or can be related to the corresponding β_{nuc}. If these quantities vary between

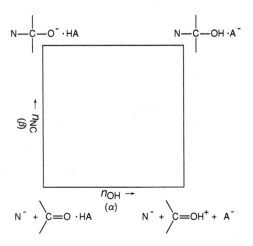

Figure 1.19 PESD for acid-catalyzed addition of a nucleophile to a carbonyl group. The bond-order coordinate can be represented by the appropriate Brönsted coefficients.

limiting values of zero and unity, then a particular α and β_{nuc} will define the position of a particular transition state within the PESD. The shift of this transition state that results from a substituent change can be predicted if the potential energy surface in the region of the saddle point can be given a specific mathematical form, for example, equation 1.20,

$$\Delta G/2.303\ RT = aX^2 + bY^2 + cXY + dX + eY + f \qquad (1.20)$$

where ΔG is the free energy of the point X,Y on the PESD relative to the reactants, with $X = \alpha$, and $Y = \beta_{nuc}$. The coordinates of the transition state are obtained by setting the first derivatives of equation 1.20 equal to zero. Changes in the position of the transition state are obtained from the derivatives of α and β_{nuc} in the appropriate reaction series, which may be estimated from the curvatures of α and β_{nuc}. In a further quantitative consideration, Grunwald developed a related algebraic treatment to reactions in which more than two intermediates can plausibly exist.[45a-c]

Other treatments based on the PESD model have led to the principles of "imbalances,"[45d] and "nonperfect synchronicity."[45e] These emphasize the idea that, as one proceeds from reactants to products, different reaction parameters, such as resonance, hydrogen bonding interactions, and solvation changes, evolve to different extents.

It is clear that the PESD method overlaps with the generalized model of Evans and Polanyi[16d] and with the VBCM model,[19] as summarized in Statement 1.2. In each of these procedures the TS is quite properly regarded as a hybrid of VB structures. However, the PESD, BEP, and VBCM methods differ in how they treat the relationship between the geometry of the TS and the mixing of the VB structures. In the BEP[16] and VBCM[19] models, it is assumed that the transition state occurs approximately at the crossing point of the reactant and product VB structures, and that the coincidence is only weakly affected by contributions from other VB structures.[19c] These assumptions imply that, in general, the charge character of the transition state does not correlate with the position of the transition state along a reaction coordinate.[19c, 35a]

On the other hand, while the PESD model also considers four (or more) VB structures,[44, 45a-c] namely reactants, products, and intermediates, the degree of mixing and, therefore, the charge character of the TS are assumed to depend directly upon the geometry of the TS.

This assumption leads to a fundamental conceptual difference between the PESD model and the BEP and VBCM models. We shall return to this difference in Chapter 6. Here we comment that the concept of the PESD is seen to offer additional dimensions for the qualitative interpretation of experimental information, but there are problems associated with the specification of the corners that describe the intermediates in Figure 1.17. In polar reactions, at least one of these corners refers to a highly ionic species. In a nucleophilic displacement reaction, this is the triple ion $(N:^- R^+ :X^-)$; in a proton transfer reaction, it is the triple ion $(B:^- H^+ :A^-)$. In the gas phase, such structures are 70–300 kcal/mol higher in energy than structures such as $(N:^- R\cdot \ \cdot X)$ and $(B:^- H\cdot \ \cdot A)$, because ionization is an energy-costly process in the gas phase.

The consequences of such behavior are illustrated in Figure 1.20, a PESD for the identity S_N2 reaction $X^- + CH_3X \rightarrow XCH_3 + X^-$ $(R = CH_3)$. Along the perpendicular mode $Q_\perp$, the transition state may correlate with either the triple ion $(X:^- CH_3^+ :X^-)$ or with the $(X:^- CH_3\cdot \ \cdot X)$ fragments. Simple thermochemical considerations[46] reveal that in the gas phase the triple ion lies higher than the fragments by an energy gap whose value is $(I_{R\cdot} - A_{X\cdot})$, that is, the ionization potential of $CH_3\cdot$ minus the electron affinity of $X\cdot$. This energy gap amounts typically to 70–190 kcal/mol, depending on X. Consequently, in the gas phase, the perpendicular mode will take the form shown in Figure 1.21, in which the identity of the loose corner I_2 is $(X:^-CH_3\cdot \ \cdot X)$, and its mirror image structure.

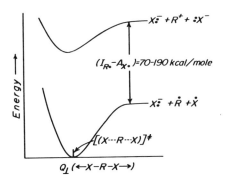

Figure 1.20 A demonstration that in gas phase the loose corner of an identity S_N2 reaction cannot be the triple ion species.

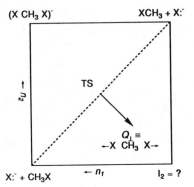

Figure 1.21 PESD for the reaction $X^- + CH_3X \rightarrow$ $XCH_3 + X^-$. The identity of the loose corner, l_2, depends on whether the diagram refers to the gas phase (Figure 1.20) or solution (Figure 1.22).

In the general case, the triple ion will be more strongly solvated in solution than the fragments $(X:^- \ R \cdot \ \cdot X)$. This is illustrated by the solvent-assisted crossing depicted in Figure 1.22, which causes the triple ion to become the corner of the PESD (Figure 1.21).[46] However, the curvature of the surface along the perpendicular mode will continue to be influenced by the neutral fragments, and changes in the energy of the carbocation R^+ may have an unpredictable effect upon the energy of the transition state. The role of the loose corner is thus no longer unambiguous.

Since the perpendicular direction l_2 in Figure 1.21 may be neutral (Figure 1.20), partly neutral, or partly triple-ionic (Figure 1.22), it follows that there is no straightforward relationship between charge distribution in the TS and the geometry (looseness) of this TS.[19c, g, 35a, 46] For example, the S_N2 transition state may be tight and yet possess high R^+ character, or vice versa. Therefore, the two aspects of the transition state must be treated separately.

Thus there is no doubt that any model must treat substituent effects on the transition structure in terms of both parallel and perpendicular effects, but care must be taken to specify thermodynamically realistic structures for the intermediate corners, and to ascertain that changes in the energy of a corner are transmitted to the

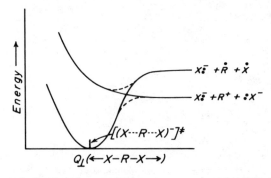

Figure 1.22 Effect of solvent on the nature of the loose corner. Dashed lines denote avoided crossing.

saddle point in a predictable manner. This caveat will apply to any process that consists of the combination of a nucleophile with an electrophile or, in general, any polar organic reaction.

1.6 THE MARCUS RELATIONSHIP AND THE CONCEPT OF THE INTRINSIC BARRIER

The PESD model does not deal directly with the reason for the variation of barriers in a particular manner. We require a procedure that incorporates the rate–equilibrium relationship of the BEP principle and, at the same time, leads to physical interpretations of linear Brönsted plots, curved Brönsted plots, and Brönsted exponents outside the range 0–1. These insights are embodied in the Marcus equation, whose organizing power may be seen in Figures 1.23 and 1.24.[47]

The Marcus equation evolved from attempts to understand the kinetics of oxidation–reduction reactions in solution. In one of the simplest of such reactions, neither the coordination number nor the coordinated ligands of the reactants are changed during the reaction. An example of such an electron transfer reaction is

$$IrCl_6^{2-} + Fe(CN)_6^{4-} \rightarrow IrCl_6^{3-} + Fe(CN)_6^{3-} \qquad (1.21)$$

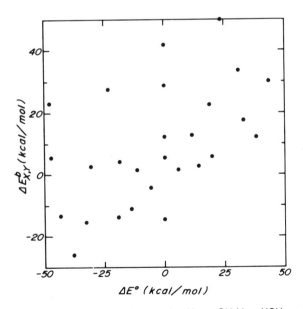

Figure 1.23 A rate–equilibrium plot for the reaction $X^- + CH_3Y \rightarrow XCH_3 + Y^-$ in the gas phase. $\Delta E_{X,Y}^b$ is the energy difference between the separated reactants and the transition state. The data refer to the 4-31G calculations of reference 47. (Copyright © 1981 by the American Chemical Society).

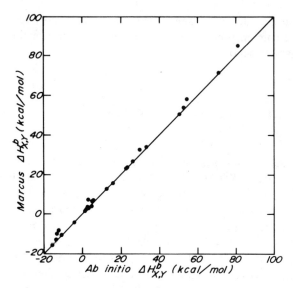

Figure 1.24   Marcus treatment of the data of Figure 1.23, using the 4-31G data of reference 47. (Copyright © 1981 by the American Chemical Society).

If the coordination shells remain intact throughout the entire reaction, the process is termed an outer-sphere electron transfer reaction.

The rate of an outer-sphere (weak overlap) electron transfer reaction can be expressed in terms of its component electron transfer reactions,[48] if it is supposed that the process can be dissected into five stages:

1. The reactants first form a precursor complex, but maintain weak overlap. The process involves electrostatic work and other nonbonded contributions. The energy of complex formation is denoted by ω^r.

2. Inner coordination and outer solvent shells are reorganized, and the critical configuration of the transition state is achieved. Defining λ as a vertical reorganization energy, the reorganization energy is 0.25λ and varies as a function of ΔE (in generalized energy units).

3. The electron is transferred in the transition state, with no changes in the nuclear configurations. The transmission coefficient is assumed to be unity.

4. The coordination and solvent shells of the newly formed products relax to their equilibrium configurations in a successor complex whose electrostatic complexation energy is ω^p.

5. The product ions move away from each other.

The result of this treatment is the Marcus equation 1.22, in which Z is a bi-

$$k = Z \exp\left(-\Delta G^{\ddagger}/RT\right)$$

$$\Delta G^{\ddagger} = \omega_r + \lambda(1 + \Delta G^{o\prime}/\lambda)^2/4 \qquad (1.22)$$

molecular collision frequency in solution, $\Delta G_R^{o\prime}$ is $\Delta G^o + \omega^P - \omega^r$, ΔG^o is the standard free energy of the reaction in the particular solvent, ω^r (or ω^P) is the work required to bring the reactants (or products) together to the mean separation distance in the activated complex (steps a, e), and the vertical reorganization energy λ for a cross reaction is the mean of that for the two-identity electron transfer reactions (equation 1.23).

$$\lambda_{12} = (\lambda_{11} + \lambda_{22})/2 \tag{1.23}$$

An equivalent form of equation 1.22 is

$$\Delta E^{\ddagger} = \Delta E_o^{\ddagger} + \Delta E/2 + (\Delta E)^2/16\Delta E_o^{\ddagger} \tag{1.24}$$

and it is recommended[48i] that free energies be employed as ΔE, and where $\Delta E_o^{\ddagger}$ is the intrinsic free-energy barrier, that is, the formal barrier when $\Delta E = 0$, given in equation 1.22 by $\lambda/4$.

Because of the statistical mechanical arguments employed in the original derivation of equation 1.22, and its specific application to solution-phase electron transfer reactions, this work was for some time known only to specialists in the field. However, following a demonstration that the Marcus equation afforded encouraging results when applied to atom transfer reactions,[49] that is, transfer of neutral particles, the original derivation was reexamined,[50] and the theory was found to be applicable not only to atom transfer, but also to proton transfer reactions.[51] The successful results of this work were then noted and extended by physical organic chemists.[52] Eventually the Marcus equations were also found to be applicable to methyl transfer reactions, in water and other solvents[53] and in the gas phase,[54] and to other chemical processes.[55]

The intense interest in the Marcus equation and the implications of the Marcus treatment that follow from these successful applications has led to numerous alternative derivations, and two of these are examined here. Although each of these less rigorous derivations is deficient in some way, collectively they lend considerable support to the notion that the rate of a transfer reaction can be analyzed as a combination of thermodynamic and intrinsic kinetic factors. It is the latter that incorporates barrier-controlling effects such as bond breaking, bond making, and overlap repulsion that are mentioned in the BEP treatment.

1.6.1 The Method of Intersecting Parabolas[56]

Figure 1.25 considers a reactant, AH, and a product, BH, of a hydrogen transfer process $AH + B \rightarrow A + BH$, to be simple harmonic oscillators having the same force constant k. If the bottom of the AH potential energy well is taken as the origin, with ΔE the energy difference the bottoms of the two wells, and d their horizontal separation on the coordinate r, we can write

$$E_{AH} = \tfrac{1}{2} kr^2 \tag{1.25}$$

$$E_{BH} = \Delta E + \tfrac{1}{2} k(r - d)^2 \tag{1.26}$$

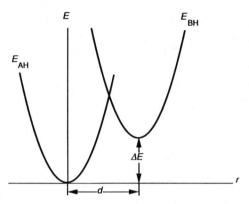

Figure 1.25 Intersecting parabolas model for a hydrogen transfer from AH to B. (Adapted, by permission, from reference 56a).

where E is a generalized energy unit. We can regard d as the distance that a hydrogen travels between its equilibrium positions in AH and HB in the complex AH $\cdots$ B. The point of intersection of the two curves as a fraction of d is obtained from the condition $E_{AH} = E_{BH}$ and occurs at

$$(r/d)_{\text{cross}} = (\Delta E + \tfrac{1}{2} kd^2)/kd^2 \qquad (1.27)$$

For an identity reaction, that is, $A = B$, $\Delta E = 0$, and $(r/d)_{\text{cross}} = 0.5$, the energy of the crossing point is $\Delta E_o^{\ddagger}$, the intrinsic barrier. Substitution of these quantities into equation 1.25 leads to

$$k = 8\Delta E_o^{\ddagger}/d^2 \qquad (1.28)$$

In the general case, the energy at the crossing point is given by

$$\Delta E^{\ddagger} = [4\Delta E_o^{\ddagger}]\,[(\Delta E + 4\Delta E_o^{\ddagger})/8\Delta E_o^{\ddagger}]^2 \qquad (1.29)$$

Rearrangement of equation 1.29 leads to equation 1.30, the Marcus equation.

$$\Delta E^{\ddagger} = \Delta E_o^{\ddagger}(1 + \Delta E/4\Delta E_o^{\ddagger})^2 \qquad (1.30)$$

Within a parabolic assumption, the barrier then consists of a kinetic parameter $\Delta E_o^{\ddagger}$ and a thermodynamic driving force ΔE. The thermodynamic driving force is seen to modify the total barrier ($\Delta E^{\ddagger}$) relative to the intrinsic barrier. Differentiation of equation 1.30 with respect to ΔE (at constant $\Delta E_o^{\ddagger}$) leads to equation 1.31, an analytical expression for a Brönsted parameter as the instantaneous slope.

$$\frac{\partial \Delta E^{\ddagger}}{\partial \Delta E} = \alpha = \frac{1}{2} + \frac{\Delta E}{8\Delta E_o^{\ddagger}} \qquad \text{(constant } \Delta E_o^{\ddagger}) \qquad (1.31)$$

Equation 1.31 states that, in a reaction series with constant $\Delta E_o^{\ddagger}$, α is a linear function of ΔE. For a reaction in the series having $\Delta E = 0$, the corresponding value of α is 0.5. As ΔE becomes negative, α decreases. When ΔE is positive, α increases. If we now substitute equation 1.28 into equation 1.27, we obtain equation 1.32 as the expression for the position of the transition state. Thus, within the

$$(r/d)_{\text{cross}} = \frac{1}{2} + \frac{\Delta E}{8 \Delta E_o^{\ddagger}} \tag{1.32}$$

restricted sense of $\Delta E_o^{\ddagger} = $ constant in equation 1.31, α is identical to $(r/d)_{\text{cross}}$ and gives the position of the transition state along the reaction coordinate; α is also a measure of the resemblance of the transition state to the reactants and products.[53d, h] For example, in the particular case of Figure 1.25, α is the fraction of the A–B distance traversed by the proton in the transition state.

1.6.2 The Derivation of Murdoch, Based on the Leffler Equation[57]

We begin with Leffler's equation $\delta E^{\ddagger} = \alpha \delta E_P + (1 - \alpha) \delta E_R$ and note that, in the limit $\delta E^{\ddagger} \rightarrow 0$, we have the Brönsted relationship $\alpha = d\Delta E^{\ddagger}/d\Delta E$, for the situation in which only ΔE varies as substituents are changed.

If α varies between 0 and 1, then $\alpha = 0.5$ when $E_R = E_P$. A stabilization of P by the quantity dE will stabilize the transition state by αdE, or $0.5dE$ and α will, in turn, decrease to some quantity α'. Stabilization of P by successive increments of dE will eventually reduce $\Delta E^{\ddagger}$ to zero at $E_P - E_R = \Delta E_{\max}$ (Figure 1.26). Since $\alpha = 0$ at $-|\Delta E_{\max}|$, and $\alpha = 1$ at $+|\Delta E_{\max}|$, α is a continuous single-valued function of ΔE.

We assume that changes in α are independent of α, that is, $d\alpha/d\Delta E$ is constant between $-|\Delta E_{\max}|$ and $+|\Delta E_{\max}|$, and that α and ΔE are linearly related. From

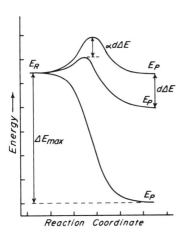

Figure 1.26 Derivation of the Marcus equation based on a consideration of barrier height as a function of reaction ergicity (Adapted, by permission, from reference 57).

the conditions for $\alpha = 0$ and $\alpha = 1$ we have equation 1.33, and subsitution of

$$\alpha = \frac{1}{2} + \frac{\Delta E}{2\Delta E_{max}} \qquad (1.33)$$

equation 1.33 into the expression for the Brönsted slope leads to equation 1.34.

$$d\Delta E^{\ddagger} = \left(\frac{1}{2} + \frac{\Delta E}{2|\Delta E_{max}|}\right) d\Delta E \qquad (1.34)$$

Integration of equation 1.34 between $\Delta E = 0$ and $\Delta E = |\Delta E_{max}|$, and using the facts that $\Delta E^{\ddagger} = \Delta E_o^{\ddagger}$ at $\Delta E = 0$ and $\Delta E^{\ddagger} = |\Delta E_{max}|$ at $\Delta E = \Delta E_{max}$, leads to equation 1.35.

$$\Delta E_o^{\ddagger} = \tfrac{1}{4}|\Delta E_{max}| \qquad (1.35)$$

Equation 1.35 shows that the intrinsic barrier of a reaction series influences the energy change for the reaction that would be required to change the Brönsted slope from 0 to 1. For example, when $\Delta E_o^{\ddagger} = 1$ kcal/mol, α will change from 0 to 1 as the catalytic strength of an acid or base changes by ca. 5 pK units at 25°C. With such a small barrier, the rate constant for a thermoneutral process will lie above the encounter-controlled limit, and such behavior would be typical of proton transfer between an oxygen or nitrogen acid–base pair. On the other hand, when $\Delta E_o^{\ddagger} = 20$ kcal/mol, a range of ca. 110 pK units would be needed to change α from 0 to 1. In this case, a 5-pK unit change in catalyst strength would change α by ca. 0.05, and the experimental Brönsted plot might well appear to be linear. It follows that the concept of the intrinsic barrier can account for the existence of both linear and curved Brönsted plots.

Integration of equation 1.34 with respect to $d\Delta E$ between the limits 0 and ΔE, and substitution of equation 1.35, leads to equation 1.36, the Marcus equation.

$$\Delta E^{\ddagger} = \Delta E_o^{\ddagger} + \frac{1}{2}\Delta E + \frac{\Delta E^2}{16\Delta E_o^{\ddagger}} \qquad (1.36)$$

As stated above, this derivation follows from the Leffler postulate that α varies between zero and unity. This is, therefore, a particular derivation for a reaction series which possesses a constant $\Delta E_o^{\ddagger}$. For such a series the Brönsted parameter is again identical to the position of the TS along a reaction coordinate that varies between zero and unity.

The two derivations presented here can leave the impression that the Marcus equation applies only to a limited set of reactions that either (a) are characterized by parabolic energy curves (Figure 1.25), or (b) obey the Leffler postulate that α varies between zero and unity. In fact, Marcus-like equations can be derived without these assumptions. Such methods include the BEBO[58] (bond energy–bond order) method,[48g] a thermodynamic approach,[59,60] and an approach based on the principle of least action.[61,62] In general, it has been found[45a,55,62,63] that, for any mathematical representation of the energy curves, the barrier exhibits a Marcus-type (Marcusian) relationship, provided that (a) the reaction coordinate is normalized to run between zero and unity, and (b) the resulting energy expression is symmetry-scaled to allow a reversal of the roles of the reactants and the products. In every derivation of the barrier expression, the importance of the intrinsic barrier is affirmed. This kinetic factor and the thermodynamic driving force are the two fundamental properties that determine the magnitude of a chemical barrier.

The Marcus equation also accounts for the existence of abnormal Brönsted parameters. To appreciate this point, we have to return to equations 1.30 and 1.31 and examine the full differential of equation 1.30, namely,[64]

$$d\Delta E^{\ddagger} = [1 - \Delta E^2 / 16 \Delta E_o^{\ddagger}]^2 d\Delta E_o^{\ddagger} + [\tfrac{1}{2}(1 + \Delta E / 4 \Delta E_o^{\ddagger})]d\Delta E \quad (1.37)$$

The form of this equation allows two conclusions:

1. If there exists a reaction series for which $\Delta E_o^{\ddagger}$ is constant ($d\Delta E_o^{\ddagger} = 0$), the Brönsted parameter of such a series will vary between zero and unity in the range $-4\Delta E_o^{\ddagger} \leq \Delta E \leq +4\Delta E_o^{\ddagger}$. In such series the Brönsted plot will be curved, and $0 \leq \alpha \leq 1$.
2. If the conditions stated above are not met in a particular experimental series, the slope of a Brönsted plot will depend upon the relative magnitudes of the two partial derivatives of equation 1.37. This slope can be larger than unity, or even negative.[64]

The consequences of (1) and (2) are the following:

a. In the most general case, the experimental Brönsted parameter may take any value, including values outside the range 0–1. The interpretation of the Brönsted parameter as a position of a transition state on a reaction coordinate is therefore unreliable in principle. Even when a normal Brönsted parameter is found ($0 \leq \alpha \leq 1$), one cannot assume that the reaction series in question obeys the condition $\Delta E_o^{\ddagger} = $ constant.[53d–i]
b. The number of reaction series that exhibit constant $\Delta E_o^{\ddagger}$ is rather small.[45a,65] Therefore, changes in observed Brönsted parameters are more likely to reflect changes in intrinsic barriers, rather than changes in the reactant-like or product-like nature of the transition state.

A unified analysis that encompasses the Marcus, PESD, and VBCM treatments is given as equation 1.38, which is the differntial of $\Delta E^{\ddagger}$, a function of ΔE, and

$$d(\Delta E^{\ddagger}) = \frac{\partial \Delta E^{\ddagger}}{\partial \Delta E} d\Delta E + \sum_j \frac{\partial \Delta E^{\ddagger}}{\partial \chi_j} \chi_j \quad (1.38)$$

also of other variables grouped together as χ_j.[35a] Equation 1.38 is general, and does not depend upon the particular choice of χ_j. For example, χ_j may represent the intrinsic barrier of the Marcus treatment, or the energies of the intermediates of the PESD model, or the degree of mixing of intermediate configurations in the VBCM model.

Since $\partial(\Delta E^{\ddagger})/\partial\Delta E$ is the definition of α, equation 1.38 may be written as

$$d\Delta E^{\ddagger} = \alpha \, d\Delta E + \sum_j \frac{\partial\Delta E^{\ddagger}}{\partial\chi_j} \, d\chi_j \qquad (1.39)$$

which shows that the change in $\Delta E^{\ddagger}$ is the sum of the changes in all of the variables, $\partial(\Delta E^{\ddagger})/\partial\chi_j$, including α. The Leffler and BEP equations (1.7b and 1.6) result from the specific choices for χ.

An experimentally derived Brönsted coefficient is obtained from the instantaneous slope of $\Delta E^{\ddagger}$ versus ΔE, and corresponds to equation 1.40. This shows that

$$\text{experimental Brönsted slope} = \frac{d\Delta E^{\ddagger}}{d\Delta E} = \alpha + \sum_j \frac{\partial\Delta E^{\ddagger}}{\partial\chi_j} \frac{\partial\chi_j}{\partial\Delta E} \qquad (1.40)$$

the Brönsted slope is comprised of a "real" α, supplemented by a blend of other effects. The result of the blend can be "normal" Brönsted behavior ($0 < \alpha < 1$) or anomalous Brönsted behavior.

It can be seen that the logic of the Marcus equation overlaps with all of the concepts so far examined. First, the Marcus treatment complements the BEP treatment by combining all kinetic factors (bond cleavage, nonbonded repulsions) into a single parameter, the intrinsic barrier. Second, by separating the intrinsic barrier from the thermodynamic driving force, the Marcus equation implicitly recognizes that the transition state contains features not present in reactants and products. In this respect, the Marcus treatment resembles the VBCM[19] and PESD[37-39] models. These features can be revealed, inter alia, through the application of the Hammett equation to a set of intrinsic barriers.[65,66] The foregoing discussion is summarized in the following statements.

Statement 1.10 A chemical barrier is determined by an intrinsic property ($\Delta E_0^{\ddagger}$) which incorporates all of the reorganization processes that typify the reaction, and by a thermodynamic driving force (ΔE) which incorporates the relative stabilities of reactants and products.

Statement 1.11 The transition state has properties that are not present in the reactants and the products. This information is in principle conveyed through the variation of the intrinsic barrier within a reaction series.

A difficulty with the Marcus approach is that, when the intrinsic barrier is not constant within a reaction series, it cannot be derived in any simple way. Since reaction series having constant $\Delta E_0^{\ddagger}$ may be rare,[45a] one cannot reliably employ extrapolation methods to derive this quantity. A number of procedures have been suggested to obtain this parameter independently.[53-55,65] In the simplest approach,

$\Delta E_o^{\ddagger}$ for a cross reaction is reasonably approximated as the average of the barriers of the two corresponding identity reactions (equation 1.41).[47,52-55,65,66]

$$\Delta E_o^{\ddagger}(XY) = \tfrac{1}{2}[\Delta E_o^{\ddagger}(XX) + \Delta E_o^{\ddagger}(YY)] \qquad (1.41)$$

In certain Hammett series,[53d, i] it is possible to derive a linear relationship between the intrinsic barriers and the thermodynamic quantities of the series. In such cases, unknown $\Delta E_o^{\ddagger}$ information can be extracted. On the other hand, equation 1.41 would not be appropriate for reactions such as cation–anion recombination, nucleophilic addition to unsaturated systems, and cycloaddition reactions generally. Such processes do not possess identity sets and an "intrinsic barrier" cannot be determined reliably from experiment.[67] Here we can appreciate the possibilities to be gained from an equation in which all of the barrier factors are treated explicitly. If such an equation could be derived, the intrinsic barrier for any reaction would become accessible from the condition $\Delta E = 0$. The discussion of Chapter 4 leads to such an equation, based on the SCD model.[46]

This capability will be developed in stages. The first stage, in Chapter 2, describes how the methods of contemporary quantum mechanics are applied to the characterization of transition states. The second stage, in Chapters 3 and 4, introduces the SCD as a qualitative quantum mechanical model that treats the barrier problem explicitly, and connects the physical organic chemical and Marcus approaches to the concepts of theoretical organic chemistry.

REFERENCES

1. J. N. Brönsted and K. J. Pedersen. Z. *Phys. Chem.* **108**, 185 (1924).

2. K. J. Pedersen. *J. Am. Chem. Soc.* **53**, 18 (1931); *J. Phys. Chem.* **38**, 581 (1934).

3. A. J. Kresge and Y. C. Tang. *J. Chem. Soc. Chem. Comm.* 309 (1980).

4. (a) J. N. Brönsted and H. C. Duus. Z. *Phys. Chem.* **117**, 299 (1925); (b) J. N. Brönsted, A. L. Nicholson, and A. Delbanco. Z. *Phys. Chem.* **169A**, 379 (1934).

5. J. N. Brönsted and E. A. Guggenheim. *J. Am. Chem. Soc.*, **49**, 2554 (1927).

6. H. M. Dawson, C. R. Hoskins, and J. E. Smith, *J. Chem. Soc.* 1884 (1929).

7. C. V. King and E. D. Bolinger. *J. Am. Chem. Soc.* **58**, 1533 (1936).

8. R. P. Bell and E. C. Baughn. *J. Chem. Soc.* 1947 (1937).

9. L. P. Hammett and H. L. Pfluger. *J. Am. Chem. Soc.* **55**, 4079 (1933).

10. G. H. Grant and C. N. Hinshelwood. *J. Chem. Soc.* 1351 (1933).

11. H. J. van Opstall. *Rec. Trav. Chim.* **52**, 901 (1933).

12. L. P. Hammett. *Chem. Rev.* **17**, 125 (1935).

13. L. P. Hammett. *Physical Organic Chemistry.* McGraw-Hill, New York, 1940.

14. (a) H. L. Pfluger. *J. Am. Chem. Soc.* **60**, 1513 (1938); (b) R. P. Bell and A. F. Trotman-Dickenson. *J. Chem. Soc.* 1288 (1949); (c) R. P. Bell and G. L. Wilson. *Trans. Faraday Soc.* **46**, 407 (1950).

15. H. Eyring. *J. Chem. Phys.* **3**, 107 (1935) and references cited therein.

16. (a) M. G. Evans and M. Polanyi. *Trans. Faraday Soc.* **34**, 11 (1938); (b) R. A. Ogg, Jr. and M. Polanyi. *Trans. Faraday Soc.* **31**, 604 (1935); (c) A. G. Evans and M. G.

Evans. *Trans. Faraday Soc.* **31,** 1401 (1935); (d) M. G. Evans and E. Warhurst. *Trans. Faraday Soc.* **34,** 614 (1938).

17. (a) R. P. Bell. *Proc. Roy. Soc. London Ser. A.* **154,** 414 (1936); (b) R. P. Bell. *The Proton in Chemistry*. Cornell University Press, Ithaca, NY, 1973, p. 207; (c) R. P. Bell and O. M. Lidwell. *Proc. Roy. Soc. London. Ser. A.* **176,** 114 (1940); (d) R. P. Bell. *J. Chem. Soc. Faraday Trans. 2.* **72,** 2088 (1976).

18. P. M. Morse. *Phys. Rev.* **34,** 210 (1932).

19. (a) A. Pross and S. S. Shaik. *J. Am. Chem. Soc.* **104,** 187 (1982); (b) A. Pross and S. S. Shaik. *J. Am. Chem. Soc.* **104,** 1129 (1982); (c) A. Pross and S. S. Shaik. *Tetrahedron Lett.* 5467 (1982); (d) A. Pross and S. S. Shaik. *Acc. Chem. Res.* **16,** 363 (1983); (e) D. J. McLennan and A. Pross. *J. Chem. Soc. Perkin Trans. 2.* 981 (1984); (f) A. Pross. *J. Org. Chem.* **49,** 1811 (1984); (g) A. Pross. *Adv. Phys. Org. Chem.* **21,** 99 (1985); (h) T. H. Lowry and K. S. Richardson, *Mechanism and Theory in Organic Chemistry*. Harper and Row, New York, 1987, pp. 359–360, 604–608.

20. J. Horiuti and M. Polanyi. *Acta Physichim. U.R.S.S.* **2,** 505 (1935).

21. R. P. Bell and O. M. Lidwell. *Proc. Roy. Soc. London. Ser. A.* **176,** 88 (1940).

22. J. N. Brönsted. *Chem. Rev.* **5,** 231 (1928).

23. G. S. Hammond. *J. Am. Chem. Soc.* **77,** 334 (1955).

24. J. E. Leffler. *Science.* **117,** 340 (1953).

25. J. E. Leffler and E. Grunwald. *Rates and Equilibria in Organic Reactions*. John Wiley, New York, 1963.

26. (a) C. D. Johnson. *Chem. Rev.* **75,** 755 (1975); C. D. Johnson and B. Stratton. *J. Chem. Soc. Perkin Trans. II.* 1903 (1988); (b) A. Pross. *Adv. Phys. Org. Chem.* **14,** 69 (1977); (c) B. Giese. *Angew. Chem. Intl. Ed. Eng.* **16,** 125 (1977); (d) C. D. Johnson. *Tetrahedron.* **36,** 3461 (1980); (e) see related discussions in C. D. Ritchie, *Acc. Chem. Res.* **5,** 348 (1972); C. D. Ritchie. *Can. J. Chem.* **64,** 2239 (1986).

27. Reference 17b, pp. 202–203 and reference cited therein.

28. M. Eigen. *Angew. Chem. Intl. Ed. Eng.* **3,** 1 (1964).

29. (a) E. A. Walters and F. A. Long. *J. Am. Chem. Soc.* **91,** 3733 (1969); (b) F. Hibbert, F. A. Long, and E. A. Walters. *J. Am. Chem. Soc.* **93,** 2829 (1971); (c) F. Hibbert and F. A. Long. *J. Am. Chem. Soc.* **94,** 2647 (1972); (d) M. M. Kreevoy and S.-W. Oh. *J. Am. Chem. Soc.* **95,** 4805 (1973).

30. (a) F. G. Bordwell and D. L. Hughes. *J. Org. Chem.* **45,** 3314 (1980); (b) D. S. Kemp and M. L. Casey. *J. Am. Chem. Soc.* **95,** 6670 (1973).

31. F. G. Bordwell and D. L. Hughes. *J. Am. Chem. Soc.* **107,** 4737 (1985).

32. (a) E. M. Arnett and R. Reich. *J. Am. Chem. Soc.* **102,** 5892 (1980); (b) I. H. Williams. *Bull. Soc. Chim. Fr. No 2.* 192 (1988); (c) H. Yamataka and S. Nagase. *J. Org. Chem.* **53,** 3232 (1988); (d) F. G. Bordwell and D. L. Hughes, *J. Am. Chem. Soc.* **108,** 7300 (1986); (e) S. Hoz, K. Yang, and S. Wolfe. *J. Am. Chem. Soc.* **112,** 1319 (1990).

33. (a) V. J. Shiner, Jr. In *Isotope Effects in Chemical Reactions*. Edited by C. J. Collins and N. S. Bowman, van Nostrand-Reinhold, Princeton, NJ, 1970, Chapter 2; (b) L. Melander and W. H. Saunders, Jr. *Reaction Rates of Isotopic Molecules*. John Wiley, New York, 1980.

34. J. M. Harris, M. S. Paley, and T. W. Prasthofer. *J. Am. Chem. Soc.* **103,** 5915 (1981).

35. (a) A. Pross and S. S. Shaik. *Nouv. J. Chim.* **13,** 427 (1989); (b) F. G. Bordwell, W. J. Boyle, Jr., J. A. Hautala, and K. C. Yee. *J. Am. Chem. Soc.* **91,** 4002 (1969); (c)

F. G. Bordwell, W. J. Boyle, Jr., and K. C. Yee. *J. Am. Chem. Soc.* **92,** 5926 (1970); (d) M. Fukuyama, P. W. K. Flanagan, F. T. Williams, Jr., L. Frainier, S. A. Miller, and H. Shechter. *J. Am. Chem. Soc.* **92,** 4689 (1970); (e) F. G. Bordwell and W. J. Boyle, Jr. *J. Am. Soc.* **93,** 511 (1971); **94,** 3907 (1972); (f) I. O. Shapiro, N. G. Zharova, Yu. I. Renneva, M. I. Terekhova, and A. I. Shatenshtein. *Theoret. Exp. Chem.* **24,** 425 (1986); (g) W. P. Jencks, M. T. Haber, D. Herschlag, and K. L. Nazaretian. *J. Am. Chem. Soc.* **108,** 479 (1986); (h) L. H. Funderburk, L. Aldwin, and W. P. Jencks, *J. Am. Chem. Soc.* **100,** 5444 (1978).

36. For a discussion of the current status see: (a) E. Buncel and H. Wilson. *J. Chem. Educ.* **64,** 475 (1987); (b) E. M. Arnett and K. Molter. *J. Phys. Chem.* **90,** 383 (1986); (c) Z. Rappoport, Ed. *Reactivity and Selectivity. Isr. J. Chem.* **26,** 303–428 (1985).

37. (a) E. R. Thornton. *J. Am. Chem. Soc.* **89,** 2915 (1967); (b) E. K. Thornton and E. R. Thornton. In *Transition States of Biochemical Processes.* Edited by R. D. Gandour and R. L. Schowen. Plenum Press, New York, 1978, Chapter 1; (c) for a purely mathematical model of these effects see J. M. Harris and J. L. Paul. *Isr. J. Chem.* **26,** 325 (1985).

38. (a) R. A. More O'Ferrall. *J. Chem. Soc. B* 274 (1970); (b) R. A. More O'Ferrall. *Notes on Organic Reactivity.* Post-Graduate Lectures, the Hebrew University, 1973.

39. W. P. Jencks. *Chem. Revs.* **72,** 705 (1972).

40. J. C. Harris and J. L. Kurz. *J. Am. Chem. Soc.* **92,** 349 (1970).

41. J. E. Critchlow. *J. Chem. Soc. Faraday Trans. 1* **68,** 1774 (1972).

42. W. J. Albery. *Progr. React. Kinet.* **4,** 355 (1967).

43. (a) E. D. Hughes, C. K. Ingold, and U. G. Shapiro. *J. Chem. Soc.* 225 (1936); (b) M. G. Evans and E. Warhurst. *Trans. Faraday Soc.* **34,** 614 (1938); (c) C. G. Swain and W. P. Langsdorf. *J. Am. Chem. Soc.* **73,** 2813 (1951); (d) S. Winstein, E. Grunwald, and H. W. Jones. *J. Am. Chem. Soc.* **73,** 2700 (1951); (e) M. J. S. Dewar. *Ann. Repts. Chem. Soc.* **48,** 121 (1951); (f) W. von E. Doering and H. H. Zeiss. *J. Am. Chem. Soc.* **75,** 4733 (1953); (g) M. G. Ettlinger and E. S. Lewis. *Texas J. Science.* **14,** 58 (1962); (h) J. F. Bunnett. *Angew. Chem., Intl. Ed. Eng.* **1,** 225 (1962); (i) A. Streitwieser, Jr. *Solvolytic Displacement Reactions.* McGraw-Hill, New York, 1962; (j) R. P. Bell, J. P. Millington, and J. M. Pink. *Proc. Roy. Soc. London, Ser. A.* **303,** 1 (1968); (k) A. J. Parker. *Chem. Revs.* **69,** 1 (1969).

44. D. A. Jencks and W. P. Jencks. *J. Am. Chem. Soc.* **99,** 7948 (1977); W. P. Jencks. *Chem. Revs.* **85,** 511 (1985).

45. (a) E. Grunwald. *J. Am. Chem. Soc.* **107,** 125 (1985); (b) *ibid.* **107,** 4710 (1985); (c) *ibid.* **107,** 4715 (1985); (d) J. M. Sayer and W. P. Jencks, *J. Am. Chem. Soc.* **99,** 464 (1977); L. Funderburk and W. P. Jencks. *J. Am. Chem. Soc.* **100,** 6708 (1978); (e) C. F. Bernasconi. In *Nucleophilicity.* J. M. Harris and S. P. McManus, Editors. ACS Advances in Chemistry Series No. 215, American Chemical Society, Washington, DC, 1987; C. F. Bernasconi. *Acc. Chem. Res.* **20,** 301 (1987).

46. S. S. Shaik. *Progr. Phys. Org. Chem.* **15,** 197 (1985).

47. S. Wolfe, D. J. Mitchell, and H. B. Schlegel. *J. Am. Chem. Soc.* **103,** 7694 (1981).

48. (a) R. A. Marcus and N. Sutin. *J. Chem. Phys.* **24,** 966, 979 (1956); (b) R. A. Marcus. *J. Chem. Phys.,* **26,** 867, 872 (1957); (c) R. A. Marcus. *Can. J. Chem.* **37,** 155 (1959); (d) R. A. Marcus. *Discuss. Faraday Soc.* **29,** 21 (1960); (e) R. A. Marcus. *J. Phys. Chem,* **67,** 853 (1963); (f) R. A. Marcus. *Ann. Rev. Phys. Chem.* **15,** 155 (1964); (g) R. A. Marcus. *J. Chem. Phys.* **43,** 679 (1965); (h) T. W. Newton. *J. Chem. Educ.* **45,** 571 (1968); (i) R. A. Marcus and N. Sutin. *Comments Inorg. Chem.* **5,** 119 (1986).

49. A. Haim and N. Sutin. *J. Am. Chem. Soc.* **88**, 434 (1966).

50. A. O. Cohen and R. A. Marcus. *J. Phys. Chem.* **72**, 4249 (1968).

51. R. A. Marcus. *J. Phys. Chem.* **72**, 891 (1968).

52. (a) M. M. Kreevoy and D. E. Konasewich. *Adv. Chem. Phys.* **21**, 243 (1971); (b) W. J. Albery, A. N. Campbell-Crawford and J. S. Curran. *J. Chem. Soc. Perkin II.* 2206 (1972); (c) A. J. Kresge, S. G. Mylonakis, Y. Sato, and V. P. Vitullo. *J. Am. Chem. Soc.* **93**, 6181 (1971); (d) A. J. Kresge. *Chem. Soc. Rev.* **2**, 475 (1973).

53. (a) W. J. Albery. *Pure Appl. Chem.* **51**, 949 (1979); (b) W. J. Albery and M. M. Kreevoy. *Adv. Phys. Org. Chem.* **16**, 87 (1978); (c) W. J. Albery. *Ann. Rev. Phys. Chem.* **31**, 277 (1980); (d) E. S. Lewis. *J. Phys. Chem.* **90**, 3756 (1986); (e) E. S. Lewis. *Bull. Soc. Chim. Fr. No. 2.* 259 (1988); (f) E. S. Lewis and D. D. Hu. *J. Am Chem. Soc.* **106**, 3292 (1984); (g) E. S. Lewis, T. A. Douglas, and M. L. McLaughlin, *Isr. J. Chem.* **26**, 331 (1985); (h) E. S. Lewis, T. A. Douglas, and M. L. McLaughlin. In *Nucleophilicity.* J. M. Harris and S. P. McManus, Editors, ACS Advances in Chemistry Series No. 215, American Chemical Society, Washington, DC, 1987; (i) M. M. Kreevoy, D. Ostovic, I. S. H. Lee, D. A. Binder, and G. W. King. *J. Am. Chem. Soc.* **110**, 524 (1988).

54. (a) M. J. Pellerite and J. I. Brauman. *J. Am. Chem. Soc.* **102**, 5993 (1980); (b) J. A. Dodd and J. I. Brauman. *J. Am. Chem. Soc.* **106**, 5356 (1984); (c) J. A. Dodd and J. I. Brauman. *J. Phys. Chem.* **90**, 3559 (1986); (d) M. J. Pellerite and J. I. Brauman. *J. Am. Chem. Soc.* **105**, 2672 (1983).

55. (a) J. R. Murdoch. *J. Am. Chem. Soc.* **102**, 71 (1980); (b) D. E. Magnoli and J. R. Murdoch. *J. Am. Chem. Soc.* **103**, 7465 (1981); (c) J. R. Murdoch and D. E. Magnoli. *J. Am. Chem. Soc.* **104**, 3792 (1982); (d) J. R. Murdoch. *J. Am. Chem. Soc.* **105**, 2159 (1983); (e) J. R. Murdoch. *J. Am. Chem. Soc.* **105**, 2660 (1983); (f) J. R. Murdoch. *J. Am. Chem. Soc.* **105**, 2667 (1983); (g) M. Y. Chen and J. R. Murdoch. *J. Am. Chem. Soc.* **106**, 4753 (1984); (h) J. Donnella and J. R. Murdoch. *J. Am. Chem. Soc.* **106**, 4724 (1984).

56. (a) G. W. Koeppl and A. J. Kresge. *J. Chem. Soc. Chem. Comm.* 371 (1973); (b) R. R. Dogonadze, A. M. Kuznetsov, and V. G. Levich. *Electrochim. Acta.* **13**, 1025 (1968); (c) D. J. McLennan. *J. Chem. Educ.* **53**, 348 (1976).

57. J. R. Murdoch. *J. Am. Chem. Soc.* **94**, 4410 (1972).

58. H. S. Johnson. *Gas Phase Reaction Rate Theory.* Ronald Press, New York, 1966, pp. 55–84.

59. (a) N. Agmon and R. D. Levine. *Chem. Phys. Lett.* **52**, 197 (1977); (b) N. Agmon and R. D. Levine. *J. Chem. Phys.* **71**, 3034 (1979); (c) N. Agmon and R. D. Levine. *Isr. J. Chem.* **19**, 330 (1980); (d) N. Agmon. *Int. J. Chem. Kinet.* **13**, 333 (1981).

60. N. Agmon. *Chem. Phys. Lett.* **45**, 343 (1977).

61. A. R. Miller. *J. Am. Chem. Soc.* **100**, 1984 (1978).

62. N. Agmon. *J. Chem. Soc. Faraday Trans. 2.* **74**, 388 (1978).

63. J. L. Kurz. *Chem. Phys. Lett.* **57**, 243 (1978).

64. R. A. Marcus. *J. Am. Chem. Soc.* **91**, 7224 (1969).

65. E. S. Lewis, M. L. McLaughlin, and T. A. Douglas. *J. Am. Chem. Soc.* **107**, 6668 (1985).

66. M. M. Kreevoy and I.-S. H. Lee. *J. Am. Chem. Soc.* **106**, 2550 (1984).

67. C. D. Ritchie, C. Kubisty, and G.-Y. Ting, *J. Am. Chem. Soc.* **105**, 279 (1983).

2

ENERGY SURFACES AND CHEMICAL REACTIONS

This chapter is concerned with the various features of the energy surfaces that are likely to be encountered in chemical reactions, the methods that exist to determine these features, and the ways in which molecules move and collide on energy surfaces. Figure 2.1 is a caricature of the landscape that might be traversed when NH_3 and O_2 react to form $NO_2 + H_2O$ (path 1), $NO + H_2O$ (path 2), or $N_2O + H_2O$ (path 3). Each of these reactions involves the exploration of a different region of the landscape and, in each case, the path proceeds via passes through the mountains.

Despite the apparent complexity of this surface, only two spatial coordinates have been used to describe the energy. Real energy surfaces actually encompass many spatial coordinates. Nevertheless, we can describe surfaces much as we would describe hilly landscapes, but keeping in mind that our thinking must extend to many dimensions.

2.1 FEATURES OF ENERGY SURFACES

An energy surface may be defined as an object that represents the changes in energy which occur as a molecular system is bent, twisted, stretched, or otherwise deformed. A very simple example of a real energy surface is seen in Figure 2.2, which describes the stereomutation of acetaldehyde enolate anion, that is, the total energy as a function of rotation about the carbon–carbon bond and inversion at the carbanionic center. The complete energy surface for this process would be much more complex than this, because the energy is a function of all the internal coordinates of the system. However, it is difficult to display an energy surface in terms of more than two or three coordinates at a time.

Figure 2.1 A caricature of movement from reactants to products within a mountainous landscape. (Reprinted, by permission, from G. C. Pimentel and R. D. Spratley. *Understanding Chemical Thermodynamics*, 1970, p. 149. Copyright © Holden-Day, Inc., San Francisco).

2.1.1 Minima, Maxima, and Saddle Points

The structure of a molecule is completely defined by a statement of the number and kinds of atoms in the molecule and of the geometrical relationships between these atoms.[1] When a molecule is in a fully relaxed state, or equilibrium structure, the bonds are not stretched and the angles are not bent or rotated from their equilibrium or lowest-energy positions. This geometry corresponds to a minimum on the energy surface for the molecule, that is, the bottom of a valley, as seen in Figure 2.2. From this point on the surface, motion in any direction leads to a higher energy.

An energy surface can exhibit more than one local minimum. Each of these refers to an equilibrium structure, and these equilibrium structures may comprise different conformations, or different structural isomers, or reactant and product molecules generally.

The valleys that contain equilibrium structures are separated from each other by ridges, mountain ranges, and peaks. In two dimensions, a movement in any direction from a peak or mountain top will proceed downhill to a lower energy. Consequently a peak is a local maximum. When there are more than two dimensions, there are different kinds of local maxima. If we define a local maximum as a point on the surface, such that the energy is a maximum along at least two

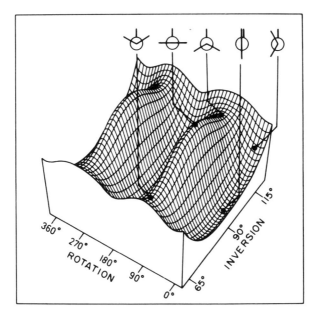

Figure 2.2 The theoretical potential energy surface for stereomutation of acetaldehyde enolate anion. The surface contains minima, saddle points, and maxima. (Reprinted, by permission, from S. Wolfe, H. B. Schlegel and I. G. Csizmadia. *Can. J. Chem.* **53**, 3365 (1975). Copyright © 1975 by the National Research Council of Canada).

mutually perpendicular directions, and is a minimum in all of the remaining independent directions, then we observe that there are two local maxima in Figure 2.2. In general, local maxima are not pertinent to the discussion of chemical reactions, except in a negative sense. If there is a peak between two valleys, it is always possible to find a lower-energy pathway connecting the valleys around the side of the peak than directly over the peak. Therefore, avoiding a local maximum will always result in a lower-energy reaction path.

The easiest way to proceed from one valley to another is via a mountain pass or col. A pass in a range of mountains is the lowest point on the ridge that separates two valleys. The pass is a minimum in one direction (along the ridge), and a maximum in the other direction (across the ridge, from one valley to the other). Because of its shape, a pass or col is also termed a saddle point. On a higher-dimensional-energy surface that might represent a chemical reaction, our problem is to find the lowest-energy path from the valley of the reactants to the valley of the products. The saddle point must be a maximum along a path connecting the valleys, and it must be a minimum in all of the remaining directions, for otherwise a lower-energy path between reactants and products would have to exist. On a higher-dimensional surface, a saddle point is, therefore, a maximum in one and only one direction, and a minimum in all of the remaining independent directions. Figure 2.2 contains four saddle points. A saddle point may be qualified further as a first-order saddle point (maximum in only one direction), or as a second-order saddle point (maximum in two directions), the latter corresponding to a local maximum.

The most important point of the foregoing discussion is summarized as Statement 2.1.

Statement 2.1 A transition structure for a reaction must be a first-order saddle point; that is, it must be a stationary point that is a maximum in only one direction and a minimum in all other perpendicular directions.

An energy surface can exhibit more than one minimum, and it can also contain more than one saddle point. Thus, for any reaction, different pathways, involving different saddle points, can connect reactants and products. Under most circumstances, the lowest-energy pathway will be followed, and the transition structure for the reaction is the highest point on this lowest-energy reaction path. The transition structure for a reaction is, therefore, one of the first-order saddle points on the energy surface. There can be only one transition structure for an elementary reaction and, since this structure is the highest point along the reaction path, its energy will determine the rate of the reaction to a considerable extent.

Some authors refer to the minima, maxima, and saddle points of energy surfaces as critical points or stationary points. From calculus, the first derivative with respect to all of the coordinates, also known as the gradient vector, is zero at a critical point. A critical point is characterized further by the signs of its second derivatives. A positive second derivative in one direction corresponds to a minimum in that direction; a negative second derivative corresponds to a maximum. To determine whether a particular critical point refers to a minimum, maximum, or saddle point, it is necessary to examine the matrix of second derivatives.

In the classical mechanical treatment of the motion of a particle on a potential energy surface, the first derivatives just described are also the negatives of the forces on the particle.[2] Therefore, at a minimum, maximum, or saddle point all of the forces on the atoms in a molecule are zero, leading to the term "stationary point." The second derivatives of the potential energy are often termed the force constants. The nature of the force constant matrix will then determine whether a stationary point is a minimum, maximum, or saddle point, as is discussed in greater detail below.

2.1.2 Intrinsic Reaction Coordinate

As we have seen, reactants and products are connected by a reaction path that passes through a saddle point on an energy surface. Since the saddle point, as well as the reactants and the products are well defined, it should be possible to connect them via a unique reaction path. For example, one could select the path of steepest descent from the saddle point to the reactants and the products. However, such a path is not unique, because the shape of an energy surface depends on the particular choice of coordinates that describe the geometry, and more than one combination of bond lengths, bond angles, and dihedral angles can be employed to represent the same structure.

For example, a three-membered ring can be defined by three bond lengths, or by two bond lengths and the subtended angle, or by a set of Cartesian coordinates; but the energy of a molecule that corresponds to a particular geometry depends only on the relative positions of the atoms, and not on the particular coordinate system employed to specify this geometry. Figures 2.3a and b refer to the same

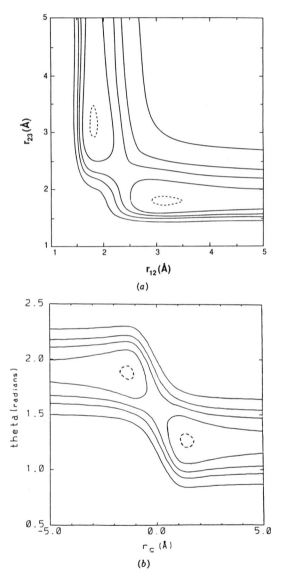

(a)

(b)

Figure 2.3 Two different views of the energy surface for the reaction $X^- + CH_3Y \rightarrow XCH_3 + Y^-$, $X = Y = Cl$. (a) $r_{12} = r(C - X)$ versus $r_{23} = r(C - Y)$. (b) Theta $= <HCX$ versus $r_C = r(C - X) - r(C - Y)$. All other coordinates of a and b are chosen so as to minimize the energy. (Reprinted, by permission, from S. C. Tucker and D. G. Truhlar. *J. Am. Chem. Soc.* **112**, 3338 (1990). Copyright © 1990 by the American Chemical Society).

process, $X^- + CH_3Y \rightarrow XCH_3 + Y^-$, but have a very different appearance, because they are based on different coordinates. The steepest descent paths will also be different.

It is possible to define an intrinsic reaction path that is independent of the coordinate system by an appeal to classical mechanics.[3,4] The movement of a classical particle on a given energy surface must be the same, regardless of how the geometry is specified. The intrinsic reaction coordinate is the path traced by a classical particle that is allowed to roll slowly from a saddle point down to the minimum for the reactants, and from the saddle point down to the minimum for the products. As soon as the particle acquires more than an infinitesimal velocity or kinetic energy, it is slowed and then allowed to continue. Since the classical equations of motion can be defined for any coordinate system, and since they must yield the same trajectory, this definition of the intrinsic reaction coordinate is unique. The classical equations of motion are simplest in mass-weighted Cartesian coordinates, because the effective mass for each coordinate is then the same. Only in such a coordinate system does the intrinsic reaction coordinate coincide with the steepest descent path. Figure 2.4 illustrates the relationship between the steepest descent path and the intrinsic reaction coordinate for an arbitrary reaction surface that refers to the process $C_2H_5F \rightarrow CH_2CH_2 + HF$.

Although intrinsic reaction coordinates like minima, maxima, and saddle points comprise geometrical or mathematical features of energy surfaces, considerable care must be exercised not to attribute excessive chemical or physical significance

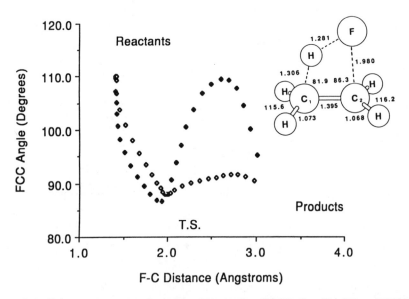

Figure 2.4 Steepest descent pathways for the reaction $CH_3CH_2F \rightarrow CH_2CH_2 + HF$ in internal coordinates (not mass weighted: open diamonds) (mass weighted: filled diamonds). (Reprinted, by permission, from C. Gonzalez and H. B. Schlegel. *J. Phys. Chem.* **94**, 5523 (1990). Copyright © 1990 by the American Institute of Physics).

to them. Real molecules have more than infinitesimal kinetic energy, and will not follow the intrinsic reaction path. Nevertheless, the intrinsic reaction coordinate provides a convenient description of the progress of a reaction, and also plays a central role in the calculation of reaction rates by variational transition state theory[5] and reaction path Hamiltonians.[6]

2.1.3 Vibrations on Energy Surfaces

Molecules are not stationary, but vibrate about their equilibrium structures, a process that corresponds to small oscillations about the minima on an energy surface. A lowering of the temperature decreases the average energy of the molecule, and also the amplitudes of the oscillations. In classical mechanics all vibrations cease at absolute zero. However, quantum mechanical oscillators[7] continue to vibrate at absolute zero, in conformity with the Heisenberg uncertainty principle. The energy associated with this vibration at $0°K$ is termed the zero point energy (ZPE). The ZPE is equal to half of the classical vibrational frequency, and can range from about 0.5 kcal/mol for a flexible bending mode or internal rotation to about 5 kcal/mol for a stiff bond stretch. Even for a small molecule, the total ZPE can amount to several tens of kcal/mol.

As chemists we have a particular interest in energy differences. Since reactants and products normally possess different numbers of different types of bonds, ZPE differences contribute to heats of reactions, typically by 0 to ± 5 kcal/mol. Further, since the bonding in transition structures differs from that in reactants and products, the ZPE will also affect activation energies, as discussed in Sections 2.6 and 2.7.

Quantum mechanical treatments of vibration[8] are usually restricted to harmonic oscillator approximation of motions on an energy surface. In this approximation, the energy surface in the vicinity of a minimum, maximum, or saddle point is expanded as a quadratic function whose coefficients are the first and second derivatives of the energy surface. At a critical point all first derivatives are zero, and the approximation to the surface can be written

$$E(\mathbf{q}) = E_o + \frac{1}{2} \sum_{ij} F_{ij}(q_i - q_i^o)(q_j - q_j^o); \quad F_{ij} = \frac{\partial^2 E}{\partial q_i \partial q_j} \quad (2.1)$$

where the q_i refer to the coordinates at the geometry of interest and the q_i^o refer to the coordinates of the stationary point. The matrix of second derivatives, $\mathbf{F}$, is termed the force constant matrix or the Hessian matrix.

The harmonic vibrational frequencies for a molecule can be computed from a knowledge of the force constant matrix, the geometry of the molecule, and the masses of the atoms by solution of the Wilson $\mathbf{F}$–$\mathbf{G}$ matrix equation[8] (equation 2.2), where $\mathbf{F}$ is the force constant matrix, $\mathbf{G}^{-1}$ is the kinetic energy matrix, and

$$\det(\mathbf{FG} - \epsilon\mathbf{I}) = 0 \quad (2.2)$$

I is the unit matrix. The eigenvalues ϵ are related to the vibrational frequencies ν (equation 2.3). The normal coordinates of vibration are those linear combinations

$$\nu = \frac{1}{2\pi} \sqrt{\epsilon} \tag{2.3}$$

of the coordinates for which the **FG** matrix equation can be solved. If **G** is a constant times the unit matrix, as is the case in mass-weighted coordinates, the normal coordinates are just the eigenvectors of **F**.

The vibrational problem can be posed in terms of internal coordinates, Cartesian coordinates, or mass-weighted coordinates. The **F** and **G** matrices will depend on the choice of coordinate system, but the final vibrational frequencies, and the geometric distortions described by the normal coordinates, are independent of the choice of coordinate system. For example, Table 2.1 illustrates the normal coordinates of vibration for H_2O in internal coordinates and in Cartesian coordinates.

At a minimum, all eigenvalues of the force constant or second derivative matrix **F** are positive, that is, the second derivative is positive for motion in any direction. At a first-order saddle point, one and only one eigenvalue is negative and all others are positive. This means that the second derivative is negative, and the energy is a maximum in one direction only. In all other perpendicular directions the second derivative is positive; that is, the energy surface is a minimum in these other directions. Because the vibrational frequencies are the square roots of the eigenvalues of the force constant matrix, a negative eigenvalue leads to an imaginary frequency. The eigenvector associated with the negative eigenvalue (the normal mode associated with the imaginary frequency) is the single direction in which the surface is a maximum, and it is termed the transition vector. For example, the transition vector for $F^- + CH_3F$ is shown in **2.1** and has an imaginary frequency of

2.1

$585i$ cm^{-1}. The transition vector is also the path of the intrinsic reaction coordinate through the saddle point. Higher-order saddle points or local maxima are characterized by two or more negative eigenvalues of the second derivative or force constant matrix.

2.1.4 Trajectories

Molecules in shallow minima vibrate with large amplitudes; more energetic molecules undergoing chemical reactions also explore large regions of the potential energy surface. Such motions are somewhat difficult to describe accurately using quantum mechanics, but can often be treated adequately by classical or semiclas-

Table 2.1 Calculated Force Constants and Normal Modes of Vibration for H₂O[a]

Cartesian Coordinate Force Constants[b]

		1	2	3	4	5	6	7	8	9
O	x	0.7155								
	y	0.0000	0.0000							
	z	0.0000	0.0000	0.4358						
H	x	-0.3577	0.0000	-0.1830	0.3953					
	y	0.0000	0.0000	0.0000	0.0000	0.0000				
	z	-0.2616	0.0000	-0.2179	0.2223	0.0000	0.2046			
H	x	-0.3577	0.0000	0.1830	-0.0375	0.0000	0.0393	0.3953		
	y	0.0000	0.0000	0.0000	0.0000	0.0000	0.0000	0.0000	0.0000	
	z	0.2616	0.0000	-0.2179	-0.0393	0.0000	0.0133	-0.2223	0.0000	0.2046

Cartesian Coordinate Normal Modes[c]

Bend 1798 cm⁻¹

O	x	0.0000
	y	0.0000
	z	-0.0740
H	x	0.3909
	y	0.0000
	z	0.5869
H	x	-0.3909
	y	0.0000
	z	0.5869

Symmetric Stretch 3813 cm⁻¹

O	x	0.0000
	y	0.0000
	z	0.0453
H	x	0.6080
	y	0.0000
	z	-0.3596
H	x	-0.6080
	y	0.0000
	z	-0.3596

Antisymmetric Stretch 3946 cm⁻¹

O	x	-0.0717
	y	0.0000
	z	0.0000
H	x	0.5693
	y	0.0000
	z	-0.4163
H	x	0.5693
	y	0.0000
	z	0.4163

Table 2.1 (Continued)

Internal Coordinate Force Constants[d]

		1	2	3
R_1	1	0.5405		
R_2	2	-0.0084	0.5405	
Θ	3	0.0428	0.0428	0.1972

Internal Coordinate Normal Modes[e]

Symmetric Stretch
3813 cm^{-1}

R_1	0.7067
R_2	0.7067
Θ	0.0339

Antisymmetric Stretch
3946 cm^{-1}

R_1	0.7071
R_2	-0.7071
Θ	0.0000

Bend
1798 cm^{-1}

R_1	-0.0884
R_2	-0.0884
Θ	0.9922

[a] At the HF/3-21G level.
[b] In hartree/bohr^2.
[c] Normalized.
[d] In hartree/bohr^2, hartree/radian bohr, or hartree/radian^2.
[e] Normalized, in bohr or radian.

sical mechanics.[9] If an initial geometry is selected, and the molecule is given an appropriate initial velocity (i.e., kinetic energy), classical mechanics can be employed to predict the trajectory of the molecule across the surface.[10] A few trajectories are illustrated in Figure 2.5 for $Cl^- + CH_3Cl$. It can be seen that none of the trajectories follows the intrinsic reaction path, that is, the bottom of the valley, nor parallels it for more than a short distance. Instead, the trajectories appear to oscillate about the intrinsic reaction path. Since the total energy (potential energy plus kinetic energy) must be conserved, each trajectory is necessarily restricted to that region of the surface whose potential energy is less than the total energy.

Of the sample trajectories in Figure 2.5, (a) and (b) reach the product valley, (c) returns to the reactant valley, and (d) is trapped for many vibrational periods in the cluster before dissociating to the products. Nevertheless, all four have the same initial energy. The difference is a consequence of the different starting positions and directions of motion. If a sufficient number of trajectories is examined, with an appropriate statistical distribution of starting geometries and velocities, the reaction rate can be approximated quite well, [10] given an accurate description of the energy surface. However, not all aspects of a reaction can be modelled by

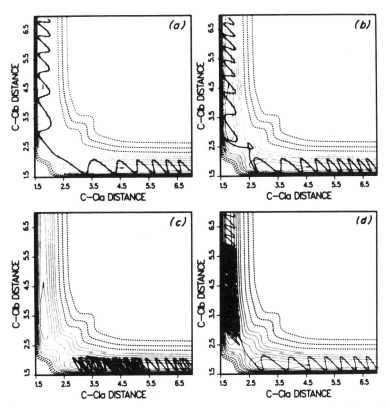

Figure 2.5 Reactive and nonreactive trajectories on the energy surface for $Cl^- + CH_3Cl$. (Reprinted, by permission, from S. R. Vande Linde and W. L. Hase, *J. Chem. Phys.* **93**, 7962 (1990). Copyright © 1990 by the American Institute of Physics).

classical trajectories. For example, although a barrier may be classically forbidden, light atoms such as hydrogen can pass through the barrier by quantum mechanical tunneling. At low temperatures, this alters the reaction rate significantly.[11]

2.1.5 Symmetry and Topology

Energy surfaces are smooth functions connecting the minima that we associate with molecular structures, and these functions must take into account the symmetries of these molecules.

Consider an energy surface with n minima. If the surface is contiguous, that is, not broken into several separate surfaces, each minimum must be connected to at least one other minimum. Since, in the region that connects two minima there must be at least one saddle point, the surface must contain at least $n - 1$ saddle points. Other relations can be derived,[12] but these either require or provide information concerning higher-order saddle points or maxima and are of less interest for the treatment of chemical reactions.

Stanton and McIver[13] have derived a number of symmetry restrictions on the transition vectors for reactions which are based on the group of symmetry operations for the transition state that either take the reactants to the products or leave the reactants and products unchanged. The three most important rules can be summarized as follows:

1. A first-order saddle point can have only one transition vector. In particular, the transition vector cannot be degenerate, that is, it cannot belong to a degenerate representation of the point group for the transition structure. If the transition vector did belong to a degenerate representation, another vector would be equivalent to it by symmetry, and would have the same negative eigenvalue of the force constant matrix. A structure having two negative eigenvalues cannot be a first-order saddle point or transition structure. Closely related to this rule, it can be shown that three valleys cannot meet at a single transition state except as a numerical accident.[13a]

Consider the S_N2 process $X^- + CH_2X_2$ depicted in structures **2.2–2.8**. The

2.2–2.8

reactants shown in **2.2** can proceed to either of the equivalent products, **2.3** or **2.4**. The process could, in principle, proceed via **2.5**, a structure of D_{3h} symmetry that connects **2.2–2.4**. However, the transition vector for **2.2** → **2.3** (C-X_1 shortening, C-X_2 lengthening) belongs to a doubly degenerate representation of D_{3h} that also includes **2.2** → **2.4**. According to rule 1, there is not one high-symmetry transition state connecting the three structures, but, rather, three lower-symmetry (C_{2v}) transition states, **2.6–2.8**.

2. The transition vector must be antisymmetric for symmetry operations that convert the reactants to the products. Consider structures **2.9–2.11**: a plane of

| 2.9 | 2.10 | 2.11 |

reflection passing through the carbon atom and perpendicular to the X–C–X axis converts the reactants (**2.9**) to the products (**2.11**). This plane of reflection is a symmetry operation for the transition structure (**2.10**), and the transition vector is antisymmetric with respect to this operation.

3. The transition vector must be symmetric for symmetry operations that leave the reactants and products unchanged. For the process **2.9** → **2.10** → **2.11**, a C_3 rotation or a reflection in a plane defined by X–C–X and any of the hydrogens leaves both the reactants and products unchanged. These are also symmetry operations of the transition state, and the transition vector is symmetric with respect to these operations.

These properties of transition vectors are summarized as Statement 2.2:

Statement 2.2   The transition vector of a first-order saddle point (a "transition state") cannot be degenerate, and must be antisymmetric with respect to symmetry operations that convert reactants to products, and symmetric with respect to operations that leave reactants and products invariant.

Other applications of symmetry operations and energy surfaces include the use of permutation groups to identify equivalent reaction paths for certain classes of reactions that comprise the rearrangement of substituents on an otherwise unchanged skeleton,[14] and the use of crystallographic space groups to treat the symmetry of energy surfaces in two or three periodic coordinates.[15]

2.2 CONSTRUCTING ENERGY SURFACES

In principle, a complete description of a reaction can be obtained if the energy surface is known.[16] For diatomic molecules, portions of the dissociation curve can be obtained from experiment under favorable circumstances using the RKR approach.[17] However, for polyatomics there is no "direct" method to obtain an experimental energy surface except in regions very close to the minima. Such surfaces must, therefore, be constructed by theoretical means. A comprehensive treatment of potential energy surfaces would encompass much of the history of quantum chemistry and reaction dynamics and is beyond the scope of this book, as is a critical evaluation of the various methods currently employed to construct energy surfaces. Such methods can be found in a number of monographs and reviews.[18] Our objective here is to familiarize the reader with the terminology, conceptual bases, and limitations of some of the more frequently used procedures. Somewhat greater emphasis will be placed on ab initio molecular orbital methods, because these ultimately are the most reliable.

2.2.1 Empirical Potential Energy Surfaces

Empirical surfaces fall into two groups: (a) methods based on some theoretical model of the underlying quantum mechanics, with parameters taken from experiment for the reactants and the products; (b) empirical functions with parameters adjusted to reproduce kinetic data. The distinction is not sharp, because adjustable parameters are usually included in (a) to improve the agreement with experiment; and the empirical functions used in (b) often have some physical basis.

The London–Eyring–Polanyi–Sato (LEPS) potential[19] is one of the better known empirical potentials for triatomic systems, for example, A + BC → AB + C. The form of the potential is based on the Heitler–London valence bond treatment of H_2 dissociation. This VB approach, extended to a triatomic system, yields the LEPS surface, equation 2.4, where Q and J are the coulomb and exchange inte-

$$
\begin{aligned}
V(r_{AB}, r_{BC}, r_{CA}) = [1/(1 + \Delta)] \{ & Q_{AB} + Q_{BC} + Q_{CA} \\
& - \{\tfrac{1}{2}[(J_{AB} - J_{BC})^2 + (J_{BC} - J_{CA})^2 + (J_{CA} - J_{AB})^2]\}^{1/2} \}
\end{aligned}
$$

(2.4)

grals for the indicated pairs of atoms, and Δ is an empirical constant. The integrals are estimated from empirical potentials for diatomics. A Morse function[20] (equation 2.5) is used for the ground state, and the corresponding anti-Morse function

$$
(Q + J)/(1 + \Delta) = (D_e) \{ \exp[-2a(r - r_e)] - 2\exp[-a(r - r_e)] \}
$$

$$
(Q - J)/(1 - \Delta) = (D_e/2) \{ \exp[-2a(r - r_e)] + 2\exp[-a(r - r_e)] \}
$$

(2.5)

(obtained by reversing the sign of the attractive term) is used for the triplet state to account for repulsive terms.

The LEPS potential has a computationally simple form, and it is free from spurious minima. It has been applied extensively to study the reactions of atoms with diatoms, and such studies have provided much insight into the dynamics of reactions.[18] For example, when there is a late barrier, the reaction rate is enhanced by a shift of some of the energy of the reactants from translation to vibration.[21]

The bond energy–bond order (BEBO) method[22] comprises a second empirical potential for the treatment of triatomic systems. In this method it is assumed that the sum of the bond orders is constant along the reaction path. For AB + C → A + BC, the energy along the path is given by equation 2.6, where $E_{AB}(r_{AB})$ and

$$V = -E_{AB}(r_{AB}) - E_{BC}(r_{BC}) + V_{AC}(r_{AC}) \qquad (2.6)$$

$E_{BC}(r_{BC})$ are the dissociation energies of the partially broken bonds and $V_{AC}(r_{AC})$ is the repulsion between A and C; V is usually taken to be the anti-Morse function obtained from the Morse potential for AC. The bond energies are assumed to be proportional to the equilibrium bond dissociation energy, times the bond order n, defined by equation 2.7, where p is an adjustable constant. The bond order is, in

$$E = D_e n^p \qquad (2.7)$$

turn, related to the bond length using Pauling's relation (equation 2.8).[23]

$$r = r_e - a \ln n \qquad (2.8)$$

The concept of *switching functions* provides a very flexible procedure for the construction of the potential energy surfaces of triatomic and polyatomic molecules.[24] In this procedure an energy surface is written in the general form of equation 2.9, where the switching functions S_1, S_2, and S_3 change the surface smoothly

$$V = S_1 V_{react} + S_2 V_{ts} + S_3 V_{prod} \qquad (2.9)$$

from that of the reactants into that of the transition state, and then into that of the products.

Another approach to the construction of global potential energy surfaces is based on a *many-body expansion* in terms of interatomic distances.[25] For a triatomic system this can be written as equation 2.10. Unfortunately, the procedures of equa-

$$V = V_{AB}(r_{AB}) + V_{BC}(r_{BC}) + V_{AC}(r_{AC}) + V_{ABC}(r_{AB}, r_{BC}, r_{AC}) \qquad (2.10)$$

tions 2.9 and 2.10 require more data than are normally available experimentally and must, therefore, rely on theoretical calculations. On the other hand, the LEPS and BEBO methods are insufficiently flexible to fit the wealth of detailed information available from computations.

Molecular mechanics is a well-established procedure for the calculation of the approximate ground-state structures of large molecules. In this procedure, the energy is written as a sum of empirical functions for each of the internal coordinates, for example, Morse functions for bond stretches, harmonic force constants for bond bending, and Fourier component analysis for torsions about bonds.[26] Some interaction force constants may also be included. Nonbonded interactions (van der Waals repulsion, London attraction) and electrostatic effects are also taken into account explicitly. Within limited classes of molecules, for which the empirical potentials are well calibrated, excellent agreement with experimental geometries can be obtained. Steric interactions for transition states and, in some cases, vibrational frequencies are also reproduced well.[26,27] For a conformational process in which the barriers arise primarily from steric interactions, molecular mechanics can provide reliable energy surfaces.

2.2.2 Semiempirical MO Methods

Molecular orbital methods can be employed to calculate the energy of a molecule in any geometry.[7] For a given nuclear configuration, the Schrödinger equation is solved to obtain the electronic energy and wavefunction. Within the Born–Oppenheimer approximation, the total energy of the molecule is the sum of the electronic energy and the nuclear–nuclear repulsion energy. However, the Schrödinger equation can be solved exactly only for a one-electron system.

Semiempirical molecular orbital methods employ an approximate form of the Schrödinger equation that can be solved more easily.[28] One important category is based on neglect of differential overlap, and includes CNDO, INDO, MINDO, MNDO, AM1, PM3, and others. In the NDO approximation, the overlap between atomic orbitals on different atoms is neglected; the remaining integrals are not computed exactly, but are assumed to be constant or are estimated through some appropriate functional form. The particular approximation used to obtain the integrals differs from method to method, and each depends on a number of parameters. These are adjusted to obtain a good average fit either to ab initio molecular orbital results (CNDO, INDO) or to experimental geometries and thermochemistries (MINDO, MNDO, AM1, PM3). Other extensions of the NDO family involve fewer approximations and calculation of additional integrals, for example, partial retention of diatomic differential overlap (PRDDO).

Semiempirical methods are particularly successful when they are applied to structures closely related to those on which their parameterization is based. However, each method has its limitation, and the failures, when they occur, can be rather spectacular. Some of the more recent methods, such as MINDO/3, MNDO, and AM1 have been applied extensively to the exploration of energy surfaces for reactions of organic molecules. Since these methods are based on the calculation of electronic wavefunctions, they can be applied to the treatment of reactions controlled by electronic effects such as orbital symmetry. However, caution is always necessary in the interpretation of the results, because the parameterization is necessarily based upon properties of energy minima rather than transition structures. To achieve a parameterization based on transition structures would require a more extensive collection of ab initio calculations than is presently available. The current semiempirical methods can, however, be useful once they have been calibrated against ab initio calculations on model reactions relevant to the problem of interest.

A semiempirical electronic structure procedure that has been used extensively by chemical dynamicists to calculate energy surfaces for triatomic systems is the diatoms-in-molecules (DIM) method.[29] This is closely related to the LEPS surfaces and uses an approximate VB model to calculate the electronic energy. The interaction energies of VB structures are calculated from an empirical bond dissociation potential for the related diatomic molecule. Since the diatomic potentials are fitted to experimental data, the correct dissociation limits are assured. The method is computationally efficient, yielding smooth surfaces. A closely related method, also based on VB theory, is a semiempirical model[30a, b] which assumes an effective Heisenberg Hamiltonian. It has been applied with success to a variety of extended π systems[30c–e] and small alkali metal clusters[30f] and affords reasonable geometries and barriers.[30b–f]

2.2.3 Ab Initio MO Methods

In ab initio MO methods[31] the full Schrödinger equation is used to treat all of the electrons of the molecule. One can, in principle, compute an energy surface to any degree of accuracy. In practice, however, approximations are necessary which restrict the complexity of the electronic wavefunction. An overview of current methods and terminology is provided in this section. Further details can be found in recent books and monographs.[31]

In the Hartree–Fock (HF) approximation, the wavefunction is an antisymmetrized product of molecular orbitals, also referred to as a Slater determinant. The molecular orbitals, in turn, are chosen as a linear combination of atomic orbitals or other suitable basis functions (LCAO–MO). The molecular orbital coefficients are adjusted to afford the lowest variational energy. In this process, each orbital is the best orbital in a *self-consistent field* of all other orbitals (therefore SCF energy and orbitals). In the restricted Hartree–Fock (RHF) method, each spatial orbital contains two electrons, one spin-up and the other spin-down. The accuracy of an SCF wavefunction depends on the number and type of basis functions used to describe the molecular orbitals. In turn, the time required to compute a wavefunction depends formally on the fourth power of the number of basis functions; the repulsion between all possible pairs of electrons must be computed, and the probability distribution of an electron involves all possible pairwise products of basis functions.

Basis sets employed for ab initio calculations can vary in size, with an increase in size normally corresponding to an increase in the reliability of the result. In a minimal basis set, each atom has one function for each pair of electrons, and, in addition, as many functions as are needed to complete the shell. For example, for carbon, these are: 1s, 2s, $2p_x$, $2p_y$, $2p_z$. A double-zeta basis set has two functions for each single function of the minimal basis set. A split-valence basis set resembles double zeta, but doubles only the functions of the valence shell. Triple-zeta basis sets are defined analogously. Any basis set can be augmented with polarization functions, that is, higher angular momentum functions than are in the valence shell (p functions on hydrogen, d functions on first- and second-row atoms). Diffuse functions can also be added for the description of anions or excited states. For calculations on molecules, the functions used to construct basis sets are normally fixed linear combinations of gaussian-type functions, whose exponents and contraction coefficients are optimized for the neurtral atoms. In all cases, considerable care must be taken that all atoms and all bonds are described equally well, so that errors caused by shortcomings of the basis set will hopefully cancel.

In the HF approximation, each electron moves in a field which is the average of all other electrons. However, in reality, each electron is affected by the instantaneous positions of all other electrons, and the motion of the electrons is correlated to allow pairs of electrons to avoid each other explicitly. When this correlation is taken into account, the HF energy is lowered, and the lowering is termed the *electron correlation energy*. If electrons remain paired during the course of a reaction, the HF energy predicts the correct shape and energy differences for an energy surface, because the correlation energy does not vary greatly. However, when the pairing of electrons is disrupted, as in bond making or bond breaking, the correlation energy is no longer constant, and the HF approximation no longer provides quantitatively correct energy differences (ΔE). Table 2.2 illustrates this point.

The HF wavefunction is composed of a single Slater determinant. Other determinants can be constructed by promotion of electrons from occupied to unoccupied orbitals to yield excited configurations. An improved wavefunction can be constructed by taking a linear combination of the reference configuration and some subset of all possible excited configurations. If the coefficients of this linear combination are computed with the variational principle, the procedure is termed configuration interaction (CI). When perturbation theory is employed, the procedure is known as many body perturbation theory (MBPT) or Möller–Plesset perturbation theory (MP).

In principle, these methods can lead to the exact energy and wavefunction, but at the expense of an infinite amount of computer time. The accuracy is determined in part by the number of excited configurations and the level of excitation included in the wavefunction. The number of configurations can range from several hundred to a few million. Of the CI calculations currently used, the degree of excitation typically includes single and double excitations (CISD), sometimes double excitations only (CID), occasionally also triple and quadruple excitations (CISDTQ), and in a few cases all possible excitations within the limit of the basis set (full CI).

Sometimes all single and double excitations from more than one reference determinant are included variationally (multireference determinant CI or MRDCI). For the perturbational approaches, the order of perturbation theory is also important. Up to fourth order is available and includes single, double, triple, and quadruple excitations (MP4SDTQ). For some excitations, the perturbational energy can be summed to infinite order. Other methods include higher excitations in some appropriately predetermined manner. These include clusters, for example, LCCD

Table 2.2 Dependence of ΔE on Basis Set and Electron Correlation[a]

Reaction	STO-3G	3-21G	6-31G*	6-31G**				Exp[b]
	HF	HF	HF	HF	MP2	MP3	MP4	
$CH_4 \rightarrow CH_3 + H$	115	87	87	87	109	110	110	113
$H_2O \rightarrow OH + H$	84	75	82	86	119	115	116	126
$CH_3CH_3 \rightarrow CH_3 + CH_3$	96	68	70	69	99	95	97	97
$CH_3F \rightarrow CH_3 + F$	66	59	69	69	113	105	108	114
$CH_2CH_2 + 2H_2 \rightarrow 2CH_4$	−91	−71	−66	−64	−61	−62	−60	−57
$CH_2O + 2H_2 \rightarrow CH_4 + H_2O$	−65	−64	−54	−58	−55	−59	−52	−59
$CH_2CH_2 + 2CH_4 \rightarrow 2CH_3CH_3$	−53	−21	−22	−22	−26	−25	−24	−20
$HCN + 2CH_4 + 2NH_3 \rightarrow 3CH_3NH_2$	−37	5	3	4	0	3	5	4
$O_2 + 2H_2O \rightarrow 2HOOH$	29	31	69	70	52	50	49	47
$NH_3 + H^+ \rightarrow NH_4^+$	259	227	219	220	220	221	220	213
$CH_2O + H^+ \rightarrow CH_3O^+$	221	183	182	187	180	183	182	183

[a]Data from references 31d and 37.
[b]Experimental enthalpy differences corrected to 0 °K minus zero-point energy.

63

(linear coupled cluster with double excitations), CCSD (coupled clusters with singles and doubles), coupled electron pair approximation (CEPA), and others.

Normally the molecular orbitals are not readjusted when the correlation energy is calculated. If the molecular orbitals and the CI coefficients are optimized concurrently, the method is known as multiconfiguration self-consistent field or MCSCF. For CASSCF (complete active space SCF) or FORS (fully optimized reaction space) MCSCF calculations, a full CI is carried out within a given set of valence orbitals. The generalized valence bond (GVB) approach is a multiconfiguration method that uses a limited set of VB-type configurations in an MCSCF calculation. Spin-coupled VB (SCVB) is another promising approach[31f] that optimizes the spin orbitals for a specific set of configurations. With any of these methods for the calculation of correlation energy, the basis set must be sufficiently large so that an adequate number of configurations can be included.

Derivatives of the energy can be calculated readily for most ab initio methods.[32] Analytical first derivatives are available for SCF, MPn, CI, CC, and MCSCF energies. These are useful in the optimization of equilibrium geometries, and essential for the location of transition structures. Analytical second derivatives are available for SCF, MP2, MCSCF, and small CI calculations. Harmonic vibrational frequencies can be computed readily from second derivatives.

Once the level of an ab initio calculation and the appropriate basis set have been selected, the energies of the stationary points on the reaction surface can be calculated. If only a few structures are of interest, these can be located directly, usually with the aid of energy derivatives. On the other hand, a study of the chemical dynamics of a reaction requires that large regions of the surface be calculated. Since ab initio calculations can be time-consuming, the best approach is to fit a suitable mathematical function to a representative collection of energies and structures. For example, the surface depicted in Figure 2.2 was based on the application of equation 2.11 to 15 calculated structures.[33] Surface fitting has developed into

$$E = E_o + c_1(\phi - 90)^2 + c_2 e^{-a(\phi - 90)^2} + c_3 \sin(\theta) \cdot (\phi - 90)$$

$$+ c_4 \sin(\theta) \cdot (\phi - 90)^3 + c_5 \sin(\theta) + c_6 \sin^2(\theta) \cdot (\phi - 90)^2$$

$$+ c_7 \sin^2(\theta) \cdot e^{-a(\phi - 90)^2} + c_8 \sin^3(\theta) \cdot (\phi - 90)$$

$$+ c_9 \sin^3(\theta) \cdot (\phi - 90)^3 + c_{10} \sin^4(\theta) + c_{11} \sin^4(\theta) \cdot (\phi - 90)^2$$

$$+ c_{12} \sin^4(\theta) \cdot e^{-a(\phi - 90)^2} \tag{2.11}$$

a high art[18,34] since it is not easy to devise a function that will have the proper asymptotic limits, avoid unwanted potholes, kinks, or bumps, and have sufficient flexibility to be fitted to the computed energies. It is not uncommon to employ 100–1000 computed points in the construction of an analytical surface.

2.3 LOCATING MINIMA ON ENERGY SURFACES

There are numerous methods for the location of minima.[35,36] Some are based only on the energy; others require energy derivatives. Some operate on only one coordinate at a time; others may change all of the coordinates concurrently. Despite these seemingly different strategies, all of the methods share certain features. The objective of an optimization algorithm is to find the minimum on an energy surface by construction of a series of points that explore a portion of the surface and proceed progressively closer to a local minimum. We assume that, at the beginning of the procedure, the details of the surface are unknown. Therefore, to begin the search, one normally adopts a simple mathematical form for the surface and adjusts this as more data are generated. Most commonly the surface is modelled by a quadratic polynomial, for example, equation 2.12, and initially no more than a crude estimate is available for the coefficients of this polynomial. The task of

$$E(\mathbf{q}) = E(\mathbf{q}^o) + \sum_i A_i(q_i - q_i^o) + \tfrac{1}{2} \sum_{ij} B_{ij}(q_i - q_i^o)(q_j - q_j^o) \quad (2.12)$$

the optimization algorithm is to adjust the model surface to fit the energies and derivatives (if available) for the structures already computed, and to use the improved surface to estimate the structure of the minimum. The different procedures vary according to the data they employ and the manner in which the model surface is adjusted to these data.

It is difficult to assess quantitatively the effectiveness of an algorithm, but desirable features include speed of convergence to the minimum, stability and reliability of the method, and the overall cost of the optimization. Energy calculations by ab initio procedures require substantial computer time. Analytical first derivatives require about the same amount of computer time as the energy. Numerical derivatives for N dimensions require N times the time to obtain the energy. Second derivatives take 5 to 10 to N times longer than first derivatives. While a method based on second derivatives may be fast in terms of the total number of steps, each step requires much more computer time than a step in an optimization method that employs only energies, or energies and gradients. Thus some compromise is usually necessary between the number of steps and the cost per step.

The ultimate test of an optimization method is its successful application to a real problem. The compendia of ab initio geometry optimizations reported in the literature and in bibliographies of ab initio wavefunctions attest to the power of gradient methods. Two particularly useful compilations are the Carnegie–Mellon Quantum Chemistry Archive,[37] which contains more than 50,000 geometry optimizations performed with the GAUSSIAN series of programs, and the Quantum Chemistry Literature Data Base, which contains ca. 20,000 entries.[38]

In this section we examine more closely a few representative methods for the location of minima. More detailed descriptions of optimization algorithms can be found in various books and review articles.[35,36] The practical aspects of geometry optimization have also been reviewed.[36b]

2.3.1 Energy-Only Algorithms

Methods that use only the energy have the widest range of applicability, but also the slowest convergence. Figures 2.6a–d illustrate two representative algorithms. The optimizations shown in this figure refer to a two-dimensional surface, and it should be remembered that real examples will be N-dimensional.

In the simplest approach, one coordinate is changed at a time and the calculations cycle over all of the coordinates.[36] This approach is also known as the axial iteration method, or the sequential univariate search method.[35] For the first coordinate, calculate the energy (1), take a step along the coordinate, calculate the energy (2), take another step, and calculate the energy again (3). There are now three points in a line, to which a parabola can be fitted. Find the minimum for the parabola and calculate the energy (4). The process is repeated for the second coordinate, and the energy (7) is calculated at the minimum of the parabola fitted to (4), (5), and (6). The process is repeated for each of the remaining coordinates. However, as is illustrated in Figure 2.6b, convergence to a minimum is not guaranteed, despite the fact that the simple two-dimensional surface is quadratic. The reason for this is that the valley of Figure 2.6b is at an angle to the coordinate axes because the coordinates are coupled ($B_{ij} \neq 0$ for $i \neq j$ in equation 2.12). A second cycle through all of the coordinates would yield a structure (13) closer to the true minimum, but it can be seen that a third cycle would be needed.

A second, more sophisticated, method is a variant of the Fletcher–Powell algorithm[39] and is illustrated in Figures 2.6c and 2.6d. The method is based on derivatives, but the implementation discussed here[39c] uses only the energy and calculates the derivatives numerically. The following steps are involved:

1. Calculate the energy at the starting geometry and at positive and negative displacement of each of the coordinates (points 1–5 in Figure 2.6c); this yields the A's and diagonal B's of equation 2.12, and amounts to a numerical calculation of the first and diagonal second derivatives.

2. Find the minimum on the model surface; if the predicted change in the coordinates is sufficiently small, the optimization is stopped.

3. Step to the minimum, calculate the energy (6), step the same distance beyond, and recalculate the energy (7).

4. Fit a parabola to (1), (6), and (7), find a minimum, and calculate the energy (8).

5. Recalculate the numerical first derivative by stepping along each coordinate (9 and 10), and readjust the model surface to fit the available data.

Steps (2)–(5) are repeated until convergence is achieved. This approach generally requires fewer steps to obtain better convergence than the first optimization

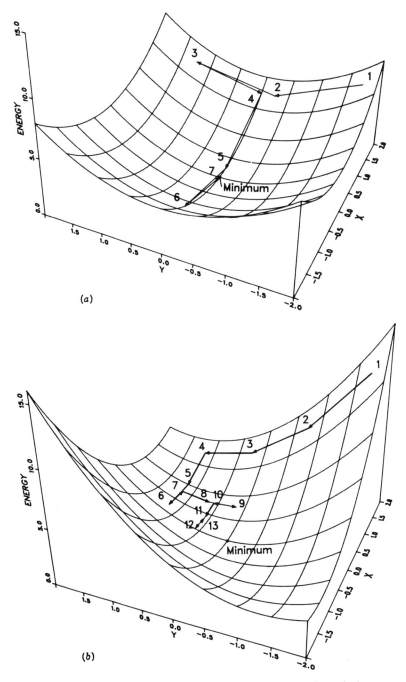

Figure 2.6 Steps in finding the minimum on a quadratic surface, using only the energy. (a) Axial iteration method on a quadratic surface without interaction terms. (b) Axial iteration method on a quadratic surface with interaction terms. (c) Fletcher–Powell method on a quadratic surface without interaction terms. (d) Fletcher–Powell method on a quadratic surface with interaction terms. (Reprinted, by permission, from reference 36a. Copyright © 1987 by Wiley-Interscience).

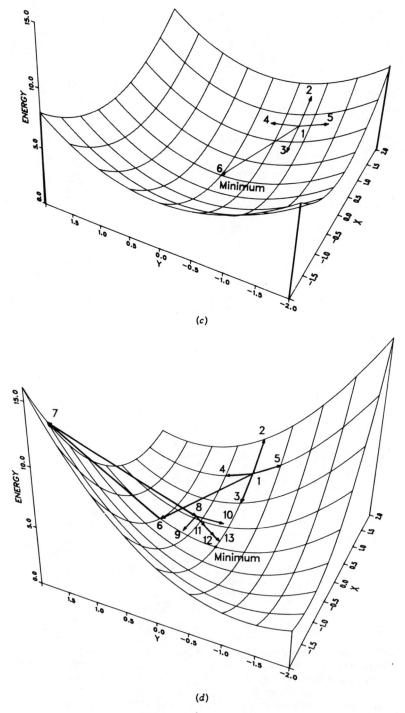

(c)

(d)

Figure 2.6 (*Continued*)

algorithm. Nevertheless, the number of steps required is still proportional to N^2, where N is the number of coordinates.

2.3.2 Gradient Algorithms

Conjugate gradient,[40] variable metric,[41] super memory gradient,[42] Fletcher–Powell,[39] Davidon–Fletcher–Powell,[39] Broyden–Fletcher–Goldfarb–Shanno,[43] optimally conditioned,[44] and Murtaugh–Sargent[45] are different kinds of optimization algorithms that make use of the first derivative of the energy[35] (preferably computed analytically). For the most part, these methods are more efficient and have better convergence properties than energy-only algorithms, but their application requires that analytical energy derivatives be available. As mentioned earlier, gradients are routinely available for SCF calculations, and in some program systems also for MCSCF, MPn, CI, GVB, and CC energies.

All gradient-based optimization methods begin with an estimate of the second derivative of the energy surface and improve the estimate as the optimization proceeds. The methods[35, 39-46] differ in the formulas used to update the second derivative (Hessian) matrix, and according to whether it is the Hessian or its inverse that is updated. A typical search is illustrated in Figure 2.7, but it must be remem-

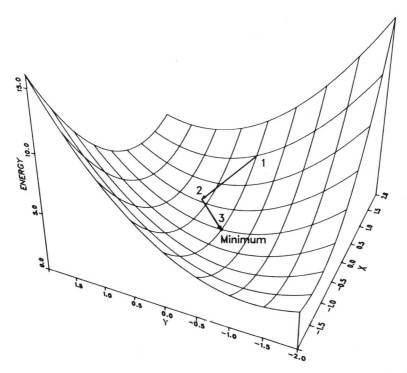

Figure 2.7 Steps in finding the minimum on a quadratic surface using a gradient algorithm. (Reprinted, by permission, from reference 36a. Copyright © 1987 by Wiley-Interscience).

bered that real optimizations take place in N dimensions. The optimization can be separated into the following steps:

1. Obtain an estimate of the Hessian, $H_{ij} = \partial^2 E / \partial q_i \partial q_j$, either as a unit matrix, as an empirical estimate, or in some other way.
2. Calculate the energy, E, and the gradient vector, $g_i = \partial E / \partial q_i$.
3. Update the Hessian (or inverse Hessian) so that the model energy surface fits the current energy and gradient as well as those from earlier steps (omit for the first point).
4. Carry out an accurate minimization along the line connecting the current point and the previous point (omitted for the first point). To save time, this is usually accomplished by fitting a cubic or quartic curve to the two energies and gradients and estimating the position of the minimum from the fitted curve, rather than by recalculation of the energy.
5. Calculate the displacement using the gradient and the updated Hessian matrix (equation 2.13). Note that this requires the inverse of the Hessian. If

$$\Delta \mathbf{q} = -\left(\frac{\partial^2 E}{\partial q^2}\right)^{-1} \frac{\partial E}{\partial \mathbf{q}} = -\mathbf{H}^{-1}\mathbf{g} \qquad (2.13)$$

the gradient and the predicted change in the geometry are sufficiently small, stop; otherwise return to Step 2.

Points (1)–(3) of Figure 2.7 illustrate this sequence. The gradient is calculated at the initial geometry (1). Point (2) is obtained from the gradient at 1 using equation 2.13 and the initial estimate of the Hessian. The gradient is calculated at 2, and point 3 is obtained by carrying out steps 3–5, as above. For an N-dimensional surface, only one gradient calculation is needed to test that a point is a minimum. When a surface is quadratic, gradient methods require $N + 2$ steps or less, and are *guaranteed* to reach the minimum, regardless of the starting geometry or the magnitude of coupling between the coordinates. Even for the nonquadratic surfaces encountered in real geometry optimizations, convergence is usually achieved in approximately N steps or less, when it is possible to make reasonable estimates of the geometry and the Hessian. The accuracy of the initial estimate of the Hessian matrix can improve the rate of convergence considerably, but will not affect the final optimized geometry, because the position of the minimum depends only on where the gradient goes to zero. Since energy plus gradient calculations require twice as much time as an energy calculation alone, each step in a gradient algorithm is more costly than in an energy-only algorithm. However, this is more than overcome by the reduced number of steps and the more stable convergence properties of gradient optimizations. In practice, as much as an order of magnitude increase in efficiency can be achieved with gradient-optimization methods compared to energy-only algorithms.

2.3.3 Second Derivatives

Analytical second derivatives are routinely available only for SCF calculations. Some programs can also compute analytical second derivatives for MP2, MCSCF, and small CI calculations. On a quadratic surface the minimum can be found in one step using the second derivatives and equation 2.14. This is the Newton-Raphson algorithm.[35] More than one step will be needed for nonquadratic surfaces, but often only the gradient, and not the second derivative matrix, has to be recalculated in subsequent steps. If it is necessary, the second derivative matrix can be recalculated every few steps, or updated, as in gradient-optimization algorithms. If second derivatives are not available, those calculated at a lower level of theory (e.g., at the SCF level) may be sufficient to assure rapid convergence to the optimized geometry.

One of the drawbacks to second derivative-based optimization methods is that the computer time required to obtain second derivatives is 5 to 10 to N times as long as a gradient calculation. Consequently, for most optimizations the gradient methods will be more efficient. However, for difficult cases, such as shallow wells and strongly coupled coordinates, analytical second derivatives may be more efficient.

2.4 LOCATING TRANSITION STRUCTURES

In some respects the search for saddle points on energy surface resembles the search for minima. In both cases the gradient at the final geometry must be zero. For a first-order saddle point, a minimum must be found for all directions but one on the energy surface. However, the saddle point must be a maximum along this one direction, which will not normally be known in advance, and must be determined as a part of the optimization. Numerous algorithms have been proposed for the location of transition structures,[47] and this section (adapted from reference 47c) surveys several of the more important methods.

2.4.1 Surface Fitting

Conceptually, the simplest approach to the location of a transition structure is to fit an analytical expression to computed energies. In this way, powerful numerical methods can be applied to find the transition structure on the fitted energy surface, and additional and expensive molecular orbital calculations can be avoided. Naturally, there are drawbacks to this strategy: (1) an acceptable functional form must be found for the many-dimensional surfaces, but there is no universally applicable way to do this, and each surface presents a different challenge; (2) a large number of energy (or energy and gradient) calculations is normally required to obtain an acceptable fit for the model surface;[48] (3) the fitted surface may not be sufficiently accurate in the region of the barrier to provide an acceptable estimate of the geometry of the transition state.

Nevertheless, certain circumstances require an analytical energy surface, for example, the study of reaction dynamics and classical trajectory calculations. An analytical expression can also be useful to illustrate the relationships between the minima, saddle points, and reaction paths in a conformational problem, or the crossing of ground and excited states.

2.4.2 Linear and Quadratic Synchronous Transit

One strategy for the location of saddle points involves certain simplifying assumptions concerning the surface. In the linear synchronous transit (LST) approach[49] it is assumed that the reaction path is a straight line connecting the reactants and the products. This is illustrated in Figure 2.8. With this assumption, the transition structure is the maximum energy point on the straight-line path. As can be seen in Figure 2.8, the LST "transition state" (1) will always be higher in energy than the true saddle point, or equal in energy if the LST passes directly through the transition state by accident or by symmetry. To improve this initial estimate of the transition structure, one can minimize the energy with respect to all of the coordinates perpendicular to the linear reaction path. The resulting point (2) is lower

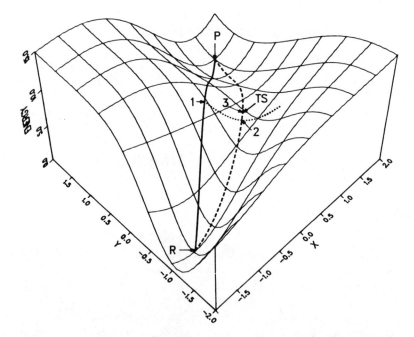

Figure 2.8 The linear synchronous transit (LST) and quadratic synchronous transit (QST) methods for finding transition structures: R, reactants; P, products; TS, true transition structure; 1, maximum on LST path (solid line); 2, minimum perpendicular to LST path; 3, maximum on QST path (dashed line). (Reprinted, by permission, from reference 36a. Copyright © 1987 by Wiley-Interscience).

in energy than the true saddle point. The reaction path can now be described by a parabola or quadratic curve that connects the reactants, 2, and the products. This is known as the quadratic synchronous transit (QST) method.[49] The maximum on the quadratic synchronous path represents a much better estimate of the energy and position of the transition state. If necessary, the process can be repeated by minimization perpendicular to the QST, followed by maximization along the new QST, until satisfactory convergence is achieved.

An advantage of the LST and QST approaches is that they require only the energy, and can be employed even when energy gradients are not available. The overall process is muct less costly than the fitting of an analytical energy surface, and it can lead to a good estimate of the transition structure under favorable circumstances. In practice, the linear interpolation of the LST approach is applied to the distance matrix of a molecule, rather than to its Cartesian or internal coordinates. This yields a reasonble approximation to the reaction path. However, if the real reaction path is too strongly curved, the LST method will provide a rather poor initial estimate of the transition structure, and the QST step may not lead to a realistic structure. This problem can be overcome by the use of structures closer to the transition state as end points of the LST and QST paths. A number of algorithms have been devised[50] that alternate between maximization along a reaction path of predetermined form, and minimization perpendicular to this path.

2.4.3 Coordinate Driving

In many reactions the change in one coordinate dominates the region of the surface between the reactants and the transition state, or between the products and the transition state, or both. For example, barriers to internal rotation are dominated by a torsion angle; also the addition of X^- or $X\cdot$ to a carbon–carbon double bond is dominated by the X–C distance. A two-dimensional surface that illustrates this point, with X the dominant coordinate, is shown in Figure 2.9a. The search for the transition state proceeds by movement along this dominant reaction coordinate, and minimization of the energy with respect to the remaining $N - 1$ coordinates. In favorable cases,[51] this series of steps will approach the transition structure, Figure 2.9a. However, the method can run into problems.[52] If the path should be curved so that a second coordinate begins to dominate, the method will fail. This failure will be signaled by an abrupt change in the geometry, since the newly dominant coordinate belongs to the set of coordinates that is being minimized. The coordinate driving method is costly, because an optimization of $N - 1$ coordinates must be carried out at each step. Moreover, the procedure does not lead to an optimized structure for the transition state, but rather to a series of points along an approximate reaction path. To improve the estimate of the transition structure, the points on the reaction path must be interpolated, or smaller and smaller steps must be taken near the saddle point. As shown in Figure 2.9b, coordinate driving could easily bypass the desired transition state if the topography of the surface is unfavorable.

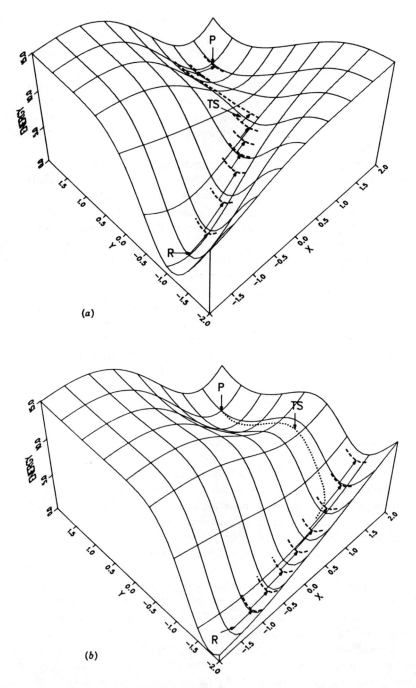

Figure 2.9 (a) The coordinate driving method for location of transition structures. A series of points is generated by stepping one coordinate (X) and optimizing the other (Y). (b) Failure of the coordinate driving method because of the presence of a long valley that does not lead to the transition structure. The dotted curve is the steepest descent path from the saddle point. (Reprinted, by permission, from reference 36a. Copyright © 1987 by Wiley-Interscience).

74

2.4.4 Walking Up Valleys

Curved reaction paths comprise a major problem for the coordinate driving method. A simple change in strategy can overcome this problem, yet not add appreciably to the computational effort. Rather than a step along a specific *coordinate*, one can take a step of a specific *length* and determine the direction that corresponds to the shallowest path up the valley. This is illustrated in Figure 2.10. Some such methods rely only on the energy,[53] or on the energy and gradients.[54] Other implementations of this strategy[55] determine the shallowest uphill path by following the appropriate vector of the second derivative matrix at each step. Another approach that shows some promise uses gradient extremals to define the best uphill steps.[56] Rather torturous reaction paths can be followed by these methods, but if the second derivative matrix must be recalculated frequently, these methods can be costly, and are limited primarily to SCF energy surfaces. In many ways, walking up a valley to the transition structure is the converse of finding the path of steepest descent from the transition structure to the reactants. However, in contrast to the steepest descent path, which is guaranteed to reach a minimum, a shallowest ascent path may not reach the desired transition structure. As in the coordinate driving method, Figure 2.9*b*, the search can lead to a dead-end valley.

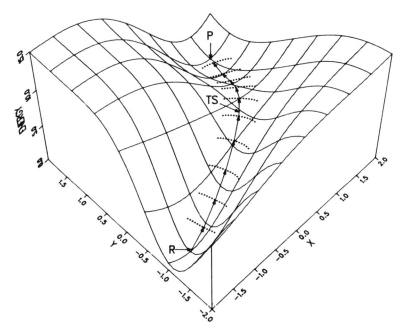

Figure 2.10 Hill climbing or walking up valleys to locate transition structures. A series of points is generated by taking a fixed-length step and finding the shallowest uphill path. (Reprinted, by permission, from reference 36a. Copyright © 1987 by Wiley-Interscience).

2.4.5   Gradient Norm Method

Energy gradients are of great utility for the location of minima. At first glance, the standard gradient methods that are employed to find minima would appear to be unsuitable for the location of transition structures, because the proof of the convergence depends on the positive definite character of the Hessian matrix (all eigenvalues positive). However, it is possible to transform a transition structure optimizaton into a minimization problem by recognition that the square of the gradient vector, or the norm of the gradient (equation 2.14), is a minimum at any

$$R = |\mathbf{g}|^2 = \sum_i \left(\frac{\partial E}{\partial q_i}\right)^2 \qquad (2.14)$$

stationary point (minimum, saddle point, or local maximum). The gradient norm method[57] changes the optimization of a saddle point to a nonlinear least-squares problem, which can be solved using Powell's algorithm.[58] The optimization converges directly to the transition structure, provided that the starting point is in the quadratic region that contains the saddle point. At the saddle point the gradient norm must be zero, and not just a minimum. If one examines the gradient norm for a surface or a cross section of a surface (Figure 2.11), it is apparent that some minima do not have a zero gradient. These correspond to shoulders on the reaction surface, and care must be taken to avoid them. Furthermore, since the radius of convergence, that is, the size of the well for the gradient norm, can be rather small, a very good guess for the transition structure may be needed to allow the gradient norm algorithm to converge.

2.4.6   Gradient Algorithms

Although it may not be possible to prove convergence for a gradient algorithm when the Hessian matrix is not positive-definite, this does not preclude the use of gradient methods for transition structure optimization.[46,59] Gradient methods employ a quadratic function to model the energy surface and locate the stationary points. A quadratic surface will have only one stationary point when the Hessian has only positive or negative eigenvalues, but no zero eigenvalues. If the Hessian has one negative eigenvalue, the stationary point located by a gradient algorithm on a quadratic surface corresponds to a first-order saddle point. With an appropriate starting geometry and a suitable estimate of the Hessian, this approach can be used to converge directly to a saddle point on the potential energy surface for a reaction. The initial geometry must be close enough to the saddle point that the quadratic approximation is valid (i.e., in the region for which the Hessian of the actual surface has one negative eigenvalue). If not, a few steps can be taken with the eigenvector following method[55] or the rational function approach[55] to reach the quadratic region of the transition state. Secondly, the initial estimate of the Hessian must have an eigenvector with a negative eigenvalue that is a reasonable approximation to the transition vector. This is usually accomplished via computation of

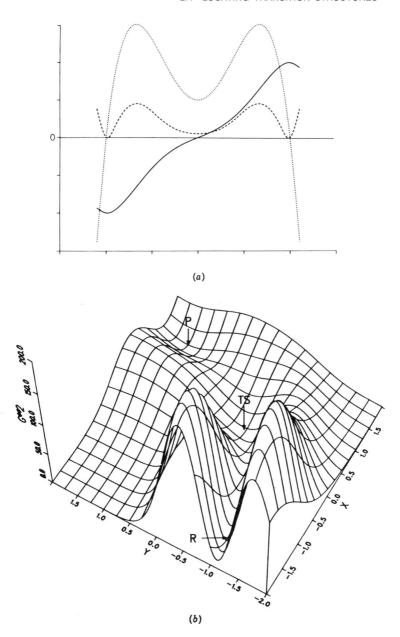

(a)

(b)

Figure 2.11 Gradient norm method for location of transition structures. (a) A one-dimensional potential energy curve (solid), its gradient (dotted), and its gradient norm squared (dashed). The gradient is zero at the minimum and at the transition state. The gradient norm squared is a minimum not only at the potential energy minimum and the transition state, but also at the shoulder. (b) The gradient norm squared for the potential energy surfaces of Figures 2.8, 2.9a, and 2.10. The gradient norm squared is a local minimum at a number of points, in addition to the minima at the reactant, product, and transition state. (Reprinted, by permission, from reference 36a. Copyright © 1987 by Wiley-Interscience).

a few elements of the initial Hessian by numerical differentiation of the gradients. Thirdly, the eigenvalue of the eigenvector resembling the transition vector must remain negative during the course of the optimization, although its magnitude and direction may change. This is achieved by placing suitable restrictions on the updating procedure for the Hessian used at each step in the optimization.[46,47] Under favorable circumstances, this procedure will converge to a transition structure in a number of steps comparable to gradient-type optimizations of equilibrium structures. Since it is usually not necessary to start the optimization with an analytical calculation of the full Hessian, the method can be more economical than walking up valleys. Nevertheless, for certain reactions, considerable care may be needed in the choice of initial geometry and the internal coordinate system.[36b]

2.4.7 Testing Transition Structures

Except for relatively simple surfaces, it is usually necessary to verify that the stationary point found in a transition structure optimization is indeed a first-order saddle point, and not a minimum or a local maximum. This requires calculation of the second derivative matrix and either diagonalization of the matrix or computation of the vibrational frequencies. For a first-order saddle point, there can be one and only one imaginary frequency or negative eigenvalue. If there are two or more imaginary frequencies, a lower-energy saddle point must exist, and can often be found by distortion of the structure along one of the eigenvectors having a negative eigenvalue, and reoptimization of the geometry. It is not sufficient to examine the approximate Hessian obtained during the transition structure optimization, because this contains no information concerning displacements that lead to lower-symmetry and, possibly, lower-energy structures. This can be summarized as Statement 2.3.

Statement 2.3 A first-order saddle point (a transition state) has a single imaginary frequency, that is, a single negative eigenvalue of the Hessian matrix. If a structure on the surface possesses two imaginary frequencies, a first-order saddle point can be reached by distortion of the structure along one of the corresponding eigenvectors, followed by geometry reoptimization.

Under some circumstances, symmetry can be used to transform a transition structure search into a minimization. If it can be determined, for example, from the Stanton–McIver symmetry rules, that the transition vector does not belong to the totally symmetric representation, then the position of the transition structure along the reaction path is fixed by symmetry. It is then necessary to optimize only the coordinates perpendicular to the reaction path. Since a transition structure must be a minimum with respect to all displacements perpendicular to the reaction path, the optimization of a transition structure is transformed into a minimization. For example, the D_{3h} transition structure for the reaction $F^- + CH_3F$ can be formed by minimization of the energy with respect to the C–F and C–H bond lengths (the transition vector has A_2'' symmetry). On the other hand, the transition state for F^-

+ CH_3Cl must be found by a full saddle-point optimization; in this case, the transition vector belongs to the A_1 representation.

Once a saddle point has been found and characterized by computation of the Hessian, it must still be tested to determine that it represents the transition structure of the desired reaction. The transition vector may indicate clearly enough that the reaction path does indeed connect the desired reactants and products. If this is not the case, it may be necessary to follow the reaction path down from the saddle point, perhaps for only a short distance, to ensure that the path leads to the correct reactants and products.

2.4.8 Areas of Difficulty for Transition Structure Optimization

The shape of a multidimensional energy surface in the region of a saddle point is more difficult to visualize than are the regions near minima. In addition, as has been discussed, a variety of strategies can be applied to the search for transition structures. It is useful to consider which procedures are most appropriate for specific kinds of problems. Several different classes of surfaces are shown in Figure 2.12.

The linear or I-shaped surface (Figure 2.12a) is the simplest, and results when one coordinate dominates the reaction path, as in torsional isomerism or pyramidal inversion. In this case almost any procedure, including LST, will be applicable.

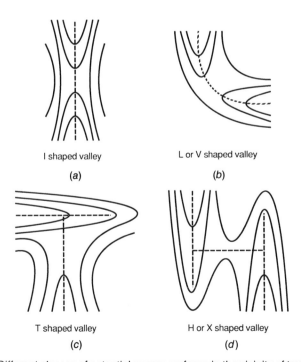

I shaped valley

(a)

L or V shaped valley

(b)

T shaped valley

(c)

H or X shaped valley

(d)

Figure 2.12 Different classes of potential energy surfaces in the vicinity of transition states.

A V- or L-shaped surface (Figure 2.12b) arises when two coordinates dominate the reaction path, as in the $S_N 2$ and other A + BC reactions. In this case the LST approach fails, but the QST method may be acceptable if the end points of the path happen to be close to the transition state. The coordinate driving approach from the reactant side will succeed if the transition state is early, but not if it is late. In the latter case, coordinate driving from the product side will succeed. Climbing up valleys, gradient norm and gradient methods will perform well on this type of surface.

If the desired saddle point is not at the end of a valley, as in I- and L-shaped surfaces, but is on the side of a main valley, then the surface is roughly T-shaped (Figure 2.12c), as is sometimes encountered in insertion reactions. In this case, coordinate driving and the method of climbing up valleys will succeed only from one direction (the stem of the T). If an attempt is made to locate the saddle point by following the main valley (the top of the T), the search will follow the valley to its end, away from the transition state. If the search is initiated in the quadratic region of the saddle point, the gradient norm and gradient methods will lead to the transition structure without problems.

A more difficult situation than the T-type surface is the H-shaped surface (Figure 2.12d), as in the weakly avoided crossing of an orbital symmetry forbidden reaction. This is observed when two long, parallel, valleys are separated by a high ridge. A pass through the ridge is a hanging valley on both sides of the ridge. This situation is more likely to occur in three or more dimensions, where the valleys do not intersect. The pass will be found near the point of closest approach, but will still be a hanging valley on both sides. In these cases, climbing up valleys and coordinate driving from either reactants or products will fail. With suitable starting geometries, any of the gradient methods should be successful.

2.5   THERMODYNAMICS AND ENERGY SURFACES

Energies and structures computed by quantum mechanics usually refer to individual, isolated molecular systems in the gas phase and at $T = 0°K$. At some point it is necessary to compare the theoretical calculations to experimentally determined thermodynamic properties. An overview of the relationships among quantities obtained from MO calculations (geometry, electronic energy, vibrational frequencies) and thermodynamic quantities (internal energy, enthalpy, entropy, and free energy) is presented in Figure 2.13. A brief, qualitative description of these relationships is given below. More quantitative discussions can be found in standard works on statistical thermodynamics.[60]

Figure 2.14a is a model energy profile for the reaction A → B. According to classical mechanics, the energy change for the reaction is just the difference in the electronic energies of A and B. However, because molecular vibrations must be treated quantum mechanically, zero-point energy must be taken into account in the energy differences, as indicated in Figure 2.14b. The internal energy of N mole-

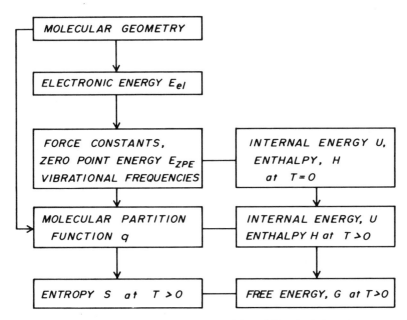

Figure 2.13 Relations among quantities obtained from MO calculations (geometry, electronic energy, vibrational frequencies), and thermodynamic quantities (partition function, enthalpy, entropy, free energy).

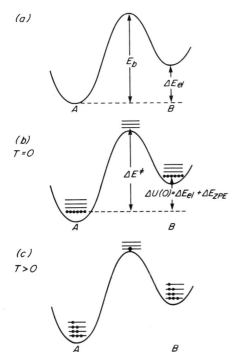

Figure 2.14 A model potential energy profile along the reaction coordinate for the process A → B. (a) Classical. (b) Quantum mechanical at $T = 0$. (c) Quantum mechanical at $T > 0$.

cules at $0°K$, $U(0)$ (equation 2.15), is the sum of the total electronic energy, E_{el},

$$U(0) = N(E_{el} + E_{ZPE}) \qquad (2.15)$$

and the zero-point vibrational energy, E_{ZPE}.

An increase in temperature leads to an increase in the internal energy. For an ensemble of molecules, such as a mole of an ideal gas, the thermal energy will be distributed among translation, rotation, and vibration (equation 2.16), where n_i is

$$U(T) = N(E_{el} + E_{ZPE}) + \sum_i n_i\epsilon_i = U(0) + \sum_i n_i\epsilon_i \qquad (2.16)$$

the number of molecules in the state with energy ϵ_i, N is the total number of molecules, and the sum is taken over *all* translational, rotational, and vibrational states and, in principle, electronic states of the N molecule system. This is illustrated in Figure 2.14c. The energies of ϵ_i are taken relative to the zero-point energy of the molecule. The number of molecules in state i is given by the Boltzmann distribution (equation 2.17),

$$n_i/N = \exp\left(-\epsilon_i/k_bT\right)\Big/\sum_j \exp\left(-\epsilon_j/k_bT\right) \qquad (2.17)$$

where k_b is Boltzmann's constant. If the energies are expressed per mole rather than per molecule, the gas constant, R, is used in place of k_b.

At a given temperature, the thermal energy can be divided in many ways among the molecules and, within each molecule, among the translational, rotational, and vibrational degrees of freedom. Equivalently, there are many ways to arrange the molecules among the various possible translational, rotational, and vibrational states. The Boltzmann distribution corresponds to the most frequently occurring arrangement. Instead of taking the distribution of molecules among available energy states, it is more convenient to write thermodynamic quantities in terms of the canonical partition function Q (equation 2.18), which is appropriate for an

$$Q = \sum_i \exp\left(-\epsilon_i/k_bT\right) \qquad (2.18)$$

ensemble of molecules at constant temperature. The sum is taken over all possible translational, rotational, and vibrational states of *all* molecules in the ensemble. For N noninteracting molecules, such as 1 mole of an ideal gas, the canonical partition function for the entire ensemble of molecules can be written in terms of the molecular partition function (equation 2.19), where the sum refers to all pos-

$$Q = q^{N/N!}; \qquad q = \sum_i \exp\left(-\epsilon_i/k_bT\right) \qquad (2.19)$$

sible translational, rotational, and vibrational states of an *individual molecule*. For translations the number of states and their energies depend on the volume and molecular mass; for rotations, on the geometry and atomic masses for the rotations; for vibrations, on the vibrational frequencies. Each of these can be computed readily from first principles for an ideal gas by standard statistical mechanical methods. The partition function is a measure of the number of states available to a molecule at a given temperature. If the states are more closely spaced, the partition function will be larger. This has an important influence on equilibrium constants and reaction rates, as discussed in Section 2.6.

The internal energy can now be written in terms of the partition function rather than as a sum over translational, rotational, and vibrational states of the molecules in the ensemble (equation 2.20). Since the partition function contains information

$$U(T) = U(0) + k_b T^2 \frac{\partial \ln Q}{\partial T} \tag{2.20}$$

concerning the number of states and the number of ways the molecules can be distributed among these states, it also contains information concerning the entropy of the system. The entropy is given by equation 2.21. The enthalpy and free energy

$$S = [U(T) - U(0)]/T + k_b \ln Q \tag{2.21}$$

can now also be defined. The change in enthalpy is the change in energy at constant pressure, and is the sum of the change in the internal energy and the change in energy due to the change in volume (equation 2.22). The change in free energy

$$\Delta H = \Delta U + \Delta(PV) \tag{2.22}$$

includes the effect of the change in entropy of the system as well as the change in enthalpy (equation 2.23).

$$\Delta G = \Delta H - T\Delta S \tag{2.23}$$

Table 2.3 collects some thermodynamic quantities available from theoretical calculations. At the level of calculation used for these examples, the errors in the computed entropies and thermal corrections to the internal energies and enthalpies are typically ± 0.5 eu and ± 0.2 kcal/mol. A more extensive discussion of the accuracies of the computed thermodynamic quantities can be found in reference 31d.

The equilibrium constant K for the reaction A $\rightleftarrows$ B is the ratio, at equilibrium, of the number of molecules of B to the number of molecules of A. In terms of $\Delta U(0)$, the internal energy difference at $T = 0°K$, and the molecular partition

Table 2.3   Thermodynamic Quantities Determined from Theoretical Calculations[a,b]

Contributions Change in Energy	$CH_4 \rightarrow$ $CH_3 + H$	$F^- + CH_3Cl$ $\rightarrow CH_3F + Cl^-$	$HC \equiv CH + 2H_2$ $\rightarrow 2CH_4$	Propene $\rightarrow$ Cyclopropane
Electronic energy ΔE_{el}	109.6	−67.3	−114.3	6.2
Zero-point energy ΔE_{ZPE}	−10.6	1.1	21.6	1.1
Thermal contribution to the internal energy $\Delta U(298) - \Delta U(0)$	1.1	−0.1	−2.5	−0.5
PV work term $\Delta(PV)$	0.6	0.0	−1.2	0.0
Entropy at $T = 298$ $\Delta S(298)$	30.2	−0.9	−51.7	−4.4
Internal energy at $T = 0$ $\Delta U(0) = \Delta E_{el} + \Delta E_{ZPE}$	99.0	−66.2	−92.7	7.3
Internal energy at $T = 298$ $\Delta U(298)$	100.1	−66.3	−95.2	6.8
Enthalpy at $T = 298$ $\Delta H = \Delta U + \Delta(PV)$	100.7	−66.3	−96.4	6.8
Free energy at $T = 298$ $\Delta G(298) = \Delta H - T\Delta S$	91.7	−66.1	−81.0	8.1

[a]Energy in kcal/mol, entropy in cal/mol-degree K.
[b]Data from references 31d and 37; ΔE computed at the MP4SDQ/6-31G** level and frequencies at the HF/6-31G** level; HF/6-31G* optimized geometries used throughout; ideal gas behavior assumed.

functions q_A and q_B, the equilibrium constant is given by equations 2.24 and 2.25.

$$K = N_B/N_A = (q_B/q_A) \exp(-\Delta U(0)/k_b T) \tag{2.24}$$

$$\Delta U(0) = [E_{el}(B) + E_{ZPE}(B)] - [E_{el}(A) + E_{ZPE}(A)] \tag{2.25}$$

Note that the energy difference includes the zero-point energy as well as the electronic energy. If the energies of A and B are the same [$\Delta U(0) = 0$], the equilibrium constant is determined by the ratio of the partition functions. In such a case, the equilibrium will favor the species having a larger partition function, that is, the species whose energy levels are more closely spaced.

2.6   KINETICS AND ENERGY SURFACES

Equilibrium constants can be determined from energy surfaces more easily than can rate constants. However, with some approximations, statistical mechanics can be used to obtain rate expressions that are applicable to some of the reactions encountered in organic chemistry. The best known of these is transition state theory, also called activated complex theory or absolute rate theory.[16]

Transition state theory begins by selection of a dividing surface that partitions the energy surface into two regions, one for the reactants and the other for the products. The dividing plane might be chosen to pass through the transition state and to be perpendicular to the reaction path, as shown in Figure 2.15. Three assumptions are then made: (1) all molecules that cross the dividing surface from the reactant side proceed to form products, and do not return to the reactant side; (2) an equilibrium distribution of reactants is always maintained; (3) the molecules at the dividing surface are in thermal equilibrium with the reactants.

The number of molecules at the dividing surface can be determined from the equilibrium expression, equation 2.26. Since zero-point energy must be taken

$$N^{\ddagger}/N_A = (q^{\ddagger}/q_A) \exp\left(-\Delta E^{\ddagger}/k_b T\right) \qquad (2.26)$$

into account, the energy difference in the exponential term is $\Delta E^{\ddagger}$, shown in Figure 2.14b, and not the classical barrier E_b in Figure 2.14a. The partition function for the transition state $q^{\ddagger}$ must be handled in a special way. One of the $3n - 6$ vibrational modes of the transition state corresponds to motion along the reaction path. This mode is best treated not as a vibration, but as a translation that carries the molecule through the dividing surface. The partition function can then be written in terms of q_{RC}, for motion along the reaction coordinate, and $\bar{q}^{\ddagger}$ for the remaining

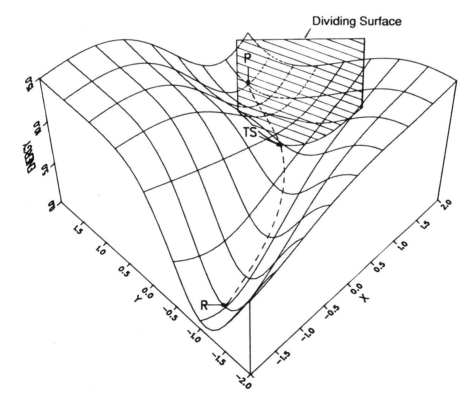

Figure 2.15 Dividing surface that separates reactants from products. Transition state theory assumes that every molecule crossing this surface from the side of reactants continues on to products.

degrees of freedom (translation, rotation, and $3n - 7$ vibrational modes) (equation 2.27). The distribution of kinetic energy for motion along the reaction path is given

$$q^{\ddagger} = q_{Rc}\, \tilde{q}^{\ddagger} \tag{2.27}$$

by the Boltzmann equation. This allows the average velocity of a molecule that travels from reactants to products to be determined. Combination of the number of molecules at the transition state (equation 2.26) with the velocity of these molecules through the transition state leads to equation 2.28, an expression for the

$$k = (k_b T/h)\, (\tilde{q}^{\ddagger}/q_A)\, \exp(-\Delta E^{\ddagger}/k_b T) \tag{2.28}$$

rate constant of the reaction k, where k_b is Boltzmann's constant and h is Planck's constant.

The rate constant according to transition state theory (TST) has the same general form as the Arrhenius rate expression, equation 2.29. Experimentally, the acti-

$$k = A \exp (-\Delta E_{act}/k_b T) \tag{2.29}$$

vation energy ΔE_{act} is obtained from a plot of $\ln k$ vs. $1/T$, with the assumption that A is independent of temperature. However, the preexponential factors and partition functions that appear in the TST rate constant do depend on the temperature. Consequently, $\Delta E^{\ddagger}$ and ΔE_{act} can differ by ca. ± 2 kcal/mol.

Transition state theory usually overestimates rate constants. One reason for this is that not all of the molecules that cross the dividing surface proceed to products. Some return to the reactants. To correct for the recrossing of the dividing surface, the rate constant is multiplied by κ, the transmission coefficient (equation 2.30).

$$k = \kappa\, (k_b T/h)\, (\tilde{q}^{\ddagger}/q_A)\, \exp (-\Delta E^{\ddagger}/k_b T) \tag{2.30}$$

The magnitude of the transmission coefficient is difficult to determine theoretically, and it is usually treated as an empirically adjustable constant, typically close to unity.

Canonical variational transition state theory is an alternative, more satisfactory, approach to the correction of some of the difficulties of transition state theory.[5] In this approach, the dividing surface does not necessarily pass through the saddle point, but is moved along the reaction path until the lowest rate is achieved. In effect, this determines the transition state at the maximum in free energy along the reaction path, rather than at the maximum in the electronic energy, and it leads to significant improvements in the prediction of reaction rates.

The assumptions of transition state theory are not necessarily valid for unimolecular reactions such as bond dissociations and isomerizations. The rate of a unimolecular reaction depends on the internal (vibrational and sometimes rotational) energy of the molecule. Because the flow of energy among translation, rotation, and vibration can be relatively slow, especially when collisions are rare, as at low pressures, the reactants and the transition state may not be in thermal equilibrium. Furthermore, some unimolecular reactions, such as bond dissociations, may not have a well-defined barrier, and the placement of the dividing surface between reactants and products may be difficult. Despite such problems, unimolecular reactions can be treated by appropriate statistical methods such as RRKM theory.[61]

RRKM theory determines the reactivity of a single activated molecule that has a specific internal energy. It then calculates the number of molecules with this energy and averages over all possible energies. It is assumed that a molecule can undergo a unimolecular reaction if its internal energy is greater than the barrier for the reaction, provided that this internal energy is localized in the appropriate coordinates. For example, dissociation of a bond requires that sufficient vibrational energy be present in the bond stretch. If the internal energy is distributed randomly, the probability that sufficient energy is localized in the required coordinates can be calculated. The rate of reaction can then be predicted for a particular activated molecule having a specific internal energy. The number of molecules with the same internal energy will depend on the rate at which molecules are activated and deactivated by collision, and on the rate at which activated molecules react to form products. To obtain an overall rate constant, the rate for an individual molecule with a given internal energy is multiplied by the number of molecules with the same internal energy, and this is integrated over all possible internal energies.

Statistical methods do not consider the details of how molecules move across energy surfaces. If quantum effects are unimportant, the trajectory of a molecule on an energy surface can be described by classical mechanics.[10] Since the accurate integration of the classical equations of motion for a polyatomic molecule on a complicated energy surface can require large amounts of computer time, the classical trajectory method is limited to processes that are complete in 10–50 psec. On the order of 10^3–10^5 trajectories may be needed for statistically meaningful results. Provided that quantum effects are not important and a sufficient number of trajectories is calculated, the classical trajectory method will simulate accurately the chemical dynamics and reaction rates associated with given energy surface.

Classical mechanics is not adequate for the description of very light atoms involved in a reaction; quantum mechanical effects, such as tunnelling, must be taken into account. A classical particle cannot cross a barrier unless its energy is greater than the barrier height. However, in quantum mechanics, tunnelling through a classically forbidden barrier is possible, depending on the mass of the particle and the thickness of the barrier. If the particle has a sufficiently small mass, for example, hydrogen in a [1,2] shift or a proton transfer, and if the barrier is sufficiently narrow, tunnelling can increase the rate significantly. This effect can be important at low temperatures where much of the reaction may proceed by tunnelling. At higher temperatures the process is dominated by the molecules that go over the barrier. A number of simple formulas for tunnelling corrections to transition state theory have been proposed.[62]

Quantum effects can also become important in the region of a surface crossing.[63] Classical motion requires that a molecule remain on a single surface but, in quantum mechanics, a molecule can hop from one surface to another if these surfaces cross, or if the crossing is only weakly avoided and the energy difference between the surfaces is small. These are nonadiabatic effects that result from a breakdown of the Born–Oppenheimer approximation, and they are sometimes important in photochemical reactions.

REFERENCES

1. E. L. Eliel. *Stereochemistry of Carbon Compounds.* McGraw-Hill, New York, 1962, p. 1.
2. H. Goldstein. *Classical Mechanics.* Addison Wesley, London, 1950.
3. K. Fukui. *J. Phys. Chem.* **74,** 4161 (1970).
4. K. Fukui. *Acc. Chem. Res.* **14,** 363 (1981).
5. D. G. Truhlar and B. C. Garrett. *Acc. Chem. Res.* **13,** 440 (1980).
6. W. H. Miller, N. C. Handy, and J. E. Adams. *J. Chem. Phys.* **72,** 99 (1980).
7. See, for example, J. P. Lowe. *Quantum Chemistry.* Academic Press, New York, 1978, or any other introductory quantum mechanics text.
8. (a) E. B. Wilson, Jr., J. C. Decius, and P. C. Cross. *Molecular Vibrations.* McGraw-Hill, New York, 1955; (b) S. Califano. *Vibrational States.* Wiley-Interscience, New York, 1976.
9. Large-scale motions usually represent higher excitations relative to the zero-point level. The correspondence principle states that the quantum mechanical picture approaches classical mechanics for high levels of excitation.
10. (a) R. N. Porter and L. M. Raff. In *Dynamics of Molecular Collisions.* Modern Theoretical Chemistry, Vol. 2, W. H. Miller, Editor, Plenum Press, New York, 1976; (b) D. L. Bunker. *Methods Comput. Phys.* **10,** 287 (1971); (c) D. G. Truhlar and J. T. Muckerman. In *Atom–Molecule Collision Theory.* R. B. Bernstein, Editor, Plenum Press, New York, 1979.
11. S. C. Tucker and D. G. Truhlar. In *New Theoretical Concepts for Understanding Organic Reactions.* J. Bertrán and I. G. Csizmadia, Editors, NATO ASI Series, Volume C267, Kluwer Publications, Dordrecht, 1989, page 291. See also reference 16a, Chapter 10.4.
12. P. G. Mezey. *Theor. Chim. Acta* **62,** 133 (1982).
13. (a) R. E. Stanton and J. W. McIver, Jr. *J. Am. Chem. Soc.* **97,** 3632 (1975); (b) J. H. Murrell and K. J. Laidler. *Trans. Faraday Soc.* **64,** 371 (1968); (c) J. H. Murrell and G. L. Pratt. *ibid.* **66,** 1680 (1970).
14. W. G. Klemperer. *J. Chem. Phys.* **56,** 5478 (1972).
15. For a review, see T. J. McLarnan. *Theor. Chim. Acta* **63,** 195 (1983).
16. (a) J. I. Steinfeld, J. S. Francisco, and W. L. Hase. *Chemical Kinetics and Dynamics.* Prentice-Hall, New Jersey 1989; (b) W. M. Smith. *Kinetics and Dynamics of Elementary Gas Reactions.* Butterworths, London, 1980; (c) J. W. Moore and R. G. Pearson. *Kinetics and Mechanism.* Wiley, New York 1981; see also reference 18.
17. (a) R. Rydberg. *Z. Physik.* **73,** 376 (1931); **80,** 514 (1933); (b) O. Klein. *Z. Physik.* **76,** 226 (1932); (c) A. L. G. Rees. *Proc. Phys. Soc. (London)* **59,** 998 (1947).
18. (a) H. S. Johnson. *Gas Phase Reaction Rate Theory.* Ronald Press, New York 1966; (b) K. J. Laidler. *Theory of Chemical Reaction Rates.* McGraw-Hill, New York 1969;

(c) G. G. Balint-Kurti. *Adv. Chem. Phys.* **30,** 137 (1975); (d) J. N. Murrell, S. Carter, S. C. Farantos, P. Huxley, and A. J. C. Varandos. *Molecular Potential Energy Functions.* Wiley, New York 1984; (e) J. N. Murrell. In *Gas Kinetics and Energy Transfer.* Specialist Periodical Reports. The Chemical Society, London, 1978, Vol. 3, Chapter 5; (f) *Potential Energy Surfaces. Faraday Disc. Chem. Soc.* **72,** (1977); (g) *Potential Energy Surfaces and Dynamics Calculations.* D. G. Truhlar, Editor, Plenum Press, New York, 1981; (h) G. C. Schatz. *Rev. Mod. Phys.* **61,** 669 (1989); (i) H. M. Duran and J. Bertrán. *Repts. Mol. Theor.* **1,** 57 (1990); (j) See also references 16 and 34.

19. S. Sato. *J. Chem. Phys.* **23,** 2465 (1955) and references cited therein.

20. P. M. Morse. *Phys. Rev.* **34,** 210 (1932).

21. J. C. Polanyi. *Acc. Chem. Res.* **5,** 161 (1972).

22. (a) H. S. Johnston and C. Parr. *J. Am. Chem. Soc.* **85,** 2544 (1963); (b) For a general overview, see reference 18a.

23. L. Pauling. *J. Am. Chem. Soc.* **69,** 542 (1947).

24. For an example, see W. L. Hase, G. Mrowska, R. J. Brudzynski, and C. S. Sloane. *J. Chem. Phys.* **69,** 3548 (1978).

25. K. S. Sorbie and J. N. Murrell. *Mol. Phys.* **29,** 1387 (1975); J. N. Murrell, K. S. Sorbie, and A. J. C. Varandar. *Mol. Phys.* **32,** 1359 (1976).

26. (a) U. Burkert and N. L. Allinger. *Molecular Mechanics.* American Chemical Society, Washington, DC, 1981; (b) O. Ermer. *Tetrahedron.* **31,** 1849 (1975); (c) N. L. Allinger, Y. H. Yuh, and J.-H. Lii. *J. Am. Chem. Soc.* **111,** 8551 (1989).

27. K. N. Houk, M. N. Paddon-Row, N. G. Rondan, Y.-D. Wu, F. K. Brown, D. C. Spellmeyer, J. T. Metz, Y. Li, and R. J. Loncharich. *Science* **231,** 1108 (1986); J.-H. Lii and N. L. Allinger. *J. Am. Chem. Soc.* **111,** 8566 (1989).

28. (a) *Semiempirical Methods of Electronic Structure Calculation.* Modern Theoretical Chemistry, Vols. 7 and 8. G. A. Segal, Editor, Plenum Press, New York, 1977; (b) J. A. Pople and D. L. Beveridge. *Approximate Molecular Orbital Theory.* McGraw-Hill, New York, 1970.

29. (a) J. C. Tully. *Adv. Chem. Phys.* **42,** 63 (1980); (b) P. J. Kuntz. In *Atom-Molecule Collision Theory.* R. B. Bernstein, Editor, Plenum Press, New York, 1979, Chapter 3.

30. (a) P. Blaise, J.-P. Malrieu, D. Maynau, and B. Ouija. *J. Mol. Struct. (THEOCHEM).* **169,** 469 (1988); J.-P Malrieu and D. Maynau. *J. Am. Chem. Soc.* **104,** 3021 (1982); (b) M. A. Robb and F. Bernardi. In *New Theoretical Concepts for Understanding Organic Reactions.* J. Bertrán and I. G. Csizmadia, Editors, ASI NATO Series, Vol. C267, Kluwer Publications, Dordrecht, 1989, page 101; (c) D. Maynau, M. Said, and J.-P. Malrieu. *J. Am. Chem. Soc.* **105,** 5244 (1983); (d) M. Said, D. Maynau, J.-P. Malrieu, and M. A. G. Bach. *J. Am. Chem. Soc.* **106,** 571 (1984); (e) H. Said, D. Maynau, and J.-P. Malrieu. *J. Am. Chem. Soc.* **106,** 580 (1984); (f) J.-P. Malrieu, D. Maynau, and J. P. Daudey. *Phys. Rev.* **B30,** 1817 (1984).

31. (a) *Methods of Electronic Structure Theory.* H. F. Schaefer, III, Editor, Plenum Press, New York, 1977; (b) *Ab initio Molecular Orbital Methods.* K. Lawley, Editor, *Adv. Chem. Phys.* Vols. 67 and 69 (1986); (c) A. Szabo and N. S. Ostlund. *Modern Quantum Chemistry: Introduction to Advanced Electronic Structure Theory.* Macmillan Publishers, New York, 1982; (d) W. J. Hehre, L. Radom, P. v. R. Schleyer, and J. A. Pople. *Ab initio Molecular Orbital Theory.* Wiley, New York 1985; (e) R. Daudel, G. Leroy, D. Peeters, and M. Sana. *Quantum Chemistry.* Wiley, New York 1983; (f) D. L. Cooper, J. L. Gerratt, and M. Raimondi. *Adv. Chem. Phys.* **59,** 319 (1987).

32. (a) P. Pulay. In *Applications of Electronic Structure Theory*. Modern Theoretical Chemistry, Vol. 4, H. F. Schaefer, III, Editor, Plenum Press, New York, 1977; (b) J. F. Gaw and N. C. Handy. *Ann. Repts. Prog. Chem. Section C.* **81c,** 291 (1985); (c) P. Pulay. *Adv. Chem. Phys.* **69,** 241 (1987); (d) *Geometrical Derivatives of Energy Surfaces and Molecular Properties*. Edited by P. Jorgensen and J. Simons. D. Reidel, NATO ASI Ser. *C166* (1986).

33. S. Wolfe, H. B. Schlegel, I. G. Csizmadia, and F. Bernardi. *Can. J. Chem.* **53,** 3365 (1975).

34. T. H. Dunning and L. B. Harding. *Theory of Chemical Reaction Dynamics*. M. Baer, Editor, CRC Press, Inc., Boca Raton, FL, 1985, Vol. 1, Chapter 1; P. J. Kuntz. *ibid.* Chapter 2.

35. For discussions of nonlinear optimization methods see, for example: (a) R. Fletcher. *Practical Methods of Optimization*. Wiley, New York 1981; (b) L. E. Scales. *Introduction to Non-linear Optimization*. MacMillan, 1985; (c) B. D. Bunday. *Basic Optimization Methods*. Edward Arnold Press, 1984.

36. (a) H. B. Schlegel. *Adv. Chem. Phys.* **67,** 249 (1987); (b) H. B. Schlegel. In *New Theoretical Concepts in Understanding Organic Reactions*. J. Bertrán and I. G. Csizmadia, Editors, NATO ASI Series, Volume C267, Kluwer Publications, Dordrecht, 1989, page 33. (c) J. C. Head and M. C. Zerner. *Adv. Quantum Chem.* **20,** 239 (1989).

37. R. A. Whiteside, M. J. Frisch, and J. A. Pople. *The Carnegie-Mellon Quantum Chemistry Archive*. 3rd Edition, Carnegie-Mellon University, Pittsburgh, PA 1983, and associated computer data base.

38. K. Ohno and K. Morokuma. *Quantum Chemistry Literature Data Base*. Elsevier, New York 1982. Yearly supplements are published in special issues of *J. Mol. Struct. (THEOCHEM)*.

39. (a) R. Fletcher and M. J. D. Powell. *Comput. J.* **6,** 163 (1963); (b) W. Davidon. Argonne National Laboratory Report ANL-5990, Argonne, IL; (c) J. S. Binkley. *J. Chem. Phys.* **64,** 5142 (1976).

40. R. Fletcher and C. M. Reeves. *Comput. J.* **7,** 149 (1964).

41. D. F. Shanno and K.-H. Phua. *Math. Programming.* **14,** 149 (1978).

42. A. Miele and J. W. Cantrell. *J. Optimiz. Theory Appl.* **3,** 459 (1969).

43. (a) C. G. Broyden. *J. Inst. Math. Appl.* **6,** 76 (1970); (b) R. Fletcher. *Comput. J.* **13,** 317 (1970); (c) D. Goldfarb. *Math. Comp.* **24,** 23 (1970); (d) D. F. Shanno. *Math. Comp.* **24,** 647 (1970).

44. W. C. Davidon. *Math Programming.* **9,** 1 (1975).

45. B. A. Murtaugh and R. W. H. Sargent. *Comput. J.* **13,** 185 (1972).

46. H. B. Schlegel. *J. Comp. Chem.* **3,** 214 (1982).

47. For recent reviews of transition structure optimization algorithms see (a) K. Müller. *Angew Chem.* **19,** 1 (1980); (b) S. Bell and J. S. Crighton. *J. Chem. Phys.* **80,** 2464 (1984); (c) H. B. Schlegel. *Adv. Chem. Phys.* **67,** 249 (1987).

48. (a) M. Sana. *Int. J. Quant. Chem.* **21,** 139 (1981); (b) D. C. Comeau, R. J. Zellmer, and I. Shavitt. In *Geometrical Derivatives of Energy Surfaces and Molecular Properties*. Edited by P. Jorgensen and J. Simons, D. Reidel, NATO-ASI Ser. *C166*, 243 (1986).

49. T. A. Halgren and W. N. Lipscomb. *Chem. Phys. Lett.* **49,** 225 (1977).

50. (a) P. Scharfenberger. *J. Comp. Chem.* **3**, 277 (1982); (b) I. Balint and M. I. Ban. *Theor. Chim. Acta* **63**, 255 (1983); (c) A. Jensen. *Theor. Chim. Acta* **63**, 269 (1983); (d) O. Tapia and J. Andres. *Chem. Phys. Lett.* **109**, 471 (1984).

51. M. J. Rothman and L. L. Lohr. *Chem. Phys. Lett.* **70**, 405 (1980).

52. U. Burkert and N. L. Allinger. *J. Comp. Chem.* **3**, 40 (1982).

53. K. Müller and L. D. Brown. *Theor. Chim. Acta* **53**, 75 (1979).

54. (a) M. V. Basilevsky and A. G. Shamov. *Chem. Phys.* **60**, 347 (1980); (b) M. J. S. Dewar, E. F. Healy, and J. J. P. Stewart. *J. Chem. Soc. Faraday Trans. 2*, **80**, 227 (1984).

55. (a) C. J. Cerjan and W. H. Miller. *J. Chem. Phys.* **75**, 2800 (1981); (b) J. Simons, P. Jorgensen, H. Taylor, and J. Ozment. *J. Phys. Chem.* **87**, 2745 (1983); (c) A. Banerjee, N. Adams, J. Simons, and R. Shepard. *J. Phys. Chem.* **89**, 52 (1985); (d) J. Baker. *J. Comp. Chem.* **7**, 385 (1986).

56. (a) D. K. Hoffman, R. S. Nord, and K. Ruedenberg. *Theor. Chim. Acta* **69**, 265 (1986); (b) P. Jorgensen, H. J. Aa. Jensen, and T. Helgaker. *Theor. Chim. Acta* **73**, 55 (1988). A point lies on a gradient extremal if the gradient at that point is an eigenvector of the Hessian at the point. Like intrinsic reaction paths, gradient extremal paths connect minima with transition states.

57. (a) J. W. McIver and A. Komornicki. *J. Am. Chem. Soc.* **94**, 2625 (1972); (b) A. Komornicki, K. Ishida, K. Morokuma, R. Ditchfield, and M. Conrad. *Chem. Phys. Lett.* **45**, 595 (1977).

58. M. J. D. Powell. *Comput. J.* **7**, 303 (1965).

59. (a) D. Poppinger. *Chem. Phys. Lett.* **35**, 550 (1975); (b) S. Bell, J. S. Crighton, and R. Fletcher. *Chem. Phys. Lett.* **82**, 122 (1981); (c) J. M. McKelvey and J. F. Hamilton, Jr. *J. Chem. Phys.* **80**, 579 (1984).

60. D. A. McQuarrie. *Statistical Thermodynamics*, Harper and Row, New York, 1973.

61. P. J. Robinson and K. A. Holbrook. *Unimolecular Reactions.* Wiley-Interscience, London, 1972.

62. R. A. Marcus and M. E. Coltrin. *J. Chem. Phys.* **67**, 2609 (1977); for leading references see B. C. Garrett and D. G. Truhlar. *J. Chem. Phys.* **74**, 1029 (1981).

63. J. C. Tully. In *Dynamics of Molecular Collisions*, Modern Theoretical Chemistry, Vol. 2, W. H. Miller, Editor, Plenum Press, New York, 1976.

3

VALENCE BOND THEORY AND THE CONCEPTUALIZATION OF POTENTIAL ENERGY SURFACES. ORIGINS OF BARRIERS, AND REACTIVITY TRENDS

In Chapter 2 we saw that the current state of the art of computational MO theory leads to very accurate data for ground states of molecules, as well as information concerning energy surfaces and the locations of saddle points and other critical regions on such surfaces. With this information we can discuss barrier heights, reaction rates, and geometric features of transition states.

Nonetheless, despite these successes, it must be admitted that the numerical results of MO calculations do not, in themselves, contain the insights that an organic chemist needs to *understand* chemistry, because the MO language does not require the concept of a bond and does not lead directly to the concept of the functional group, or to a description of the making and breaking of bonds.

The Bell–Evans–Polanyi (BEP) principle (Chapter 1) comprised a first attempt to understand why barriers exist on reaction surfaces, what governs the structure of a transition state, and what factors control the height of the barrier. A central feature of the BEP principle is the statement that a *reaction profile is the result of a crossing of reactant- and product-like energy curves*, Statement 1.1).

This concept of curve crossing is not found in qualitative approaches based on MO theory, with the exception of the Woodward–Hoffmann "forbidden" reactions.[1] In such reactions the barrier is treated as the result of an *avoided* crossing of reactant- and product-like energy curves, which becomes transparent[1b] when conservation of orbital symmetry is invoked. This postulate reveals crossings of occupied and unoccupied molecular orbitals for forbidden, but not for "allowed" reactions, so that reasons for the existence of barriers in the latter case are not clear.[2]

In this chapter we argue that insights involving curve crossing follow more naturally when VB arguments are applied to the breaking and/or making of bonds. Since these processes are localized events, the VB method seems to be the appropriate conceptual tool, because the central tenet of this theory is the concept of the localized bond.

Accordingly, we present here a method for the construction of qualitative energy surfaces from building blocks consisting of VB configurations. These configurations can comprise classical localized VB configurations based on atoms, or partly localized configurations based on fragment molecular orbitals. This strategy will allow the insights of VB theory to be supplemented by insights and language borrowed from MO theory.

Because of our use of such a mixed language, a first objective of this chapter is to draw attention to the interrelationship between the MO and VB procedures.[3] Following the establishment of these common features, we introduce a qualitative procedure, founded on the notion of avoided crossings, which will provide the basis for all of our subsequent analyses of quantitative reactivity trends.

3.1 COMPARISON OF THE MO AND VB METHODS

3.1.1 The Hydrogen Molecule

The hydrogen molecule contains the simplest example of a covalent bond, and it can be used to illustrate some of the essential features of MO and VB wavefunctions.[3,4] The bonding and antibonding orbitals of the molecule can be written in terms of the 1s orbitals of the hydrogen atoms,

$$\phi_\sigma = \frac{1}{\sqrt{2 + 2S_{12}}}(\chi_1 + \chi_2); \qquad \phi_{\sigma*} = \frac{1}{\sqrt{2 - 2S_{12}}}(\chi_1 - \chi_2) \qquad (3.1)$$

where the atomic orbitals (AO's) χ_1 and χ_2 are the 1s functions on atoms 1 and 2. The overlap between χ_1 and χ_2 is S_{12}, and the $(2 \pm 2S_{12})^{-1/2}$ factors ensure that the molecular orbitals are properly normalized. In the MO description, the ground state has a doubly occupied σ orbital. This can be written as a Slater determinant,[5]

$$\psi^{MO}(\sigma^2) = |\phi_\sigma\bar{\phi}_\sigma| = \frac{1}{2 + 2S_{12}}|(\chi_1 + \chi_2)(\bar{\chi}_1 + \bar{\chi}_2)| \qquad (3.2)$$

which is also depicted in Figure 3.1. No bar over an orbital in equation 3.2 refers to an alpha electron (↑), and a bar refers to a beta electron (↓).

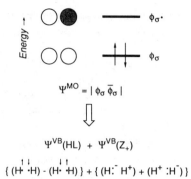

Figure 3.1 The σ^2 MO configuration and its VB constituents for H_2. The dots on the VB structures refer to electrons and vertical arrows to spins.

$$\Psi^{VB}(HL) + \Psi^{VB}(Z_+)$$

$$\{\,(H\!\cdot\ \cdot H) - (H\!\cdot\ \cdot H)\,\} + \{\,(H\!:\ \ H^+) + (H^+\ :H^-)\,\}$$

The Slater determinant in terms of MO's can be expanded into Slater determinants of AO's,[3a-c]

$$\psi^{MO}(\sigma^2) = \frac{1}{2 + 2S_{12}} |\chi_1\bar{\chi}_1 + \chi_1\bar{\chi}_2 + \chi_2\bar{\chi}_1 + \chi_2\bar{\chi}_2|$$

$$= \frac{1}{2 + 2S_{12}} \{|\chi_1\bar{\chi}_2| + |\chi_2\bar{\chi}_1| + |\chi_1\bar{\chi}_1| + |\chi_2\bar{\chi}_2|\}$$

$$= \frac{1}{2 + 2S_{12}} \{|\chi_1\bar{\chi}_2| - |\bar{\chi}_1\chi_2| + |\chi_1\bar{\chi}_1| + |\chi_2\bar{\chi}_2|\} \qquad (3.3)$$

In the second term of equation 3.3 the orbitals have been interchanged, to place them in the standard order. This requires the sign to be reversed, because the terms of the equation are Slater determinants, and an interchange of any two columns of a determinant changes its sign.

The combination of the first and second terms of equation 3.3 is the Heitler–

$$\psi^{VB}(HL) = \frac{1}{\sqrt{2 + 2S_{12}^2}} \{|\chi_1\bar{\chi}_2| - |\bar{\chi}_1\chi_2|\} \qquad (3.4)$$

London (HL) configuration of the VB wavefunction, and is depicted in Figure 3.1 below the MO wavefunction. The third and fourth terms of equation 3.3 describe zwitterionic VB configurations, and these are depicted alongside the HL configuration in Figure 3.1.

$$\psi^{VB}(Z_1) = |\chi_1\bar{\chi}_1| \qquad (3.5)$$

$$\psi^{VB}(Z_2) = |\chi_2\bar{\chi}_2| \qquad (3.6)$$

For a symmetric bond, as is present in the hydrogen molecule, it is convenient to take positive and negative combinations of Z_1 and Z_2, and Z_+ and Z_- are now orthogonal.

$$\psi^{VB}(Z_+) = \frac{1}{\sqrt{2 + 2S_{12}^2}} \{|\chi_1\bar{\chi}_1| + |\chi_2\bar{\chi}_2|\} \tag{3.7}$$

$$\psi^{VB}(Z_-) = \frac{1}{\sqrt{2 - 2S_{12}^2}} \{|\chi_1\bar{\chi}_1| - |\chi_2\bar{\chi}_2|\} \tag{3.8}$$

The MO wavefunction can be expressed as an equal mixture of the HL and Z_+ VB configurations (equation 3.9).

$$\psi^{MO}(\sigma^2) = \frac{\sqrt{2 + 2S_{12}^2}}{2 + 2S_{12}} \{\psi^{VB}(HL) + \psi^{VB}(Z_+)\} \tag{3.9}$$

Three excited states must be considered, in addition to the σ^2 ground state, in the MO picture. These are the singlet and triplet states that correspond to an orbital occupancy $\sigma\sigma^*$, and a singlet of type σ^{*2}. As with σ^2, expansion of the σ^{*2} MO configuration leads to a linear combination of the HL and Z_+ VB configurations (equation 3.10). These are depicted in Figure 3.2. The singlet $\sigma\sigma^*$ configuration

$$\psi^{MO}(\sigma^{*2}) = |\phi_{\sigma^*}\bar{\phi}_{\sigma^*}| = \frac{1}{2 - 2S_{12}} |(\chi_1 - \chi_2)(\bar{\chi}_1 - \bar{\chi}_2)|$$

$$= \frac{1}{2 - 2S_{12}} |\chi_1\bar{\chi}_1 - \chi_1\bar{\chi}_2 - \chi_2\bar{\chi}_1 + \chi_2\bar{\chi}_2|$$

$$= \frac{1}{2 - 2S_{12}} \{|\chi_1\bar{\chi}_1| + |\chi_2\bar{\chi}_2| - |\chi_1\bar{\chi}_2| + |\bar{\chi}_1\chi_2|\}$$

$$= \frac{-\sqrt{2 + 2S_{12}^2}}{(2 - 2S_{12})} \{\psi^{VB}(HL) - \psi^{VB}(Z_+)\} \tag{3.10}$$

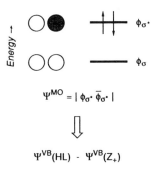

$$\Psi^{MO} = |\phi_{\sigma^*}\bar{\phi}_{\sigma^*}|$$

$$\Downarrow$$

$$\Psi^{VB}(HL) - \Psi^{VB}(Z_+)$$

$$\{(H\!\!\uparrow\ \cdot\!H) - (H\!\!\cdot\ \uparrow\!H)\} - \{(H\!:^- H^+) + (H^+ :H^-)\}$$

Figure 3.2 The σ^{*2} MO configuration and its VB constituents for H_2.

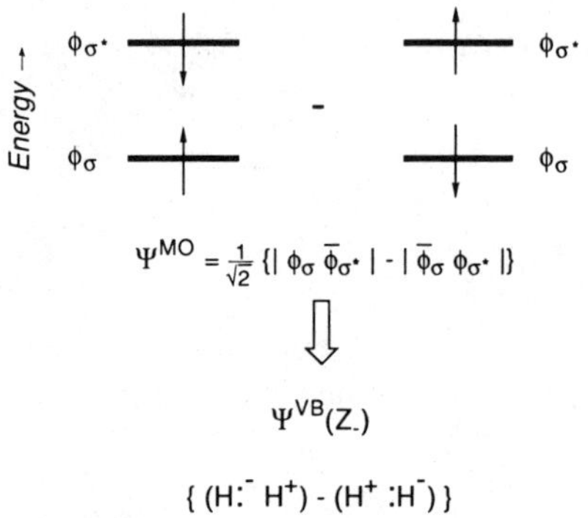

Figure 3.3 The $\sigma^1\sigma^{*1}$ singlet MO configuration and its VB constituents for H_2.

$$\Psi^{MO} = \frac{1}{\sqrt{2}}\{|\phi_\sigma \bar{\phi}_{\sigma^*}| - |\bar{\phi}_\sigma \phi_{\sigma^*}|\}$$

$$\Downarrow$$

$$\Psi^{VB}(Z_-)$$

$$\{(H:^- H^+) - (H^+ :H^-)\}$$

can also be expanded (equation 3.11), and it corresponds to the Z_- combination of the VB configurations Z_1 and Z_2, as depicted in Figure 3.3. An analogous treatment of the triplet

$$^1\psi^{MO}(\sigma\sigma^*) = \frac{1}{\sqrt{2}}\{|\phi_\sigma \bar{\phi}_{\sigma^*}| - |\bar{\phi}_\sigma \phi_{\sigma^*}|\}$$

$$= \frac{1}{2\sqrt{2}-2S_{12}^2}\{|(\chi_1 + \chi_2)(\bar{\chi}_1 - \bar{\chi}_2)| - |(\bar{\chi}_1 + \bar{\chi}_2)(\chi_1 - \chi_2)|\}$$

$$= \frac{1}{\sqrt{2}-2S_{12}^2}\{|\chi_1\bar{\chi}_1| - |\chi_2\bar{\chi}_2|\}$$

$$= \psi^{VB}(Z_-) \tag{3.11}$$

$\sigma\sigma^*$ configuration leads to the triplet analogue of the HL configuration, which differs from $\Psi^{VB}(HL)$ in the sign between the determinants,

$$^3\psi^{MO}(\sigma\sigma^*) = \frac{1}{\sqrt{2}}\{|\phi_\sigma \bar{\phi}_{\sigma^*}| + |\bar{\phi}_\sigma \phi_{\sigma^*}|\}$$

$$= \frac{1}{2\sqrt{2}-2S_{12}^2}\{|(\chi_1 + \chi_2)(\bar{\chi}_1 - \bar{\chi}_2)| + |(\bar{\chi}_1 + \bar{\chi}_2)(\chi_1 - \chi_2)|\}$$

$$= \frac{-1}{\sqrt{2}-2S_{12}^2}\{|\chi_1\bar{\chi}_2| + |\bar{\chi}_1\chi_2|\} \tag{3.12}$$

There are two other triplet sublevels, $|\sigma\sigma^*|$ and $|\bar{\sigma}\bar{\sigma}^*|$, which have the same total spin as the triplet state of equation 3.12 but differ in the orientation of their spin. All three VB components of the triplet are depicted in Figure 3.4.

A wavefunction constructed from a single VB configuration or from a single MO determinant represents only a first approximation of the correct many-electron wavefunction. A better wavefunction can be constructed by taking a linear com-

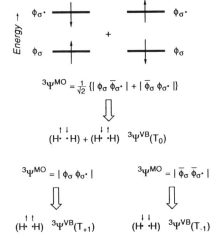

$$^3\psi^{MO} = \frac{1}{\sqrt{2}} \{|\,\phi_\sigma\,\bar{\phi}_{\sigma^*}\,| + |\,\bar{\phi}_\sigma\,\phi_{\sigma^*}\,|\}$$

$$\Downarrow$$

$$(\text{H}\!\cdot \cdot\!\text{H}) + (\text{H}\!\cdot \cdot\!\text{H}) \quad ^3\psi^{VB}(\text{T}_0)$$

$$^3\psi^{MO} = |\,\phi_\sigma\,\phi_{\sigma^*}\,| \qquad\qquad ^3\psi^{MO} = |\,\bar{\phi}_\sigma\,\bar{\phi}_{\sigma^*}\,|$$

$$\Downarrow \qquad\qquad\qquad \Downarrow$$

$$(\text{H}\!\cdot \cdot\!\text{H}) \quad ^3\psi^{VB}(\text{T}_{+1}) \qquad (\text{H}\!\cdot \cdot\!\text{H}) \quad ^3\psi^{VB}(\text{T}_{-1})$$

Figure 3.4 The $\sigma^1\sigma^{*1}$ triplet MO configuration and its VB constituents for H_2. Subscripts (0, ± 1) of the triplet sublevel refer to the spin components in the z direction and the x,y plane, respectively. Only the zero sublevel is shown for the MO configuration, while all sublevels are shown in the VB configurations.

bination of MO determinants or VB configurations. This is a configuration interaction (CI) wavefunction,[5]

$$\Psi = C(\sigma^2)\psi^{MO}(\sigma^2) + C(\sigma\sigma^*)\psi^{MO}(\sigma\sigma^*) + C(\sigma^{*2})\psi^{MO}(\sigma^{*2}) \qquad (3.13)$$

The coefficients $C(\sigma^2)$, $C(\sigma\sigma^*)$, and $C(\sigma^{*2})$ are selected so as to minimize the variational energy, $E = \langle\Psi|\hat{H}|\Psi\rangle/\langle\Psi|\Psi\rangle$. In the present example, $C(\sigma\sigma^*)$ is zero by symmetry. For the ground state of the hydrogen molecule, $C(\sigma^2) = 0.994$ and $C(\sigma^{*2}) = -0.113$, using the STO-3G basis set and a bond length of 0.74 Å.

An alternative, but entirely equivalent, wavefunction can be constructed from a linear combination of singlet VB configurations,

$$\Psi = D(\text{HL})\psi^{VB}(\text{HL}) + D(Z_-)\psi^{VB}(Z_-) + D(Z_+)\psi^{VB}(Z_+) \qquad (3.14)$$

The coefficients of this wavefunction are again found by minimization of the variational energy. With the STO-3G basis set, $D(\text{HL}) = 0.788$, $D(Z_+) = 0.226$, and $D(Z_-) = 0$ by symmetry. It is important to note that since the Ψ^{MO} can be expressed in terms of the Ψ^{VB}, the coefficients for the VB configurations could also have been determined by substitution of equations 3.9 and 3.10 into equation 3.13.

It must also be emphasized that the *same* many-electron wavefunction is obtained regardless of whether one takes a linear combination of MO determinants or of VB configurations, *provided that all possible MO determinants and all possible VB configurations are used*. These two equivalent approaches for the construction of the wavefunction are depicted in Figure 3.5. In this figure, $\Psi^{VB}(Z_-)$ cannot mix with $\Psi^{VB}(\text{HL})$ or $\Psi^{VB}(Z_+)$ because of symmetry. Equivalently, $\Psi^{MO}(\sigma\sigma^*)$ has a different symmetry than $\Psi^{MO}(\sigma^2)$ and $\Psi^{MO}(\sigma^{*2})$.

The radical cation and radical anion of H_2 also merit consideration here, because these will represent building blocks of the VB configurations of more complex systems. The radical cation is a one-electron species, and its ground-state

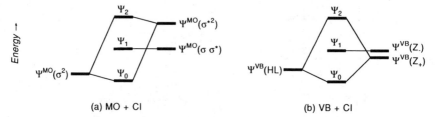

Figure 3.5 Construction of the ground and excited-state wavefunctions for H_2: (a) starting from the MO configurations based on the σ and σ^* molecular orbitals; (b) starting from the VB configurations. Provided that all MO determinants or all VB configurations are included in the configuration interaction calculation, the same total wavefunctions are obtained.

wavefunction is just the σ orbital, which can also be expressed as two VB configurations:

$$\psi_0 = \psi^{MO}(\sigma) = |\phi_\sigma| = \frac{1}{\sqrt{2 + 2S_{12}}} |(\chi_1 + \chi_2)|$$

$$= \frac{1}{\sqrt{2 + 2S_{12}}} \{\psi^{VB}(H^{\cdot}\ H^+) + \psi^{VB}(H^+\ {}^{\cdot}H)\} \qquad (3.15)$$

In the same way, the σ^* orbital is the wavefunction for the excited state, Ψ_1:

$$\psi_1 = \psi^{MO}(\sigma^*) = |\phi_{\sigma*}| = \frac{1}{\sqrt{2 - 2S_{12}}} |(\chi_1 - \chi_2)|$$

$$= \frac{1}{\sqrt{2 - 2S_{12}}} \{\psi^{VB}(H^{\cdot}\ H^+) - \psi^{VB}(H^+\ {}^{\cdot}H)\} \qquad (3.16)$$

The radical anion, H_2^-, has a doubly occupied σ orbital and a singly occupied σ^* orbital in the MO description:

$$\psi_0 = \psi^{MO}(\sigma^2\sigma^*) = |\phi_\sigma \bar{\phi}_\sigma \phi_{\sigma*}|$$

$$= \frac{1}{(2 + 2S_{12})\sqrt{2 - 2S_{12}}} |(\chi_1 + \chi_2)(\bar{\chi}_1 + \bar{\chi}_2)(\chi_1 - \chi_2)|$$

$$= \frac{1}{(1 + S_{12})\sqrt{2 - 2S_{12}}} \{|\chi_1\chi_2\bar{\chi}_2| - |\chi_1\bar{\chi}_1\chi_2|\}$$

$$= \frac{1}{\sqrt{2 + 2S_{12}}} \{\psi^{VB}(H^{\cdot}\ H\!:^-) - \psi^{VB}(H\!:^-\ {}^{\cdot}H)\} \qquad (3.17)$$

This MO wavefunction and its VB components are depicted in Figure 3.6. The normalized VB configurations for the radical anion correspond to a hydride ion

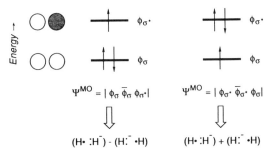

Figure 3.6 The $\sigma^2\sigma^{*1}$ and $\sigma^1\sigma^{*2}$ configurations of H_2^- and their VB constituents.

and a hydrogen atom:

$$\psi^{VB}(\text{H}^{\cdot}\ \text{H:}^-) = \frac{1}{\sqrt{1 - S_{12}^2}}\ |\chi_1\chi_2\bar{\chi}_2|$$

$$\psi^{VB}(\text{H:}^-\ {}^{\cdot}\text{H}) = \frac{1}{\sqrt{1 - S_{12}^2}}\ |\chi_1\bar{\chi}_1\chi_2| \tag{3.18}$$

The excited state of the radical anion contains a doubly occupied σ^* orbital and a singly occupied σ orbital.

$$\psi_1 = \psi^{MO}(\sigma\sigma^{*2}) = |\phi_{\sigma^*}\bar{\phi}_{\sigma^*}\phi_{\sigma}|$$

$$= \frac{1}{(2 - 2S_{12})\sqrt{2 + 2S_{12}}}\ |(\chi_1 - \chi_2)(\bar{\chi}_1 - \bar{\chi}_2)(\chi_1 + \chi_2)|$$

$$= \frac{1}{(1 - S_{12})\sqrt{2 + 2S_{12}}}\ \{|\chi_1\chi_2\bar{\chi}_2| + |\chi_1\bar{\chi}_1\chi_2|\}$$

$$= \frac{1}{\sqrt{2 - 2S_{12}}}\ \{\psi^{VB}(\text{H}^{\cdot}\ \text{H:}^-) + \psi^{VB}(\text{H:}^-\ {}^{\cdot}\text{H})\} \tag{3.19}$$

The relationship between the MO and VB descriptions is shown in Figure 3.6. It is seen that the radical anion and its excited state correspond to the negative and positive linear combinations of the same VB structures. The bonding combination is the negative combination. This will be the case for VB functions whenever the matrix element that connects the VB configurations[6] possesses a negative sign, as is the case for the configurations of equation 3.18.

3.1.2 A–X, A Model Heteronuclear Sigma Bond

A more typical example of a σ bond would involve atoms of different electronegativity, as in the model system A–X. If χ_1 is a hybrid orbital on A, the less electronegative atom, and χ_2 is a hybrid orbital on X, the more electronegative atom, then the σ orbital can be written as

$$\phi_\sigma = a\chi_1 + b\chi_2; \qquad b > a > 0, \qquad a^2 + b^2 + 2abS_{12} = 1 \qquad (3.20)$$

The MO coefficient b is larger than a, leading to greater electron density on the more electronegative center. In the MO description, the ground state has a doubly occupied σ orbital, and the σ bond is a mixture of the HL and zwitterionic configurations. However, since $b > a$, the contribution of A^+X^-, that is, $\Psi^{VB}(Z_2)$, is larger than that of A^-X^+, that is, $\Psi^{VB}(Z_1)$.

$$\psi^{MO}(\sigma^2) = |\phi_\sigma \overline{\phi_\sigma}| = |(a\chi_1 + b\chi_2)(a\overline{\chi}_1 + b\overline{\chi}_2)|$$

$$= ab\{|\chi_1\overline{\chi}_2| - |\overline{\chi}_1\chi_2|\} + a^2|\chi_1\overline{\chi}_1| + b^2|\chi_2\overline{\chi}_2|$$

$$= ab\sqrt{2 + 2S_{12}^2}\ \psi^{VB}(HL) + a^2\psi^{VB}(Z_1) + b^2\psi^{VB}(Z_2) \qquad (3.21)$$

The antibonding orbital σ^* can be written as

$$\phi_{\sigma^*} = b^*\chi_1 - a^*\chi_2; \qquad b^* > a^* > 0 \qquad (3.22)$$

In this case, there are three conditions on the values of a, b, a^*, and b^*:

$$a^2 + b^2 + 2abS_{12} = 1$$

$$a^{*2} + b^{*2} - 2a^*b^*S_{12} = 1$$

$$ab^* - a^*b + (bb^* - aa^*)S_{12} = 0 \qquad (3.23)$$

The first two are the normalization conditions for σ and σ^*, and the third is the orthogonality condition $\langle \sigma | \sigma^* \rangle = 0$. These relations can be used to evaluate a^* and b^* in terms of a and b.

$$a^* = \frac{a + bS_{12}}{\sqrt{1 - S_{12}^2}}, \qquad b^* = \frac{b + aS_{12}}{\sqrt{1 - S_{12}^2}} \qquad (3.24)$$

We observe that $b^* > a^*$, so that the electron density is larger at the less electronegative center.[7]

The doubly excited MO determinant can be expressed as:

$$\psi^{MO}(\sigma^{*2}) = |\phi_{\sigma^*}\overline{\phi_{\sigma^*}}| = |(b^*\chi_1 - a^*\chi_2)(b^*\overline{\chi}_1 - a^*\overline{\chi}_2)|$$

$$= b^{*2}|\chi_1\overline{\chi}_1| + a^{*2}|\chi_2\overline{\chi}_2| - a^*b^*\{|\chi_1\overline{\chi}_2| - |\overline{\chi}_1\chi_2|\}$$

$$= -a^*b^*\sqrt{2 + 2S_{12}^2}\ \psi^{VB}(HL) + b^{*2}\psi^{VB}(Z_1) + a^{*2}\psi^{VB}(Z_2) \qquad (3.25)$$

The singly excited determinant is dominated by the zwitterionic VB configurations, with only a small contribution from the HL configuration:

$$\begin{aligned}
{}^1\psi^{MO}(\sigma\sigma^*) &= \frac{1}{\sqrt{2}}\{|\phi_\sigma\bar{\phi}_{\sigma*}| - |\bar{\phi}_\sigma\phi_{\sigma*}|\} \\
&= \frac{1}{\sqrt{2}}\{|(a\chi_1 + b\chi_2)(b^*\bar{\chi}_1 - a^*\bar{\chi}_2)| \\
&\quad - |(a\bar{\chi}_1 + b\bar{\chi}_2)(b^*\chi_1 - a^*\chi_2)|\} \\
&= \sqrt{2}ab^*|\chi_1\bar{\chi}_1| - \sqrt{2}a^*b|\chi_2\bar{\chi}_2| \\
&\quad + \frac{1}{\sqrt{2}}(bb^* - aa^*)\{|\chi_1\bar{\chi}_2| - |\bar{\chi}_1\chi_2|\} \\
&= \sqrt{2}ab^*\psi^{VB}(Z_1) - \sqrt{2}a^*b\psi^{VB}(Z_2) \\
&\quad + \sqrt{1 + S_{12}^2}(bb^* - aa^*)\psi^{VB}(HL)
\end{aligned} \tag{3.26}$$

The CI wavefunction for the ground state can now be written as

$$\begin{aligned}
\Psi &= C(\sigma^2)\psi^{MO}(\sigma^2) + C(\sigma\sigma^*)\psi^{MO}(\sigma\sigma^*) + C(\sigma^{*2})\psi^{MO}(\sigma^{*2}) \\
&= D(HL)\psi^{VB}(HL) + D(Z_1)\psi^{VB}(Z_1) + D(Z_2)\psi^{VB}(Z_2)
\end{aligned} \tag{3.27}$$

and the relationship between the MO–CI and VB wavefunctions is depicted in Figure 3.7.

The analogy to the hydrogen molecule (Figure 3.5) is clear, but we must recognize that in the general case zwitterionic configurations are not equal in energy, and we therefore do not generate linear combinations from them, as in Figure 3.5. Instead, these configurations are mixed separately into the HL configuration.

For the radical cation of A–X, the ground-state wavefunction Ψ_0 is the σ orbital, and the excited state wavefunction Ψ_1 is the σ^* orbital:

$$\begin{aligned}
\psi_0 = \psi^{MO} = |\phi_\sigma| &= |(a\chi_1 + b\chi_2)| \\
&= a\psi^{VB}(A^{\cdot}\,X^+) + b\psi^{VB}(A^+\,{}^{\cdot}X)
\end{aligned} \tag{3.28}$$

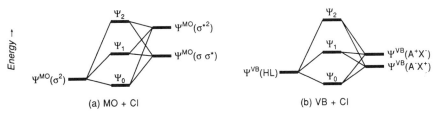

Figure 3.7 Construction of the ground and excited states of A–X, (a) starting from the MO configurations based on the σ and σ^* molecular orbitals; (b) starting from VB configurations.

$$\psi_1 = \psi^{MO} = |\phi_\sigma| = |(b^*\chi_1 - a^*\chi_2)|$$
$$= b^*\psi^{VB}(A^\cdot \, X^+) - a^*\psi^{VB}(A^+ \, {}^\cdot X) \qquad (3.29)$$

In the ground state of the radical anion, the σ orbital is doubly occupied and σ^* is singly occupied:

$$\psi_0 = \psi^{MO}(\sigma^2\sigma^*) = |\phi_\sigma \overline{\phi}_\sigma \phi_{\sigma^*}|$$
$$= |(a\chi_1 + b\chi_2)(a\overline{\chi}_1 + b\overline{\chi}_2)(b^*\chi_1 - a^*\chi_2)|$$
$$= (b^2b^* + aa^*b)|\chi_1\chi_2\overline{\chi}_2| - (a^2a^* + abb^*)|\chi_1\overline{\chi}_1\chi_2|\}$$
$$= (b^2b^* + aa^*b)\sqrt{1 - S_{12}^2}\; \psi^{VB}(A^\cdot \, X{:}^-)$$
$$- (a^2a^* + abb^*)\sqrt{1 - S_{12}^2}\; \psi^{VB}(A{:}^- \, {}^\cdot X) \qquad (3.30)$$

Since $b, b^* > a, a^*$ the $(A^\cdot \, X{:}^-)$ configuration is weighted more heavily than $(A{:}^- \, {}^\cdot X)$.

In the excited state of the radical anion, the σ^* orbital is doubly occupied, σ is singly occupied, and $(A{:}^- \, {}^\cdot X)$ has the larger coefficient:

$$\psi_1 = \psi^{MO}(\sigma\sigma^{*2}) = |\phi_{\sigma^*}\overline{\phi}_{\sigma^*}\phi_\sigma|$$
$$= |(b^*\chi_1 - a^*\chi_2)(b^*\overline{\chi}_1 - a^*\overline{\chi}_2)(a\chi_1 + b\chi_2)|$$
$$= (aa^{*2} + a^*bb^*)|\chi_1\chi_2\overline{\chi}_2| + (bb^{*2} + aa^*b^*)|\chi_1\overline{\chi}_1\chi_2|\}$$
$$= (aa^{*2} + a^*bb^*)\sqrt{1 - S_{12}^2}\; \psi^{VB}(A^\cdot \, X{:}^-)$$
$$+ (bb^{*2} + aa^*b^*)\sqrt{1 - S_{12}^2}\; \psi^{VB}(A{:}^- \, {}^\cdot X) \qquad (3.31)$$

Figure 3.8 shows the relationship between the MO and VB wavefunctions.

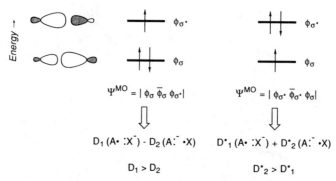

Figure 3.8 The $\sigma^2\sigma^{*1}$ and $\sigma^1\sigma^{*2}$ configurations and their VB constituents for the radical anion of A–X. The D's are coefficients.

3.1.3 Relationships Between General MO and VB Wavefunctions[3,6]

For a system that contains more than two or three electrons, or more than two or three basis functions, it is a tedious process to expand a MO wavefunction in terms of VB configurations by explicit multiplication of all of the terms. However, there is a direct procedure to obtain the coefficient of any VB configuration from the MO coefficients. In the MO approach, the most general form of a wavefunction is the CI wavefunction, which, as we have seen, can be written as a linear combination of determinants of molecular orbitals:

$$\Psi = \Sigma_i C_i \psi_i^{MO} \tag{3.32}$$

The determinants, ψ^{MO}, are antisymmetrized products of spin molecular orbitals,

$$\psi^{MO} = |\phi_1\phi_2 \cdots \phi_{n\alpha}\overline{\phi}_1\overline{\phi}_2 \cdots \overline{\phi}_{n\beta}| = n!^{-1/2}\Sigma_i(-1)^{p_i}P_i(\phi_1(1)\phi_2(2) \cdots) \tag{3.33}$$

where n_α and n_β are the number of alpha (spin up) and beta (spin down) electrons, respectively, and $n = n_\alpha + n_\beta$. The ψ^{MO} are normalized and orthogonal, and the P_i are permutation operators that act on the electrons. These P_i have a sign $(-1)^{p_i}$ that depends on the number of twofold permutations that are performed on $\phi_1(1)\phi_2(2) \cdots \phi_n(n)$ to obtain the permuted product number i in equation 3.33. The ψ_i^{MO} are all normalized and mutually orthogonal.

As in the MO approach, the most general form of the VB wavefunction is a linear combination of configurations,

$$\Psi = \Sigma_i D_i \psi_i^{VB} \tag{3.34}$$

each of which is a normalized, antisymmetrical, product (i.e., determinant) of basis functions.

$$\psi^{VB} = N|\cdots \chi_i \cdots \chi_j \cdots \chi_k \cdots| \tag{3.35}$$

It is often necessary to take linear combinations of antisymmetrized products, to ensure that the VB configuration has the correct spin multiplicity.[8] The normalization coefficients can be written as determinants of overlaps between basis functions.

$$N = |\cdots S_{ij} \cdots|^{-1/2} \tag{3.36}$$

Although the ψ_i^{VB} are normalized, the overlap between different configurations is not zero.

Consider a MO wavefunction with three occupied spin orbitals, each a linear combination of five basis functions.

$$\phi_1 = c_{\mu 1}\chi_\mu + c_{\nu 1}\chi_\nu + c_{\kappa 1}\chi_\kappa + c_{\sigma 1}\chi_\sigma + c_{\tau 1}\chi_\tau$$

$$\phi_2 = c_{\mu 2}\chi_\mu + c_{\nu 2}\chi_\nu + c_{\kappa 2}\chi_\kappa + c_{\sigma 2}\chi_\sigma + c_{\tau 2}\chi_\tau$$

$$\phi_3 = c_{\mu 3}\chi_\mu + c_{\nu 3}\chi_\nu + c_{\kappa 3}\chi_\kappa + c_{\sigma 3}\chi_\sigma + c_{\tau 3}\chi_\tau \tag{3.37}$$

If we had a particular interest in the VB configuration $|\chi_\mu \chi_\kappa \chi_\sigma|$, we could take the coefficient of χ_μ in ϕ_1 *times* the coefficient of χ_κ in ϕ_2 *times* the coefficient of χ_σ in ϕ_3. An additional contribution to $|\chi_\mu \chi_\kappa \chi_\sigma|$ comes from the product of χ_μ in ϕ_2, χ_κ in ϕ_1, and χ_σ in ϕ_3. Since the configurations are determinants, and the sign must be changed whenever any two columns are interchanged (e.g., $|\chi_\kappa \chi_\mu \chi_\sigma| = -|\chi_\mu \chi_\kappa \chi_\sigma|$), the additional contribution is equal to *minus* the coefficient of χ_κ in ϕ_1 *times* the coefficient of χ_μ in ϕ_2 *times* the coefficient of χ_σ in ϕ_3. All other permutations must also be included, with the appropriate sign. This corresponds to a determinant of coefficients.[3a,b]

$$\tilde{D}_{\mu\kappa\sigma} = \begin{vmatrix} c_{\mu 1} & c_{\kappa 1} & c_{\sigma 1} \\ c_{\mu 2} & c_{\kappa 2} & c_{\sigma 2} \\ c_{\mu 3} & c_{\kappa 3} & c_{\sigma 3} \end{vmatrix}$$

$$= c_{\mu 1}c_{\kappa 2}c_{\sigma 3} - c_{\mu 1}c_{\kappa 3}c_{\sigma 2} - c_{\mu 2}c_{\kappa 1}c_{\sigma 3}$$

$$+ c_{\mu 2}c_{\kappa 3}c_{\sigma 1} + c_{\mu 3}c_{\kappa 1}c_{\sigma 2} - c_{\mu 3}c_{\kappa 2}c_{\sigma 1} \tag{3.38}$$

Using equation 3.36, the normalization coefficient for $|\chi_\mu \chi_\kappa \chi_\sigma|$ is

$$N_{\mu\kappa\sigma} = \begin{vmatrix} 1 & S_{\kappa\mu} & S_{\sigma\mu} \\ S_{\kappa\mu} & 1 & S_{\sigma\kappa} \\ S_{\sigma\mu} & S_{\sigma\kappa} & 1 \end{vmatrix}^{-1/2} \tag{3.39}$$

and the coefficient for the normalized VB configuration is

$$D_{\mu\kappa\sigma} = \tilde{D}_{\mu\kappa\sigma}/N_{\mu\kappa\sigma} \tag{3.40}$$

Since the alpha and beta spin orbitals are orthogonal, this determinant and the normalization coefficient can be factored into its alpha and beta components.[3a,c]

$$D = (\tilde{D}^\alpha/N^\alpha)(\tilde{D}^\beta/N^\beta) \tag{3.41}$$

This method can be generalized to any number of electrons and basis functions, and is equally applicable to VB configurations based on atomic orbitals or fragment orbitals, as is illustrated in the following section.

3.1.4 A Three-Electron Problem: Hydrogen Molecule Plus Hydrogen Atom

The simplest example of a bimolecular reaction is provided by the interaction of hydrogen atom with hydrogen molecule. The molecular orbitals for this system are shown qualitatively in Figure 3.9. An ROHF/STO-3G calculation on the linear transition state for this reaction yields

$$\phi_1 = 0.35120\ \chi_1 + 0.53601\ \chi_2 + 0.35120\ \chi_3$$

$$\phi_2 = 0.77580\ \chi_1 + 0.0 \qquad \chi_2 - 0.77580\ \chi_3$$

$$\phi_3 = -0.90133\ \chi_1 + 1.37874\ \chi_2 - 0.90133\ \chi_3 \tag{3.42}$$

where the χ's are 1s functions on atoms 1, 2, and 3, and the H–H distance is 0.89 Å. The spin-restricted Hartree–Fock ground state ψ_0 has two electrons in ϕ_1 and one electron in ϕ_2. Eight excited configurations, ψ_1-ψ_8, can be generated by promotion of one or more electrons to higher-lying orbitals.

$$\psi_0^{MO} = |\phi_1\phi_2\bar{\phi}_1|, \qquad \psi_1^{MO} = |\phi_1\phi_2\bar{\phi}_2|, \qquad \psi_2^{MO} = |\phi_1\phi_2\bar{\phi}_3|$$

$$\psi_3^{MO} = |\phi_1\phi_3\bar{\phi}_1|, \qquad \psi_4^{MO} = |\phi_1\phi_3\bar{\phi}_2|, \qquad \psi_5^{MO} = |\phi_1\phi_3\bar{\phi}_3|$$

$$\psi_6^{MO} = |\phi_2\phi_3\bar{\phi}_1|, \qquad \psi_7^{MO} = |\phi_2\phi_3\bar{\phi}_2|, \qquad \psi_8^{MO} = |\phi_2\phi_3\bar{\phi}_3| \tag{3.43}$$

In the VB description, three electrons must be distributed among three basis functions, and this can be done in a total of nine ways:

$$\psi_1^{VB} = \frac{1}{\sqrt{1 - S_{12}^2}}\ |\chi_1\chi_2\bar{\chi}_3|$$

$$\psi_2^{VB} = \frac{1}{\sqrt{1 - S_{13}^2}}\ |\chi_1\bar{\chi}_2\chi_3|$$

$$\psi_3^{VB} = \frac{1}{\sqrt{1 - S_{23}^2}}\ |\bar{\chi}_1\chi_2\chi_3|$$

$$\psi_4^{VB} = \frac{1}{\sqrt{1 - S_{12}^2}}\ |\chi_1\chi_2\bar{\chi}_2|$$

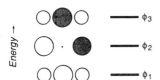

ϕ_3

ϕ_2

ϕ_1

Figure 3.9 Qualitative representation of the molecular orbitals of H_3.

$$\psi_5^{VB} = \frac{1}{\sqrt{1 - S_{13}^2}} |\chi_1\chi_3\bar{\chi}_3|$$

$$\psi_6^{VB} = \frac{1}{\sqrt{1 - S_{13}^2}} |\chi_1\bar{\chi}_1\chi_3|$$

$$\psi_7^{VB} = \frac{1}{\sqrt{1 - S_{23}^2}} |\chi_2\bar{\chi}_2\chi_3|$$

$$\psi_8^{VB} = \frac{1}{\sqrt{1 - S_{23}^2}} |\chi_2\chi_3\bar{\chi}_3|$$

$$\psi_9^{VB} = \frac{1}{\sqrt{1 - S_{12}^2}} |\chi_1\bar{\chi}_1\chi_2| \tag{3.44}$$

The first three of these configurations resemble the Heitler–London configurations of the hydrogen molecule (equation 3.4). The negative combination of VB configurations ψ_1 and ψ_2 will be given by χ_1 *times* ψ(HL) for atoms 2 and 3; the negative combination of ψ_2 and ψ_3 will be given by χ_3 *times* ψ(HL) for atoms 1 and 2. Configurations ψ_4, ψ_5 and ψ_6, ψ_7 are related to the zwitterionic configurations of the hydrogen molecule (equations 3.5 and 3.6). The configurations ψ_4 and ψ_7–ψ_9 are charge transfer configurations, and correspond to structures of the type $[H^+ (H_2^- H)^-]$.

It follows that the VB configurations for the H_3 transition state could be constructed from the VB configurations for H and H_2 (equations 3.4, 3.5, and 3.12), together with various configurations arising from charge transfer (equation 3.17). Since the final wavefunction is independent of the orbitals selected to construct the configurations, these could be written in terms of atomic basis functions (χ_1, χ_2, χ_3), molecular orbitals (ϕ_1, ϕ_2, ϕ_3), or fragment orbitals (σ_{12}, σ_{12}^*, χ_3 or σ_{23}, σ_{23}^*, χ_1).

Figure 3.10 shows the VB type configurations that can be generated by distribution of electrons among these fragment orbitals (FO's) for the H_3 transition state. We could equally well have chosen our fragment orbitals to be χ_1 and σ_{23}, σ_{23}^*, or χ_2, σ_{13}, σ_{13}^*. The specific choice of FO's, or the choice between VB with FO's or AO's, will depend upon the specific features to be illustrated.[3b,6]

The ground-state molecular orbital determinant, or any of the excited state determinants, can be expanded in terms of VB configurations by, for example, insertion of equation 3.42 into equation 3.43, multiplying out all of the factors, and collecting the contributions from all of the equivalent terms. Alternatively, the general scheme outlined in Section 3.1.3 could be used to expand the MO wavefunction.[3a–c] The following result is obtained for the Hartree–Fock ground state

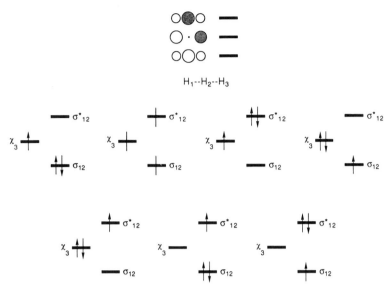

Figure 3.10 Molecular orbitals and fragment VB configurations (H_3 + H_1H_2) for the H_1---H_2---H_3 transition state. There are two alternative representations depending on the definitions of the fragments (H_1 + H_2H_3; H_2 + H_1H_3). For the second configuration, from left to right there are two VB configurations with three electrons in three orbitals and a total doublet spin.

determinant:

$$
\begin{aligned}
\psi_0^{MO} &= \quad 0.1207\ \psi_1^{VB} - 0.2879\ \psi_2^{VB} + 0.1207\ \psi_3^{VB} \\
&\quad + 0.1840\ \psi_4^{VB} + 0.1886\ \psi_5^{VB} - 0.1886\ \psi_6^{VB} \\
&\quad - 0.1841\ \psi_7^{VB} + 0.1207\ \psi_8^{VB} - 0.1207\ \psi_9^{VB} \\
&= \quad 0.1441(\psi_1^{VB} - \psi_2^{VB}) - 0.1440(\psi_2^{VB} - \psi_3^{VB}) - 0.0233(\psi_1^{VB} + \psi_3^{VB}) \\
&\quad + 0.1841\ \psi_4^{VB} + 0.1886\ \psi_5^{VB} - 0.1886\ \psi_6^{VB} \\
&\quad - 0.1841\ \psi_7^{VB} + 0.1207\ \psi_8^{VB} - 0.1207\ \psi_9^{VB}
\end{aligned}
\tag{3.45}
$$

The best many-electron wavefunction that can be constructed with a given basis set is, as we have seen, a linear combination of the ground state and *all* excited state determinants. The same wavefunction is obtained by taking the best linear combination of the atomic orbital-based VB configurations. With the STO-3G basis set and $R(HH) = 0.89$ Å, the final result is

$$
\begin{aligned}
\Psi &= \quad 0.9862\ \psi_0^{MO} + 0.0\ \psi_1^{MO} + 0.06111\ \psi_2^{MO} \\
&\quad + 0.0\ \psi_3^{MO} + 0.1208\ \psi_4^{MO} + 0.0\ \psi_5^{MO} \\
&\quad + 0.0597\ \psi_6^{MO} + 0.0\ \psi_7^{MO} + 0.0746\ \psi_8^{MO}
\end{aligned}
$$

$$= \quad 0.2159 \ \psi_1^{VB} - 0.5105 \ \psi_2^{VB} + 0.2159 \ \psi_3^{VB}$$

$$+ \ 0.0914 \ \psi_4^{VB} + 0.0926 \ \psi_5^{VB} - 0.0926 \ \psi_6^{VB}$$

$$- \ 0.0914 \ \psi_7^{VB} + 0.0661 \ \psi_8^{VB} - 0.0661 \ \psi_9^{VB}$$

$$= \quad 0.2553(\psi_1^{VB} - \psi_2^{VB}) - 0.2553(\psi_2^{VB} - \psi_3^{VB}) - 0.0394(\psi_1^{VB} + \psi_3^{VB})$$

$$+ \ 0.0914 \ \psi_4^{VB} + 0.0926 \ \psi_5^{VB} - 0.0926 \ \psi_6^{VB}$$

$$- \ 0.0914 \ \psi_7^{VB} + 0.0661 \ \psi_8^{VB} - 0.0661 \ \psi_9^{VB} \qquad (3.46)$$

3.2 CONSTRUCTION OF POTENTIAL ENERGY CURVES FOR CHEMICAL REACTIONS

Having demonstrated that a delocalized wavefunction can be expressed as a linear combination of VB configurations, we can begin the construction of potential energy curves from VB-type building blocks, using as a prototype the three-electron process of the preceding section.

3.2.1 The Hydrogen Atom Exchange, H + H–H → H–H + H

The process $H^{\bullet}$ + H–H → H–H + $H^{\bullet}$ belongs to a class of reactions, including radical addition to an unsaturated system and atom abstraction, in which a net electron spin is shifted from one atom to another. Such reactions may be described as *spin-transfer* transformations.[6,9]

The Covalent HL Curves. Along the linear reaction coordinate, the principal configurations that describe the H_3 complex are the HL forms **3.1** and **3.2**. For conciseness of presentation, the HL bond is drawn as two dots connected by a line,

$$H_C{\bullet} \quad H_A{\bullet}\!-\!{\bullet}H_B \qquad\qquad H_C{\bullet}\!-\!{\bullet}H_A \quad {\bullet}H_B$$

$$\textbf{3.1} \qquad\qquad\qquad\qquad \textbf{3.2}$$

where the line denotes the singlet pairing of the two electrons (this pairing is specified by arrows in Figure 3.1). As seen in Figure 3.11a, the two HL configurations interchange in energy along a reaction coordinate defined as R (reactants) → P (products). At the R extremum, A–B is short, A–C is long, and $\psi(\textbf{3.1})$ is the lowest-energy configuration.

This configuration increases in energy in the $R \to P$ direction, because the A–B bond is broken, and replaced by C $\cdots$ A repulsion. The existence of such

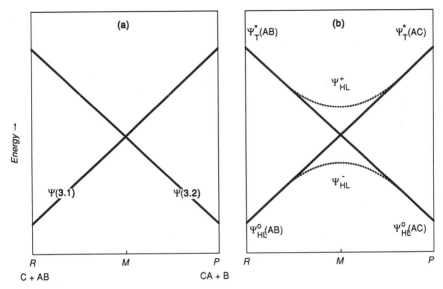

Figure 3.11 (a) Variation of the energies of **3.1** and **3.2** along an $R \to P$ reaction coordinate. (b) Consequences of the avoided crossing of the VB configurations **3.1** and **3.2**.

repulsion follows from the form of the HL wavefunction, equation 3.47. The first

$$\psi^{HL}(A-B) = N\{|\chi_C \chi_A \bar{\chi}_B| - |\chi_C \bar{\chi}_A \chi_B|\}$$

$$N'|\chi_C \bar{\chi}_A \chi_B| = \tfrac{1}{2}\{\psi_{HL}(C-A) + \psi_T(C-A)\}$$

$$\psi_{HL}(C-A) = N\{|\chi_C \bar{\chi}_A \chi_B| - |\bar{\chi}_C \chi_A \chi_B|\}$$

$$\psi_T(C-A) = N\{|\chi_C \bar{\chi}_A \chi_B| + |\bar{\chi}_C \chi_A \chi_B|\} \tag{3.47}$$

term (determinant) describes a triplet interaction between A and C; in the second term, the interaction between A and C is approximately one-half singlet and one-half triplet. The two terms together lead to ca. 75% of triplet repulsion between A and C.

It follows that, at the P extremum of the reaction coordinate, $\psi(\mathbf{3.1})$ becomes an excited configuration of $\psi(\mathbf{3.2})$, the HL wavefunction of the A–C bond. Moreover, this excited configuration is close in character to the triplet state of the A–C bond plus an uncoupled radical on H_B. An analogous but symmetric argument applies to the $\psi(\mathbf{3.2})$ curve in the direction $P \to R$.

The two curves cross at the geometry at which A is equidistant from B and C. At this point (M of Figure 3.11a), the two HL configurations $\psi(\mathbf{3.1})$ and $\psi(\mathbf{3.2})$ are degenerate, and will avoid the crossing by *configuration mixing*,[6,10] to yield $\psi_{HL}{}^+$ and $\psi_{HL}{}^-$. The latter has the lower energy, because the matrix element

$\langle \Psi(3.1)|\hat{\mathbf{H}}|\Psi(3.2)\rangle$ has a negative sign; in such a case the bonding combination is Ψ_{HL}^{-}.[6]

The consequences of the avoided crossing are seen in Figure 3.11b, and we observe that this comprises a reorganization mechanism,[3b,6] which enables the reaction complex to change its electronic character from reactant-like to product-like via a transition state that is a resonance hybrid of the two. The avoided crossing therefore reflects the interchange of bonds that occurs during the chemical reaction, and its relationship to the BEP treatment is apparent.

Secondary Configuration Mixing and the State Correlation Diagram.
Figure 3.11b is a state correlation diagram derived from Figure 3.11a by allowing mixing between the configurations. At the R extremum, the anchor points are $\Psi_{HL}^{0}(AB)$, the HL configuration at the equilibrium bond length of the A–B bond, coupled to Ψ_C to give an overall doublet, and $\Psi_T^{*}(AB)$, the vertical triplet excited state of this bond at the same bond length. At the P extremum, the anchor points are $\Psi_{HL}^{0}(CA)$, coupled to Ψ_B, and $\Psi_T^{*}(CA)$.

The upper anchor points at the two extrema will comprise 75% triplet of the short bond, if we insist on site specificity of the spin coupling, or a full triplet if site specificity is not imposed.[9g] For example, if we require that $\Psi_T^{*}(AB)$ involves singlet spin coupling across the C $\cdots$ A linkage, $\Psi_T^{*}(AB)$ is given by equation 3.48, and is 75% triplet across A $\cdots$ B. On the other hand, if we remove the

$$\psi_T^{*}(AB) = N\{|ab\bar{c}| - |\bar{a}bc|\} \qquad a = \chi_A, \text{ etc.} \tag{3.48}$$

requirement for site specificity, arguing that the C $\cdots$ AB distance is infinite at the R extremum, $\Psi_T^{*}(AB)$ will be given by equation 3.49.

$$\psi_T^{*}(AB) = N\{2|ab\bar{c}| - |a\bar{b}c| - |\bar{a}bc|\} \tag{3.49}$$

This describes a triplet A $\cdots$ B spin coupled to a radical center C, half through site A and half through site B.

The process is completed by mixing of secondary configurations into the covalent curves.[6] These secondary configurations are zwitterionic structures,

$$Z_1 = H_C^{\cdot} \; H_A^{+} \; :H_B^{-} \tag{3.50a}$$

$$Z_2 = H_C^{\cdot} \; H_A:^{-} \; H_B^{+} \tag{3.50b}$$

$$Z_3 = H_C:^{-} \; H_A^{+} \; {}^{\cdot}H_B \tag{3.50c}$$

$$Z_4 = H_C^{+} \; H_A:^{-} \; {}^{\cdot}H_B \tag{3.50d}$$

$$Z_5 = H_C:^{-} \; H_A^{\cdot} \; H_B^{+} \tag{3.50e}$$

$$Z_6 = H_C^{+} \; H_A^{\cdot} \; :H_B^{-} \tag{3.50f}$$

and their behavior along the reaction coordinate is depicted in Figure 3.12a.

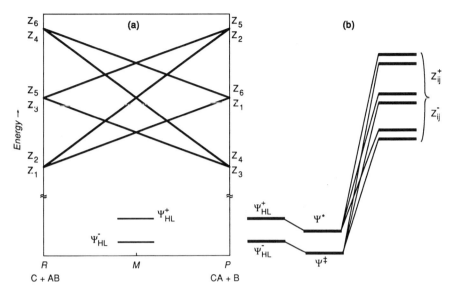

Figure 3.12 (a) Behavior of the Z configurations for the $H\cdot + H_2$ reaction along an $R \rightarrow P$ reaction coordinate. (b) The VB mixing diagram and the resulting twin states at the transition state. (Adapted, by permission, from reference 6, Figure 22. Copyright © 1989 by Kluwer Publications).

Consider the crossing of Z_1 and Z_3: at the R extremum, Z_1 is stabilized relative to Z_3, because of the electrostatic attraction between A and B when A–B is short. These stabilities are reversed at the P extremum. Analogous arguments apply to the pairs Z_2, Z_4 and Z_5, Z_6, so that the Z configurations cross, in pairs, above the crossing point of the HL curves at the midpoint of the reaction coordinate.

In each case there is an avoided crossing that results from the mixing of pairs of structures via electron shifts and, by taking positive and negative linear combinations of the Z_{ij} pairs, we can understand their mixing patterns with Ψ_{HL^+} and Ψ_{HL^-}.[6] The VB mixing diagram that results is seen in Figure 3.12b. The lower HL combination mixes with the negative combination of each Z_{ij} pair, and the higher HL combination mixes with the positive combination of this Z_{ij} pair to yield $\Psi^\ddagger$ and Ψ^*. These are, respectively, the transition state and its vertical excited state, which may, therefore, be expressed as equations 3.51a and 3.51b:

$$\Psi^\ddagger \approx \{(\text{H}^{\cdot}\ \text{H–H}) - (\text{H–H}\ ^{\cdot}\text{H})\} + \lambda\{(\text{H}:^{-}\ \text{H}^{\cdot}\ \text{H}^{+}) + (\text{H}^{+}\ \text{H}^{\cdot}\ \text{H}:^{-})\} \quad (3.51a)$$

$$\Psi^* \approx \{(\text{H}^{\cdot}\ \text{H–H}) + (\text{H–H}\ ^{\cdot}\text{H})\} + \lambda(\text{H}:^{-}\ \text{H}^{\cdot}\ \text{H}^{+}) - (\text{H}^{+}\ \text{H}^{\cdot}\ \text{H}:^{-})\} \quad (3.51b)$$

By convention, the VB structure H–H in equations 3.51a,b represents a Lewis structure which is a mixture of the HL configuration and its zwitterionic counterparts $\text{H}^{+}:\text{H}^{-}$ and $\text{H}:^{-}\text{H}^{+}$. With this interpretation, the transition state $\Psi^\ddagger$ *is seen to contain contributions from configurations that are absent from the Lewis structures of the reactant and the product.*

The final step in the construction of the state correlation diagram involves the mixing of the Z configurations with the HL configurations at other points along the reaction coordinate. At the extrema, a positive combination of a pair of Z configurations mixes into the HL configurations (Ψ_{HL}^0 in Figure 3.11b): $Z_1 + Z_2$ at the R extremum, and $Z_3 + Z_4$ at the P extremum. There is little mixing between Z configurations and the upper anchor points of the extrema, because the latter are predominantly (ca. 75%) triplet states of the ground state bonds,[6] coupled to the isolated atom to give an overall doublet, as already discussed. At other points along the reaction coordinate, all of the Z configurations mix into each of the covalent curves to form state curves with mixed VB character (cf. equation 3.46).

The final result can be summarized by Figure 3.13. This is a *state correlation diagram* (SCD), and it is built up from two curves that are anchored at the ground and triplet excited states of the reactants and the products.[6,10] The two curves intersect along the reaction coordinate, and the barrier and transition state of the spin transfer reaction are a consequence of the avoided crossing of the curves. Clearly, the derivation of the SCD comprises an alternative derivation of Statement 1.1. However, as we shall see, the form of an SCD permits qualitative and quantitative predictions that are not possible in the BEP treatment.

3.2.2 The Hydrogen Ion Exchange Reaction, $H^- + H–H \rightarrow H–H + H^-$

This reaction is the prototype of cationic transfer between two centers, for example, R^+ transfer between X^- and Y^- in an S_N2 reaction, or H^+ transfer between X^- and Y^- in a proton transfer reaction.[3b,6,11,12]

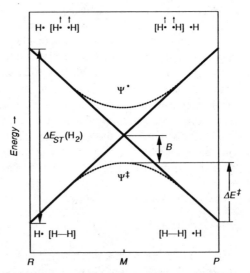

Figure 3.13 A state correlation diagram (SCD) for the process $H\cdot + H–H \rightarrow H–H + \cdot H$. The energy gap of the diagram is proportional to ΔE_{ST}, the singlet-to-triplet excitation of the short bond; B is the degree of avoided crossing, $\Delta E^{\ddagger}$ is the barrier, and $\Psi^{\ddagger}$ is the transition state. The excited states involve a triplet $H\cdot \cdot H$ spin coupled to $H\cdot$. (Adapted, by permission, from reference 6, Figure 23. Copyright © 1989 by Kluwer Publications).

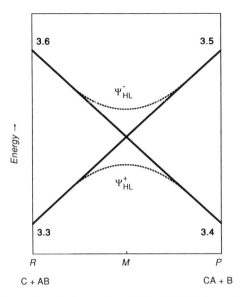

Figure 3.14 Crossing and avoided crossing (dashes) of the HL configurations **3.3–3.6** for the reaction $H_C:^- + H_A–H_B \rightarrow H_C–H_A + :H_B^-$. (Adapted, by permission, from reference 6, Figure 25. Copyright © 1989 by Kluwer Publications).

The Covalent HL Curves.   The HL structures that describe H_3^- are **3.3** and **3.4,** and their intersection along an $R \rightarrow P$ reaction coordinate is depicted in Figure 3.14, which also illustrates the avoided crossing that occurs in this case.

$$H_C:^- + (H_A\cdot —\cdot H_B) \qquad (H_C\cdot —\cdot H_A) + :H_B^-$$
$$\text{\textbf{3.3}} \qquad\qquad\qquad \text{\textbf{3.4}}$$

There is a change in the energy of Ψ (**3.3**) along the reaction coordinate, because the HL bond is broken and, in addition, at the P extremum **3.3** becomes **3.5,** which

$$(H_C:^- \overline{\;\cdot H_A\;}) + \overline{\;}\cdot H_B$$
$$\text{\textbf{3.5}}$$

contains a repulsive three-electron interaction across the $C \cdots A$ linkage, and the spins are paired along the infinitely long $A \cdots B$ linkage. In the same way, **3.4** becomes **3.6** in the reverse direction.

$$H_C\overline{\cdot\;} + (\overline{\;H_A\cdot} :H_B^-)$$
$$\text{\textbf{3.6}}$$

Secondary Configuration Mixing and the State Correlation Diagram.
The secondary configurations fall into two categories: the long-bond structure **3.7**

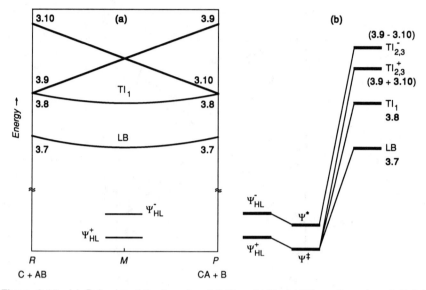

Figure 3.15 (a) Behavior of the long bond (LB) and tri-ionic (TI) configurations **3.7–3.10** along an $R \rightarrow p$ reaction coordinate. (b) The VB mixing diagram and the resulting twin states at the crossing point. (Adapted, by permission, from reference 6, Figure 26. Copyright © 1989 by Kluwer Publications).

(LB), and the tri-ionic structures **3.8–3.10** (TI_1–TI_3). Figure 3.15a shows

$$H_C{}^\cdot \overline{\quad H_A{:}^- \quad} {}^\cdot H_B \qquad H_C{:}^- \quad H_A{}^+ \quad {:}H_B{}^-$$

$$\text{(LB)} \qquad\qquad\qquad \text{(TI}_1\text{)}$$

$$\mathbf{3.7} \qquad\qquad\qquad\qquad \mathbf{3.8}$$

$$H_C{:}^- \quad {:}H_A{}^- \quad H_B{}^+ \qquad H_C{}^+ \quad H_A{:}^- \quad {:}H_B{}^-$$

$$\text{(TI}_2\text{)} \qquad\qquad\qquad \text{(TI}_3\text{)}$$

$$\mathbf{3.9} \qquad\qquad\qquad\qquad \mathbf{3.10}$$

how the energies of these configurations vary along the reaction coordinate.

In **3.7** and **3.8**, an interaction across $A \cdots B$ is replaced by the identical interaction across $C \cdots A$ as the reaction proceeds. Variation of the energies of these configurations along the reaction coordinate is, therefore, expected to be small. On the other hand, **3.9** must increase in energy because of the four-electron overlap repulsion that develops across $C \cdots A$; **3.10** exhibits analogous behavior in the opposite direction. At the midpoint of the reaction coordinate, there is a crossing of **3.9** and **3.10**, above the crossing point of the two HL configurations.

At the crossing point M, the mixing of the secondary configurations into the two HL combinations leads to the transition state and its vertical excited state, as depicted in Figure 3.15b. These states may be expressed as equations 3.52.

$$\Psi^{\ddagger} \approx \{(H{:}^{-}\ H{-}H) + (H{-}H\ {:}H^{-})\} - \lambda_1(LB) + \lambda_2(TI_1) + \lambda_3\{(TI_2) + (TI_3)\}$$

$$\text{(3.52a)}$$

$$\Psi^{*} \approx \{(H{:}^{-}\ H{-}H) - (H{-}H\ {:}H^{-})\} + \lambda\{(TI_2) - (TI_3)\} \qquad \text{(3.52b)}$$

Again we find that the transition state contains contributions from configurations that are absent from the Lewis structures of the reactants and the products.

The final SCD is shown in Figure 3.16, which consists of two curves, anchored at the ground and *vertical charge transfer* states of reactants and products.

3.2.3 Generalizations of State Correlation Diagrams and Their Interpretation

The SCD's of the $H + H_2$ and $H^- + H_2$ reactions can be represented by the general diagram of Figure 3.17.[3b,6,11b,12] Such a diagram is intended to convey the fact that there is no direct correlation between the ground state wavefunctions of reactants and products, because spin pairing is not the same in Ψ_R and Ψ_P. A transformation from Ψ_R to Ψ_P will require the intervention of two unique species, Ψ_{RP}^{*} and Ψ_{PR}^{*}, that are vertical excited states of Ψ_R and Ψ_P, respectively, but with the spin pairing of Ψ_P and Ψ_R. This similarity in spin pairing allows the correlations $\Psi_{RP}^{*} \to \Psi_P$ and $\Psi_{PR}^{*} \to \Psi_R$. It is convenient to think of these unique species as "*electronic images*" or "*image states.*"[3b,6,11b]

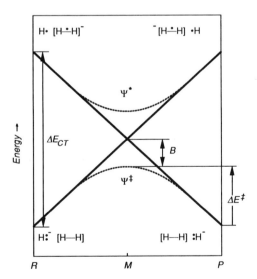

Figure 3.16 A state correlation diagram (SCD) for the process $H{:}^{-} + H{-}H \to H{-}H + {:}H^{-}$. The gap of the diagram is ΔE_{CT}, the vertical charge transfer energy; B is the degree of avoided crossing, $\Delta E^{\ddagger}$ is the barrier, and $\Psi^{\ddagger}$ is the transition state. The odd electrons in the charge transfer states are singlet paired. (Adapted, by permission, from reference 6, Figure 27. Copyright © 1989 by Kluwer Publications).

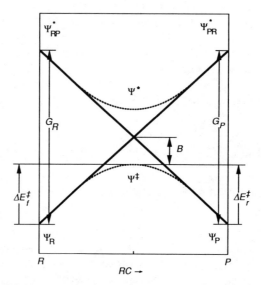

Figure 3.17 An SCD for a general chemical reaction. The G's are energy gaps at the two extrema of the reaction coordinate (RC). (Adapted, by permission, from reference 6, Figure 28. Copyright © 1989 by Kluwer Publications).

A most important feature of the treatment of any reaction in terms of these principles will be the specification of the identities of the image states. The SCD of Figure 3.17 will apply to any chemical step in which at least one bond is exchanged, or to a step involves a formal redox process, for example, $R^+ + :X^- \rightarrow$ R–X, and one way to identify the image states is illustrated in structures **3.11–3.14**, which employ fragment orbital-based VB representations of the anchor states.[3b,6]

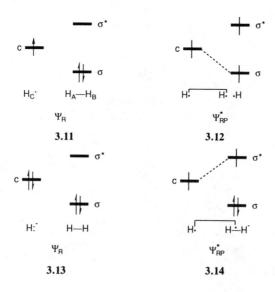

Structures **3.11** and **3.13** represent the Ψ_R of the spin transfer and hydride exchange reactions, respectively; in each case, the electrons are paired in the separate reactants. In the Ψ_{RP}^* representations, **3.12** and **3.14**, there is in each case a bond-pair (a dash connecting the orbitals, or a line connecting the dots) which becomes the C–A bond at the P extremum of the reaction coordinate, and which couples the electrons of the reactants, the coupling being shown by dashes that connect orbitals. In addition, within each Ψ_{RP}^*, placing one electron in a σ^* orbital leads to cleavage of A–B at the P extremum.

Thus, to identify image states, we must reorganize electrons among the orbitals of the reactants, so as to generate *one intermolecular bond-pair for each new bond that is formed during the reaction*. Different kinds of chemical reactions can then be classified in terms of the excitations needed to satisfy the correct number of bond-pairs in the image states: a triplet excitation will be required when there is no change in the formal oxidation states of the reaction centers, as in spin-transfer; a charge transfer excitation will be required whenever there are changes in the formal oxidation states of the reaction centers, as in a reaction between a nucleophile and an electrophile. To deduce the formal oxidation state of a group or an atom, the electrons in a Lewis bond are distributed equally between the bonded groups or atoms. These remarks are summarized as Statements 3.1 and 3.2.

Statement 3.1   The image states of spin transfer reactions involving the exchange of one bond are vertical triplet states, obtained by triplet excitation of the electron pair in the bond that is broken during the transformation. Two spins are then recoupled to a singlet across the infinitely long linkage.

Statement 3.2   The image states of electrophile–nucleophile reactions involving the exchange of one bond are vertical charge transfer states, obtained by transfer of a single electron from the nucleophile to the vacant orbital of the bond that is broken during the transformation. The total spin remains singlet.

For reactions in which more than one bond is exchanged, for example, elimination and cycloaddition, each bond exchange will, in general, require a single excitation. Depending on the nature of the process, this excitation may be triplet, or charge transfer, or a combination of the two. The possibilities are summarized in the following statements:

Statement 3.3   When a reaction proceeds via bond exchanges without changes in the formal oxidation states of the reacting centers, only triplet excitations are necessary to obtain the image states, one such excitation for each bond that is broken.

Statement 3.4   When a reaction proceeds via both bond exchange and changes in the oxidation states of reaction centers, charge transfer excitations of each pair of centers undergoing formal redox will lead to their image states, and all other excitations will be triplet in nature. The total number of excitations must equal the number of bonds that are broken.

Regiochemical Pathways. In the general case, the radical abstraction reaction of C· with A–B can lead to C–A or C–B. How are these pathways distinguished in the SCD procedure?

At the R extremum of the reaction coordinate, the anchor state of the reactant is given by **I** in Figure 3.18. This is characterized by the orbitals and electrons of the A–B bond and the singly occupied orbital of C, and is present in both pathways. Since, in each case, one bond is broken, one triplet excitation will be required (Statement 3.3), and this is the $\sigma \rightarrow \sigma^*$ excitation **I** $\rightarrow$ **II**. A single bond-pair (indicated by dashes) can now be formed from the three unpaired electrons of **II** in two ways, corresponding to the image states **III** and **IV**.

These image states are degenerate at the R extremum, because the bond-pair interaction is zero for the separated reactants. However, as we proceed from R to P along the reaction coordinate, maximization of the c–σ and c–σ^* interactions will be different in the two reactions: **III** correlates with the product state in the reaction that maximizes the interaction of the radical with the σ orbital of AB. If atom A is more electronegative than atom B, the σ orbital will have the larger coefficient on atom A (Section 3.12), and the attack will occur at this site. On the other hand, **IV** correlates with the product state in the reaction that maximizes the interaction of the radical with σ^*, and this will lead to attack at B.

The same analysis can be given for the addition of a radical to a multiple bond, except that σ,σ^* become π,π^*. It will be noted that the orbital pair of the molecule undergoing reaction is a HOMO–LUMO pair, so that the image states correspond conceptually to regiochemical pathways controlled by the coefficients in the HOMO (**III**) or in the LUMO (**IV**).[13] In terms of Figure 3.17, the SCD's of the two

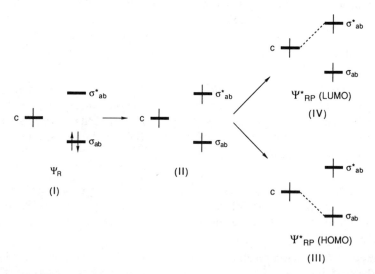

Figure 3.18 The Ψ_R and Ψ_{RP}^* states of the state correlation diagrams that describe the attack of C· upon A or B of A–B. The dashed lines indicate the modes of singlet pairing. (Adapted, by permission, from reference 6, Figure 30. Copyright © 1989 by Kluwer Publications).

reactions will be the same at the R extremum, but the gap, G_P, at the P extremum may well be different.

SCD's for Stepwise Reactions. When the mechanism of the reaction has the form $R \rightarrow I \rightarrow P$, with I an intermediate, the SCD will be comprised of three main curves, as depicted in Figure 3.19. In this figure, two of the curves describe the transformation $R \rightarrow P$ as before (Statements 3.1–3.4), and the third is based on the excited states Ψ_{RI}^* and Ψ_{PI}^*, which are images of I. The $S_N 1$[11] and $S_N V$[14] mechanisms are examples of processes described by Figure 3.19. Elimination mechanisms have been discussed analogously;[15] and the SiH_5^- intermediate has been found to arise from the same type of three-curve crossing.[16] Other systems also conform to this picture.[17]

The Avoided Crossing: Resonance Energy of the Transition State. An important quantity in Figure 3.17 is B, the stabilization that results from the mixing of the two state curves at the crossing point. Each of these curves is localized, because each describes a single bond-pairing pattern, so that the stabilization due to mixing (B) can be regarded as the quantum mechanical resonance energy (QMRE) of the transition state.[6,9d–g,18,19]

Deformation and the Origins of Barriers. Figure 3.17 states that the transition state will be found at that point on the reaction coordinate at which Ψ_P and Ψ_{RP}^* cross. For crossing to occur, it is necessary that the system overcome a fraction of the vertical energy gap G_R, which exists at the R extremum. As depicted in Figure 3.20, this is accomplished via bond and/or angular distortions, as well as nonbonded interactions.

These *deformations* serve to destabilize the ground state and, concomitantly, to

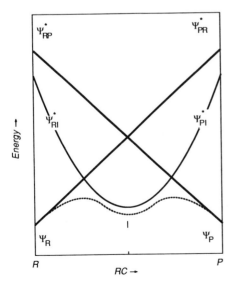

Figure 3.19 State correlation diagram for a stepwise process $R \rightarrow I \rightarrow P$, where I is an intermediate. The avoided crossing is shown only on the lower surface.

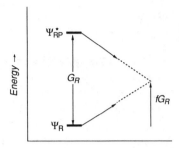

Figure 3.20 The crossing of Ψ_R and Ψ_{RP}^* is achieved by deformation of the system. The total deformation energy (bond distortion and nonbonded repulsion) is fG_R. (Adapted, by permission, from reference 6, Figure 33. Copyright © 1989 by Kluwer Publications.)

stabilize the excited state. The height of the crossing point is then determined by the total deformation energy needed to achieve resonance between the ground and excited states. Expressed in terms of ground-state deformations, this can be written as equation 3.53, that is, the deformation energy is proportional to the energy gap

$$\Delta E_{\text{def}}(R) = f\, G_R; \quad f < 1 \tag{3.53}$$

that must be overcome in order to bring two different bonding schemes into resonance. The barrier in the forward direction can, therefore, be expressed as equation 3.54, and equation 3.55 can be written for the reverse reaction.

$$\Delta E_f^{\ddagger} = \Delta E_{\text{def}}(R) - B = f\, G_R - B \tag{3.54}$$

$$\Delta E_{\text{def}}(P) = f'\, G_P = f\, G_R - \Delta E$$

$$\Delta E_r^{\ddagger} = f'\, G_P - B \tag{3.55}$$

Equations 3.53–3.55 are of great importance, because they describe the activation process of a chemical reaction in terms of understandable properties of reactants, products, and transition states. For example, from equation 3.53 we may write $f = \Delta E_{\text{def}}/G$, which states that f determines the deformation energy required to overcome a given energy gap G. As we shall see in Section 3.3.2, f can be analyzed in terms of the *steepness* or *shallowness* of the intersecting curves.

3.3 AN ALGEBRAIC TREATMENT OF BARRIERS BASED ON TWO-CURVE STATE CORRELATION DIAGRAMS

The factors that determine the height of a barrier are the energy gaps at the reactant and product extrema of the reaction coordinate (G_R and G_P), the overall energy change for the reaction ($\Delta E = E_P - E_R$), the curvatures of the two intersecting curves (f_R and f_P), and the quantum mechanical resonance energy (B). Each of these quantities can be analyzed to yield qualitative and, in some cases, semi-quantitative information regarding barriers.

3.3.1   Definition of the Reaction Coordinate of an SCD

A first problem in the development of a detailed expression for the reaction barrier in terms of the SCD model is to provide a definition of a reaction coordinate (RC). Ideally, the RC might consist of a series of geometric configurations along the path of steepest descent (Chapter 2). However, this definition is not very convenient, because it includes parameters whose values vary from equilibrium values to infinity.

An RC having a finite length can be defined from the bond length–bond order relationship,[20]

$$r_0 - r_n = a \ln n; \qquad n = \exp\left[-(r_0 - r_n)/a\right] \tag{3.56}$$

in which a is a constant, n is the *bond order*, and r_n and r_0 are bond lengths at a general distance and at equilibrium. Equation 3.56 serves as a useful coordinate transformation to a logarithmic scale that provides a finite length for the reaction coordinate. For example, $n = 0$ for $r_n \gg r_0$, and $n = 1$ for $r_n = r_0$.

The bond orders at three critical points along the RC of the identity exchange reaction X + A-X → X-A + X are shown in Figure 3.21, along with Δn ($= n_1 - n_2$). This RC is defined, in terms of Δn, by equation 3.57, and it varies between zero and unity.

$$RC = \tfrac{1}{2}(\Delta n + 1); \qquad \Delta n = n_2 - n_1 \tag{3.57}$$

In the *TS*, individual bond orders are $n^{\ddagger}$, whose value need not be equal to 0.5. The assumption that $n^{\ddagger} = 0.5$ would imply that the total bond order ($\Sigma n = $ constant) is conserved along the reaction coordinate. In such a case, the RC could be defined adequately in terms of only one of the bond orders, and would vary between zero and unity. However, this assumption is not necessary, and the definition of equation 3.56 will be employed hereafter. For the case of a nonidentity reaction, X + A-Y → X-A + Y, this means that the RC will still vary between zero and unity, despite the fact that the *TS* no longer is found at $RC = 0.5$. Clearly, no physical significance should be attached to reaction coordinates that vary between zero and unity in the manner described.

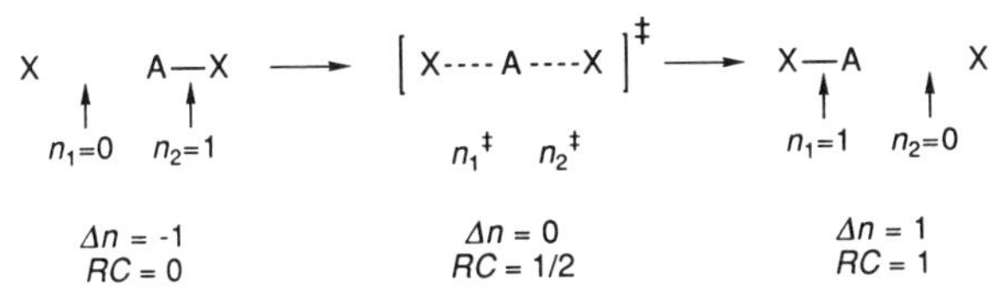

Figure 3.21   Definition of a reaction coordinate (RC) that varies from 0 to 1 in terms of bond orders n_i.

3.3.2 Intrinsic Barriers of Identity Reactions

Obviously, the existence of an identity reaction will facilitate the experimental determination of an intrinsic barrier. Figure 3.22 illustrates the situation in this case.

The crossing point and *TS* are found at $RC = 0.5$. The height of the crossing point is some fraction f of the energy gap G and the *TS* is stabilized relative to the crossing point by the quantum mechanical resonance energy B, leading to equation 3.58, as we have seen.

$$\Delta E^{\ddagger} = f G - B; \quad f < 1 \tag{3.58}$$

The magnitude of f will depend upon the behavior of the individual curves along the reaction coordinate. As seen in Figure 3.23*a*, a steep descent from the higher to the lower anchor points will reflect a process having a small f; on the other hand, Figure 3.23*b* illustrates a shallow descent of the curves, which will reflect a process having a large f.

Two factors especially can be expected to influence the magnitude of f in any reaction, namely, electron delocalization and steric effects. In Figure 3.23, the excited states descend in energy because there is a bond coupling interaction of all bond-pairs that become covalent bonds. However, when the structure of the system allows bond-pair electrons to be delocalized, their coupling interaction will be weak, and a shallow descent of the excited state should be expected (higher f). Likewise, any steric effect that inhibits bond coupling will increase the magnitude of f. Based on these considerations, it is convenient to think of f as a *bond-coupling delay index*.

An additional point may be noted at this stage. If B is constant within a series

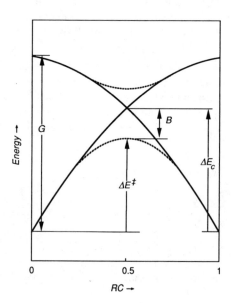

Figure 3.22 Important features of the state correlation diagram for an identity exchange reaction.

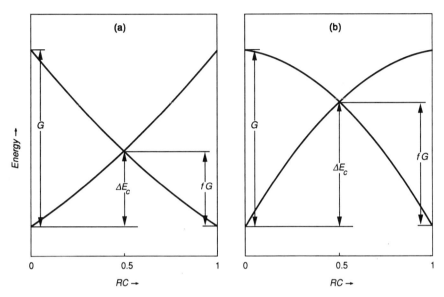

Figure 3.23 (a) Steep descent of curves, corresponding to a process characterized by a small f. (b) Shallow descent of curves, corresponding to a process characterized by a large f.

(family) of reactions, equation 3.58 indicates that there should be a linear structure–reactivity relationship between $\Delta E^{\ddagger}$ and the gap, G, as shown in Figure 3.24. The different families of reactions are *characterized by different values of f.*

The different curvatures in Figure 3.23 can be expresed mathematically, with the assumption that the individual curves are quadratic functions of RC. It has been shown, for example, that a Morse curve can be written as a quadratic function when it is expressed in terms of a bond-order reaction coordinate.[21] The energy of

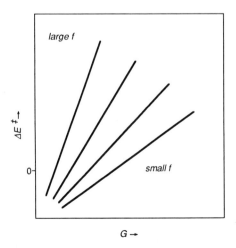

Figure 3.24 Nature of the structure–reactivity relationships expected to exist for identity reactions. Each line represents a "family" having a common f and B.

the reactant curve of Figure 3.23a can be written as equation 3.59, where x is the

$$E_A = (4f - 1)Gx + (2 - 4f)Gx^2 \qquad (3.59)$$

distance along the reaction coordinate RC. For the reactants ($RC = 0$), substitution of $x = 0$ into equation 3.59 leads to $E_A = 0$; for the products, $x = 1$ and $E_A = G$, as required. In the same way, the energy of the product curve of Figure 3.23a is given by equation 3.60. For an identity reaction, the intersection of E_A and E_B

$$E_B = G + (4f - 3)Gx + (2 - 4f)Gx^2 \qquad (3.60)$$

occurs at $x = 0.5$. The coefficients of x and x^2 in equations 3.59 and 3.60 are chosen so that the energy at the crossing point is $\Delta E_c = fG$. While this offers no special insights in the case of an identity reaction, equations 3.59 and 3.60 are of considerable utility in the analysis of nonidentity reactions.

Figure 3.25 illustrates how f and, therefore, the height of the crossing point varies as the curvature varies. The condition $f = 0.25$ leads to two intersecting parabolas. The condition $f = 0.5$ leads to two straight lines.

3.3.3 Intrinsic Barriers of Thermoneutral Nonidentity Reactions

Thermoneutral reactions can also exist in reaction series that do not possess identity sets. Because of the Marcus relationship (Chapter 1), the barriers of such reactions must also be regarded as intrinsic barriers. This situation is treated in Figure 3.26, which is described by two curvature indices, f_R and f_P, and two gaps, G_R and G_P.

As detailed in the Appendix to this chapter, a quadratic expansion of the reac-

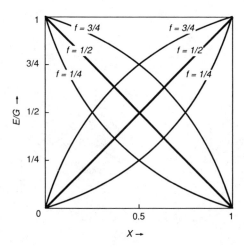

Figure 3.25 An illustration of how f varies with curvature.

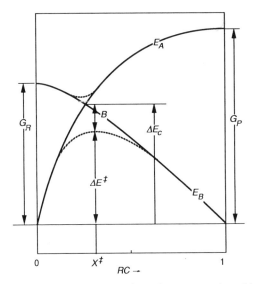

Figure 3.26   State correlation diagram for a thermoneutral nonidentity reaction.

tant and product curves leads to equations 3.61 and 3.62. The position of the

$$E_A = (4f_P - 1) G_P x + (2 - 4f_P) G_P x^2 \qquad (3.61)$$

$$E_B = G_R + (4f_R - 3) G_R x + (2 - 4f_R) G_R x^2 \qquad (3.62)$$

crossing point, $X^\ddagger$, is obtained from the condition $E_A(RC) = E_B(RC)$. Solving for RC leads to a quadratic equation; an approximate value for $X^\ddagger$ (equation 3.63) can

$$x^\ddagger = \frac{1}{2} - \frac{f_P G_P - f_R G_R}{G_R + G_P} \qquad (3.63)$$

be obtained by expansion of E_A and E_B around $x = 0.5$, with retention of only the linear terms. Note that a noncentral TS is predicted for the nonidentity thermoneutral process if $f_R G_R \neq f_P G_P$.

To find the height of the crossing point of Figure 3.26, we have to substitute equation 3.63 into equation 3.61 or 3.62, and the assumption that the curves are linear in the region $X = 0.5$ leads to equation 3.64 and thence to equation 3.65.

$$\Delta E_C \approx (f_R + f_P) \frac{G_R G_P}{G_R + G_P} \qquad (3.64)$$

$$\Delta E^\ddagger \approx (f_R + f_P) \frac{G_R G_P}{G_R + G_P} - B \qquad (3.65)$$

These equations highlight the manner in which the gaps and curvatures of the SCD combine to determine the height of the crossing point, and also the height of the barrier. There are many examples of nonidentity thermoneutral reactions, whose SCD's must be described by different curves. For example,[22] the recombination reaction of pyronin cation with phenoxide ion in water solvent exhibits $K_{eq} \cong 1$.

With certain simplifications, this approach reduces to the Marcus treatment of electron transfer reactions, in terms of intersecting parabolas.[23] In the latter case, equations 3.66 and 3.67 express the energies of the curves, for a "reaction coor-

$$E_A(x) = E_A^{(0)} + x^2 G \tag{3.66}$$

$$E_B(x) = E_B^{(0)} + (x - 1)^2 G \tag{3.67}$$

dinate" (x) that varies between zero and unity, and has "force constants" given by equation 3.68. The energy minima of the curves are $E_A = E_A^{(0)}$ at $x = 0$, and

$$\frac{\partial^2 E_A}{\partial x^2} = \frac{\partial^2 E_B}{\partial x^2} = 2G \tag{3.68}$$

$E_B = E_B^{(0)}$ at $x = 1$. Since the curvatures ("force constants") are thus assumed to be constant, the variation of the energy with x will lead to a crossing point ΔE_c, relative to $E_A^{(0)}$, given by equation 3.69, for $E_A^{(0)} = E_B^{(0)}$. If we compare equations 3.65 and 3.69,

$$\Delta E^{\ddagger} = \Delta E_C = \tfrac{1}{4} G \tag{3.69}$$

we find that the Marcus treatment corresponds to the case $B = 0$ and $f_R = f_P = 0.25$.

Statement 3.5 summarizes the foregoing remarks.

Statement 3.5 For any thermoneutral reaction that can be described by the generalized SCD of Figure 3.26, the intrinsic barrier will be determined by the gaps of the diagram and the bond-coupling delay indices of the two curves.

3.3.4 Nonthermoneutral Reactions, the BEP Principle and Its Breakdown

An SCD of the type shown in Figure 3.27 characterizes a nonthermoneutral reaction. In such cases the barrier can be expressed in terms of G_P, the excitation energy from ground state to excited state at the P extremum of the RC. Alternatively, the gaps might be defined so as to take ΔE, the energy change for the reaction, into account. In the latter case, the gaps are G_R and G_R^P, where

$$G_R^P = G_P + \Delta E; \quad \Delta E > 0 \tag{3.70}$$

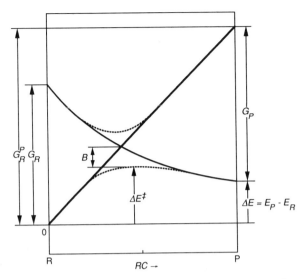

Figure 3.27 State correlation diagram for a nonthermoneutral reaction.

Approximate expressions for the barrier can be derived in terms of G_P or G_R^P. An example of such expressions, which take ΔE into account, is given in equations 3.71 and 3.72. The two equations are interconvertible through the relationship G_P

$$\Delta E_f^\ddagger \approx \left[f_R + f_P + (1 - f_R - f_P) \frac{\Delta E}{G_R} \right] \frac{G_R G_R^P}{G_R + G_R^P - \Delta E} - B \quad (3.71)$$

$$\Delta E_f^\ddagger \approx \left[f_R + f_P + (1 - f_R - f_P) \frac{\Delta E}{G_R} \right] \frac{G_R(G_P + \Delta E)}{G_R + G_P} - B \quad (3.72)$$

$= G_R^P - \Delta E$, and it will be noted that both of these equations reduce to equation 3.65 in the thermoneutral process ($\Delta E = 0$). Since ΔE is defined as $E_P - E_R$, making the process more exoergic will make ΔE more negative and, with all other barrier factors unchanged, *will lower the barrier*. This is the BEP principle, but derivation of the principle in this way reveals its limits of applicability (Statement 3.6).

Statement 3.6 The BEP principle will be obeyed whenever an increase in exergonicity of a reaction is not accompanied by opposing changes in the energy gaps and/or bond-coupling delay indices.

In the general case of a nonidentity reaction, the barrier is a function of six variables, as shown in equation 3.73. A full understanding of structure–reactivity

$$\Delta E^\ddagger = \Delta E^\ddagger(f_R, f_P, G_R, G_P, \Delta E, B) \quad (3.73)$$

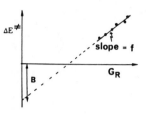

Figure 3.28 A linear free-energy plot using equation 3.54. The slope of the line and its intercept are f and B, respectively. (Reprinted, by permission, from reference 6, Figure 34. Copyright © 1989 by Kluwer Publications).

relationships would require a seven-dimensional space, a situation that could easily lead to despair.

We can, however, make progress using equation 3.54, by introduction of the well-established physical organic chemical concept of "reaction families.[6,18,19] Let us assume, for example, that among an ensemble of reactions which obey a single mechanism there are families characterized by constant f and B. The existence of such families might be identified by experimental relationships of the type shown in Figure 3.28.

If such relationships were found, and the linearity could be shown to have the interpretation suggested, two properties of the transition state would thereby have been determined experimentally: one is the deformation energy of the reactants and products in the geometry of the transition state, from fG_R and the energy change for the reaction (equations 3.54 and 3.55); the other is B.

As we shall find in Chapters 5 and 6, it is sometimes possible to use experimental information to estimate all of the parameters of an SCD, to give physical meaning to these parameters and, thereby, to extract much information regarding the reactivity patterns and geometries of transition states.

APPENDIX

Derivation of Barrier Equations for a Two-Curve SCD

For the partial SCD depicted in Figure 3.29 for the thermoneutral case, the energies of the two curves as a function of the reaction coordinate x can be written

$$E_A(x) = G_P(4f_P - 1)x + G_P(2 - 4f_P)x^2 \tag{3A.1}$$

$$E_B(x) = G_R + G_R(4f_R - 3)x + G_R(2 - 4f_R)x^2 \tag{3A.2}$$

where, in each case, $E(x = 0.5) = fG$. The position of the crossing point, $x^{\ddagger}$, is obtained from the condition $E_A(x) = E_B(x)$. This leads to

$$[G_R(2 - 4f_R) - G_P(2 - 4f_P)]x^2$$
$$+ [G_R(4f_R - 3) - G_P(4f_P - 1)]x + G_R = 0 \tag{3A.3}$$

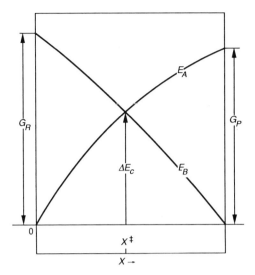

Figure 3.29 A portion of a state correlation diagram used to derive an algebraic expression for a thermoneutral reaction.

which has solutions

$$x^{\ddagger} = \frac{-b - \sqrt{b^2 - 4ac}}{2a}$$

$$a = G_R(2 - 4f_R) - G_P(2 - 4f_P);$$

$$b = G_R(4f_R - 3) - G_P(4f_P - 1); \qquad c = G_R \qquad (3A.4)$$

The desired solution is in the range of $0 \le x \le 1$. An approximate expression for $x^{\ddagger}$ is obtained by rewriting equations 3A.1 and 3A.2 in terms of $x - 0.5$,

$$E_A(x) = f_P G_P + G_P(x - \tfrac{1}{2}) + G_P(2 - 4f_P)(x - \tfrac{1}{2})^2 \qquad (A3.5)$$

$$E_B(x) = f_R G_R - G_R(x - \tfrac{1}{2}) + G_R(2 - 4f_R)(x - \tfrac{1}{2})^2 \qquad (A3.6)$$

Dropping the quadratic terms from equations 3A.5 and 3A.6 and equating the resulting expressions for E_A and E_B leads to equation 3A.7. The $x^{\ddagger}$(approx) will be acceptable whenever $f_R G_R$ and $f_P G_P$ do not differ greatly.

$$x^{\ddagger}(\text{approx}) - \tfrac{1}{2} = (f_R G_R - f_P G_P)/(G_R + G_P) \qquad (3A.7)$$

An estimate of ΔE_c, the height of the crossing point of Figure 3.29, is obtained by substitution of a $x^{\ddagger}$(approx) into the linear portion of equation 3A.6. This gives

$$\Delta E_C(\text{approx}) = (f_R + f_P)G_R G_P/(G_R + G_P) \qquad (3A.8)$$

The final expression for the barrier in the thermoneutral case is

$$\Delta E^{\ddagger}(\text{approx}) = (f_R + f_P)G_R G_P/(G_R + G_P) - B \qquad (3A.9)$$

For the case of a nonthermoneutral process depicted in Figure 3.30, the expressions for the individual curves are

$$E_A(x) = (G_P + \Delta E)(4f_P - 1)x + (G_P + \Delta E)(2 - 4f_P)x^2 \qquad (3A.10)$$

$$E_B(x) = G_R + (G_R - \Delta E)(4f_R - 3)x + (G_R - \Delta E)(2 - 4f_R)x^2 \qquad (3A.11)$$

and, proceeding as before, we arrive at

$$E_A(x) = f_P(G_P + \Delta E) + (G_P + \Delta E)(x - \tfrac{1}{2})$$
$$+ (G_P + \Delta E)(2 - 4f_P)(x - \tfrac{1}{2})^2 \qquad (3A.12)$$

$$E_B(x) = \Delta E + f_R(G_R - \Delta E) - (G_R - \Delta E)(x - \tfrac{1}{2})$$
$$+ (G_R - \Delta E)(2 - 4f_R)(x - \tfrac{1}{2})^2 \qquad (3A.13)$$

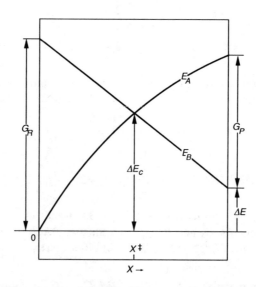

Figure 3.30. A portion of a state correlation diagram used to derive an algebraic expression for a nonthermoneutral reaction.

which leads to

$$x^{\ddagger}(\text{approx}) - \frac{1}{2} = \frac{f_R(G_R - \Delta E) - f_P(G_P + \Delta E) + \Delta E}{G_R + G_P} \qquad (3A.14)$$

and

$$\Delta E^{\ddagger}(\text{approx}) = \frac{(f_R + f_P)(G_R - \Delta E)(G_P + \Delta E) + \Delta E(G_P + \Delta E)}{G_R + G_P} - B$$

$$= \left[f_R + f_P + (1 - f_R - f_P) \frac{\Delta E}{G_R} \right] \frac{G_R(G_P + \Delta E)}{G_R + G_P} - B.$$

$$(3A.15)$$

as the final expression for the barrier.

REFERENCES

1. (a) R. B. Woodward and R. Hoffmann. *Angew. Chem.*, *Intl. Ed. Engl.* **8**, 781 (1969); (b) H. C. Longuet-Higgins and E. W. Abrahamson, *J. Am. Chem. Soc.* **87**, 2045 (1965).

2. However, see the method of natural MO correlations: (a) A. Sevin and P. Chaquin. *Chem. Phys.* **95**, 49 (1985); (b) A. Devaquet, A. Sevin, and B. Bigot. *J. Am. Chem. Soc.* **100**, 2009 (1978).

3. (a) P. C. Hiberty and C. Leforestier. *J. Am. Chem. Soc.* **100**, 2012 (1978); (b) S. S. Shaik. *J. Am. Chem. Soc.* **103**, 3691 (1981); (c) P. Karafiloglu and J.-P. Malrieu. *Chem. Phys.* **104**, 383 (1986); (d) M. A. Fox and F. A. Matsen, *J. Chem. Educ.* **62**, 477 (1985); F. A. Matsen. *Acc. Chem. Res.* **11**, 387 (1978); (e) B. Oujia, M.-B. Lepetit, D. Maynau, and J.-P. Malrieu. *Phys. Rev.* **A39**, 3289 (1989).

4. (a) L. Salem and C. Rowland, *Angew. Chem.*, *Intl. Ed. Engl.* **11**, 92 (1972); (b) L. Salem. *Electrons in Chemical Reactions: First Principles*. Wiley-Interscience, New York, 1982, pp. 6, 72-75.

5. A. Szabo and N. S. Ostlund. *Modern Quantum Chemistry*. MacMillan Publishing Co., New York, 1982.

6. S. S. Shaik. In *New Theoretical Concepts for Understanding Organic Reactions*. J. Bertrán and I. G. Csizmadia, Editors, NATO ASI Series, Vol. C267, Kluwer Publications, Dordrecht, 1989, p. 165.

7. T. A. Albright, J. K. Burdett, and M.-H. Whangbo. *Orbital Interactions in Chemistry*. John Wiley, New York, 1985, p. 29.

8. For a general consideration of VB wavefunctions see R. McWeeny and B. T. Sutcliffe, *Methods of Molecular Quantum Mechanics*. Academic Press, London, 1969, Chapters 3 and 6.

9. (a) S. S. Shaik and R. Bar. *Nouv. J. Chim.* **8**, 411 (1984); (b) S. S. Shaik and P. C. Hiberty. *Am. Chem. Soc.* **107**, 3089 (1985); (c) S. S. Shaik, P. C. Hiberty, G. Ohanes-

sian, and J.-M. Lefour. *Nouv. J. Chim.* **9**, 385 (1985); (d) S. S. Shaik, P. C. Hiberty, J.-M. Lefour, and G. Ohanessian. *J. Am. Chem. Soc.* **109**, 363 (1987); (e) S. S. Shaik, P. C. Hiberty, G. Ohanessian, and J.-M. Lefour. *J. Phys. Chem.* **92**, 5086 (1988); (f) S. S. Shaik and E. Canadell. *J. Am. Chem. Soc.* **111**, 4306 (1989); (g) P. Maitre, P. C. Hiberty, G. Ohanessian, and S. S. Shaik, *J. Phys. Chem.* **94**, 4089 (1990).

10. For lucid discussions of avoided crossings see (a) L. Salem, C. Leforestier, G. Segal, and R. Wetmore, *J. Am. Chem. Soc.* **97**, 479 (1975); (b) L. Salem. *Science* **191**, 822 (1976); (c) A. Devaquet. *Pure Appl. Chem.* **41**, 455 (1975).

11. (a) A. Pross and S. S. Shaik. *Acc. Chem. Res.* **16**, 363 (1983); (b) S. S. Shaik. *Progr. Phys. Org. Chem.* **15**, 197 (1985); (c) A. Pross, *Adv. Phys. Org. Chem.* **21**, 99 (1985).

12. O. K. Kabbaj, F. Volatron, and J.-P. Malrieu. *Chem. Phys. Lett.* **147**, 353 (1988).

13. (a) V. Bonancic-Koutecky, V. Koutecky, and L. Salem. *J. Am. Chem. Soc.* **99**, 842 (1977); (b) J. M. Poblet, E. Canadell, and T. Sordo. *Can. J. Chem.* **61**, 2068 (1983).

14. D. Cohen, R. Bar, and S. S. Shaik. *J. Am. Chem. Soc.* **108**, 231 (1986).

15. A. Pross and S. S. Shaik. *J. Am. Chem. Soc.* **104**, 187 (1982).

16. G. Sini, P. C. Hiberty, and S. S. Shaik. *J. Chem. Soc. Chem. Commun.* 772 (1989).

17. A. Demoliens, O. Eisenstein, P. C. Hiberty, J.-M. Lefour, G. Ohanessian, S. S. Shaik, and F. Volatron. *J. Am. Chem. Soc.* **111**, 5623 (1989).

18. (a) S. S. Shaik. *J. Org. Chem.* **52**, 1563 (1987); (b) S. S. Shaik, P. C. Hiberty, J. M. Lefour, and G. Ohanessian. *J. Am. Chem. Soc.* **109**, 363 (1987).

19. E. Buncel, S. S. Shaik, I.-H. Um, and S. Wolfe. *J. Am. Chem. Soc.* **110**, 1275 (1989).

20. (a) H. S. Johnston. *Gas Phase Reaction Rate Theory.* Ronald Press, New York, 1966; (b) H. S. Johnston and C. Parr. *J. Am. Chem. Soc.* **85**, 2544 (1963); (c) N. Agmon. *Chem. Phys. Lett.* **45**, 343 (1977); (d) J. D. Dunitz, *Phil. Trans. Roy. Soc. London.* **272B**, 99 (1975).

21. E. Shusterovich. *Acc. Chem. Res.* **21**, 183 (1988).

22. C. D. Ritchie. In *Nucleophilicity.* J. M. Harris and S. P. McManus, Editors, Advances in Chemistry Series, American Chemical Society, Washington, DC, 1986.

23. R. A. Marcus and N. Sutin. *Comments Inorg. Chem.* **3**, 119 (1986).

4

REACTIVITY FACTORS FOR
S$_N$2 REACTIONS

In Chapter 3 we began the introduction of the so-called state correlation diagram (SCD) model of reaction barriers, which is based on a VB treatment of the crossing of reactant and product curves along a reaction coordinate. The model is expressed formally in terms of energy gaps, reaction ergicity, curvatures, and quantum mechanical resonance energies. The first three of these reactivity factors are accessible experimentally or can be estimated in some way. The fourth, the quantum mechanical resonance energy, requires a separate discussion, which is deferred to Chapter 5. Our main objective here is to assemble the experimental data concerning energy gaps and curvatures, where these exist, and to suggest ways to obtain the required information when the experimental data are not available. This analysis has been carried out for the case of the S$_N$2 reaction, the prototype of a process that allows the insights of the SCD model to be applied in some detail. As we shall see, these insights can be developed not only for the gas phase, but also for some of the solvents in which the majority of the experimental work is performed.

Figure 4.1 is the SCD which describes the S$_N$2 reaction Y:$^-$ + R–X → Y–R + :X$^-$.[1] The ground anchor points of the diagram are the ion–dipole complexes **4.1** and **4.2**, which have been found to exist in the gas phase[2,3] and also in

4.1	**4.2**

certain solvents.[4,5] The excited anchor points are the corresponding vertical charge transfer states. The gaps of the diagram therefore refer to vertical electron transfer energies.

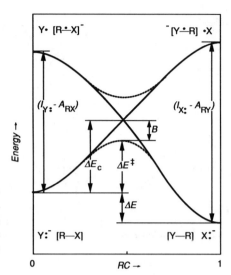

Figure 4.1 State correlation diagram (SCD) for an S_N2 process. The ground state anchor points refer, for convenience, to the geometries of the ion–molecule complexes; the upper anchor points are vertical charge transfer states in which the odd electrons are singlet coupled.

The figure asserts that the process comprises the movement of a *single* electron coupled to bond reorganization. In the gas phase, the movement of the single electron is coupled to the distortions associated with the C–X/Y–C bond interchange which, in turn, are coupled to the umbrella motion that inverts the configuration at the central carbon atom. In solution, the movement of the single electron is coupled to these molecular distortions, and also to the reorganization of solvent molecules around the reacting centers.[1d,e,f]

4.1 THE ELECTRON TRANSFER ENERGY GAPS—VERTICAL IONIZATION POTENTIALS AND VERTICAL ELECTRON AFFINITIES

4.1.1 Vertical Ionization Potentials of Nucleophiles in the Gas Phase

Gas phase ionization potentials $[I_{Y:}(g)]$ have been determined experimentally for many anions,[6] and the data for a number of the common nucleophiles are collected in Table 4.1. We note that $I_{Y:}(g)$ increases systematically as the nucleophilic center moves from left to right along a row of the periodic table (entries 1, 5, 8, 10). However, down a column, the variation is much smaller (entries 1–4, 5–7).

4.1.2 Vertical Ionization Potentials of Nucleophiles in Solution

These quantities, termed $I_{Y:}(s^*)$, are available for water solvent from photoemission spectra of anions,[7] and in water and other solvents from electronic transitions associated with charge transfer to solvent (CTTS).[8] Other values can be estimated from equation 4.1,[1d,e,f;9]

$$I_{Y:}(s^*) \approx I_{Y:}(g) + (1 + \rho)S_{Y:} \tag{4.1}$$

Table 4.1 Ionization Potentials of S_N2 Nucleophiles (Y:$^-$) in the Gas Phase [$I_{Y:}$ (g)] and in Solution [$I_{Y:}$ (s)*]^{a,b}

			$I_{Y:}$ (s)*	
Entry	Y:$^-$	$I_{Y:}$ (g)	H_2O	DMF
1	F$^-$	78	240	213
2	Cl$^-$	83	204	183
3	Br$^-$	78	186	170
4	I$^-$	71	166	156
5	HO$^-$	44	195	168
6	HS$^-$	54	177	156
7	HSe$^-$	51	159	147
8	H$_2$N$^-$	19	—	—
9	H$_2$P$^-$	29	—	—
10	CH$_3^-$	2	—	—
11	SiH$_3^-$	32^c	—	—
12	CN$^-$	89	203	181

^{a}All values are in kcal/mol.
^{b}$I_{Y:}$ (g) values are from Table A4.1 (see Appendix) and reference 6; $I_{Y:}$ (s)* values in H_2O for entries 2–5 are from reference 7. The remainder of the $I_{Y:}$ (s)* data can be calculated using $S_{X:}$ (see equation 4.15). Note that $\rho(H_2O) = 0.56$; $\rho(DMF) = 0.48$.^{1d,e,f,9}
^{c}From reference 29.

where $S_{Y:}$ is the free energy of desolvation of Y:$^-$ (see Section 4.1.4), and ρ is termed the solvent reorganization factor. This factor is related to the ability of the solvent superstructure to undergo change, and it has been expressed in terms of the optical (ϵ_{op}) and static (ϵ_s) dielectric constants.^{7,9,10}

$$\rho = \frac{\epsilon_s - \epsilon_{op}}{\epsilon_{op}(\epsilon_s - 1)} \tag{4.2}$$

In equation 4.2, ϵ_{op} is determined from the refractive index of the solvent,^{9b} according to $\epsilon_{op} = n^2$.

Some values of $I_{Y:}(s^*)$ are collected in the second and third columns of Table 4.1. We observe that, along a row, the gas phase trends are maintained in solution but, down a column, the ionization potentials now seem to decrease progressively and significantly.

4.1.3 Gas Phase Vertical Electron Affinities (A_{RX})

The adiabatic electron affinity (EA) of a molecule M is defined as the negative of the standard energy change for the reaction

$$M + e^- \rightarrow M^- \quad \Delta E = -EA \tag{4.3}$$

A positive EA means that the anion is stable, while a negative EA indicates that the anion is unstable with respect to electron loss.

The subject has often been reviewed,[6, 11] and more than 30 different methods for *EA* determinations have been developed. Unfortunately, the extensive compilations of data that are thus available are not helpful to us, because in most cases (reference 11a is an exception) these compilations refer to *adiabatic* quantities, that is, values corresponding to differences in the energy minima of the charged and neutral species at their respective equilibrium geometries. This point is illustrated in Figure 4.2, which shows the relationship between vertical and adiabatic electron affinities for an unstable anionic state $(RX)^-$. As defined in this figure, the vertical quantity A_{RX} is negative, so that a decrease in the magnitude of A_{RX} corresponds to an increasing instability of $(RX)^-$.

There are some experimental strategies that can be used to obtain qualitative estimates of vertical electron affinities and trends in A_{RX} within series of related molecules. One such strategy is based on Figure 4.3.[12] We are dealing here with excitation from the nonbonding *np* orbital on X of R–X; the difference between the term value of the Rydberg excited state ionization potential E^+, and the $np \rightarrow$ σ^* excitation energy E^*, written $\mathscr{A}(np, \sigma^*)$, can be regarded as the ionization potential of an electron in a σ^* orbital. Since inspection of Figure 4.3 shows that $\mathscr{A}$ decreases as the σ^* energy increases, there is an inverse relationship between the magnitude of $\mathscr{A}$ and the magnitude of A_{RX}, that is, A_{RX} decreases as $\mathscr{A}$ increases.

For the series CH_3X (X = F, Cl, Br, I), the $\mathscr{A}$ values are 77, 92, 100, and 110 kcal/mol, respectively,[12] and are seen to vary inversely with the CH_3–X bond dissociation energies (D_{R-X}). Indeed, a linear relationship is observed.[12] We should, therefore, expect that A_{RX} becomes less negative as X is varied down a column of the periodic table.

A second experimental procedure which allows a more direct estimate of vertical electron affinities is based on measurements of electron attachment energies.

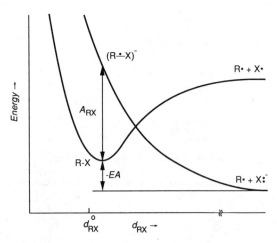

Figure 4.2 The relationship between vertical (A_{RX}) and adiabatic (*EA*) electron affinities for an unstable radical anion, $(R \pm X)^-$

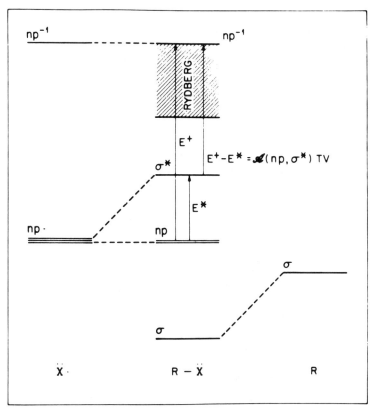

Figure 4.3 Definition of the *A*-band term value in alkyl halides (Reprinted, by permission, from reference 12. Copyright © 1985 by the National Research Council of Canada).

These are obtained from electron–molecule scattering experiments[13–17] in which temporary anion states are formed, with lifetimes 10^{-12}–10^{-17} sec. These temporary states have negative electron affinities.

Electron transmission spectroscopy (ETS)[14c] is a scattering technique that studies the transparency of a molecular gas to an electron beam of sharp energy as a function of the electron energy. The transparency is inversely proportional to the probability (cross section) of electron scattering, and the total cross section is enhanced significantly by a resonant capture of the electron. Since the temporary anion decays rapidly, electron capture is observed as a broad band ("resonance") in the scattering cross section. The spectrum is normally recorded as a derivative of transmitted current as a function of electron energies.

The attachment energy is the energy at which the derivative spectrum exhibits a fluctuation, and it can be interpreted as the negative electron affinity of a temporary state which contains an excess electron in one of the virtual orbitals.[15, 16] When the lifetime is $\geq 10^{-14}$ sec, it is possible to observe a progression of peaks, corresponding to vibrational progression of the temporary state. For lifetimes $< 10^{-14}$ sec, one observes a wide vibrationless resonance. The midpoint of this latter resonance is taken as the *vertical attachment energy*, and this quantity is the negative of the vertical electron affinity.

There are some problems with the method, which limit the quantitative significance of the numerical values.[13] For example, dissociative σ^* states are not always observed[15-17] and, when they are, very broad resonances are seen. For example,[17] the resonance for CH_3F stretches from ca. -5 to -8 eV, and its midpoint (-6.1 eV[16]) has been taken as the vertical electron affinity. On the other hand, theoretical studies[13] suggest that four closely spaced anionic states contribute to this broad resonance, with the σ_{CF}^* state the lowest lying, at -4.6 eV. In the cases of CH_3Br and CH_3I, σ_{CX}^* states are not observed[18] but, using substituent effects in other bromo and iodo compounds,[16, 17] values of -2.8 eV (CH_3Br) and -1.2 eV (CH_3I) can be estimated.

Despite the problems associated with the method, ETS spectroscopy remains the most reliable and uniform experimental procedure to obtain electron affinities of dissociative states and to obtain qualitative information regarding substituent effects and trends.

It is useful to try to complement this information with qualitative numerical data based on the thermochemical cycle summarized in equations 4.4a–d,[1a, b, d]

$$(R \pm X)^- \rightarrow R\cdot + X:^- \qquad \Delta E_1 = D_{R \pm X} \qquad (4.4a)$$

$$X:^- \rightarrow X\cdot + e^- \qquad \Delta E_2 = A_{X\cdot} \qquad (4.4b)$$

$$R\cdot + X\cdot \rightarrow R-X \qquad \Delta E_3 = -D_{R-X} \qquad (4.4c)$$

$$\overline{(R \pm X)^- \rightarrow R-X + e^- \qquad \Delta E = A_{RX}} \qquad (4.4d)$$

The vertical electron affinity of the molecule R–X can then be expressed as

$$A_{RX} = A_{X\cdot} - D_{R-X} + D_{R \pm X} \qquad (4.5)$$

The first step of this cycle is the dissociation of the radical anion (starting from the geometry of the neutral R–X) into its most stable fragments. The second step is the ionization of $X:^-$, and its energy change is the electron affinity of $X\cdot$, which is also the adiabatic ionization potential of $X:^-$. The energy change for the third step is the negative of the dissociation energy of R–X. The unknown quantity of equation 4.5 is $D_{R \pm X}$, which refers to the vertical process depicted in Figure 4.4. An empirical estimate of $D_{R \pm X}$ can be obtained from simple VB considerations. As we bring $R\cdot$ and $X:^-$ fragments together to the equilibrium distance (d_{RX}^o) of the ground state molecule, three-electron overlap repulsion (*OR*) leads to a progressive increase in energy (curve a). Concurrently, mixing in the higher-energy $(R:^- \cdot X)$ configuration (equation 4.6) stabilizes $(R\cdot :X^-)$ by the quantity ΔE_{int},

$$(R \pm X)^- = a(R\cdot :X^-) + b(R:^- \cdot X); \; a^2 + b^2 + 2ab\langle R\cdot :X^- | R:^- \cdot X \rangle = 1$$

$$(4.6)$$

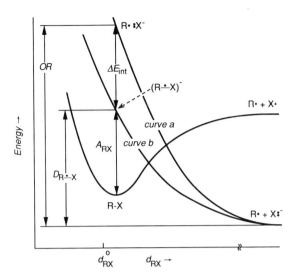

Figure 4.4 Definition of $D_{R \pm X}$: the Morse curve describes the formation of R–X from its fragments. Curve a represents the approach of R· and :X⁻; Curve b refers to the formation of $(R \pm X)^-$ as a result of interaction with the higher-lying curve for $(R:^- \cdot X)$ (not shown).

with formation of the radical anion $(R \pm X)^-$. As we have seen, this latter species is normally characterized by a negative dissociation energy $D_{R \pm X}$ (curve b).

Based on Figure 4.4, we may write

$$OR = D_{R \pm X} + \Delta E_{int} \qquad (4.7)$$

The absolute magnitude of OR corresponds to the energy of a Heitler–London bond,[1d] and it can be estimated, from the Pauling relationship for the strength of a C–X bond, in terms of the geometric mean of the covalent bond energies,

$$OR = -m \sqrt{D_{RR} D_{XX}} \qquad (4.8)$$

where

$$m = (d_R^{cov} + d_X^{cov})/d_{RX}^o \qquad (4.9)$$

and m is a correction factor suggested by Sanderson.[19] Further details can be found in the Appendix to this chapter. In equation 4.9, d^{cov} is a covalent radius and d_{RX}^o is the bond length of R–X.

The interaction term ΔE_{int} can be approximated by the standard neglect of overlap expression for the interaction of two configurations,[20]

$$\Delta E_{int} = \tfrac{1}{2} \Delta E - \tfrac{1}{2} \sqrt{\Delta E^2 + 4h_{RX}^2} \qquad (4.10)$$

where ΔE is the difference in the energies of the $(R:^- \cdot X)$ and $(R \cdot :X^-)$ VB configurations, that is,

$$\Delta E = A_{X \cdot} - A_{R \cdot} \tag{4.11}$$

and h_{RX} is approximated by $m(D_{RR}D_{XX})^{1/2}$ (see Appendix).

Based on equations 4.7–4.11, we propose to estimate $D_{R \pm X}$ as

$$D_{R \pm X} = OR - \Delta E_{int}$$

$$= -m\sqrt{D_{RR}D_{XX}} - \tfrac{1}{2}(A_{X \cdot} - A_{R \cdot})$$

$$+ \tfrac{1}{2}\sqrt{(A_{X \cdot} - A_{R \cdot})^2 + 4m^2(D_{RR}D_{XX})} \tag{4.12}$$

Table 4.2 illustrates the calculation of this quantity for the methyl halides, and substitution into equation 4.5 of values estimated in this way leads to the gas phase vertical electron affinities of R–X collected in Table 4.3.

We note immediately that A_{RX} becomes more positive as X is varied down a column of the periodic table; this is the trend we expected earlier from our analysis of experimental $\alpha(np, \sigma^*)$ values. We also note that A_{RX} becomes more positive as X is varied from left to right along a row of the periodic table.

In general, Table 4.3 suggests that, in harmony with equation 4.5, when X has a low electron affinity $(A_{X \cdot})$, or forms a strong bond to R (high D_{R-X}), the vertical electron affinity will be more negative.

Table 4.4 contains a more complete listing of calculated and experimental gas phase vertical electron affinities. The trends are similar, so that the observation made in the preceding paragraph appears also to be consistent with experiment.

4.1.4 Vertical Electron Affinities in Solution

Experimental data concerning solution electron affinities are sparse;[21] *vertical* electron affinities in solution $[A_{RX}(s^*)]$ are unknown and have to be estimated.

Table 4.2   Data Employed to Calculate $D_{R \pm X}$ for the Methyl Halides

Entry	X	m^a	$(D_{RR}D_{XX})^{1/2 b}$	$A_{X \cdot} - A_{R \cdot}{}^c$	$D_{R \pm X}{}^d$
1	F	1.079	58.5	76	−27
2	Cl	0.99	72.2	82	−29
3	Br	1	64.6	76	−28
4	I	1	57.2	69	−25

$^a d_R^{cov} = 0.77$ Å; $d_F^{cov} = 0.72$ Å; $d_{Cl}^{cov} = 0.99$ Å; $d_{Br}^{cov} = 1.15$ Å; $d_I^{cov} = 1.34$ Å.
$^b D_{RR} = 90$ kcal/mol (from D. F. McMillen and D. M. Golden. *Annu. Rev. Phys. Chem.* **33**, 493 (1982); $D_{FF} = 38$ kcal/mol; $D_{ClCl} = 58$ kcal/mol; $D_{BrBr} = 46.4$ kcal/mol; $D_{II} = 36.4$ kcal/mol (from the *CRC Handbook of Chemistry and Physics*).
^{c} Data from Table 4.1.
^{d} Rounded off.

Table 4.3 A_{RX} Values[a] for CH$_3$X Molecules Calculated with Equations 4.5 and 4.12

Entry	X	A_X[b]	D_{R-X}[b]	$D_{R \cdot X}$	A_{RX}[b]
1	F	78	108	−27.0	−57
2	Cl	83	84	−29.0	−30
3	Br	78	71	−28.0	−21
4	I	71	56	−25.0	−10
5	CH$_3$CO$_2$	78	83	−25.4	−30
6	HO	44	91	−18.0	−65
7	CH$_3$O	37	81	−15.0	−61
8	HS	54	74	−21.4	−42
9	CH$_3$S	43	77	−18.0	−52
10	NC	89	122	−45.5	−68
11	HCC	68	117	−28.8	−78
12	H$_2$N	19	81	−8.5	−70
13	H	17	104	20.4	−66

[a] All values are in kcal/mol, and have been rounded off.
[b] The data source has been collected in Appendix 1.

This quantity can be obtained from the thermochemical cycle,

$$(R\text{–}X)(s) \rightarrow (R\text{–}X)(g) \qquad \Delta E_1 = S_{RX} \qquad (4.13a)$$

$$(R\text{–}X)(g) \rightarrow R\cdot(g) + X\cdot(g) \qquad \Delta E_2 = D_{R\text{-}X} \qquad (4.13b)$$

$$X\cdot(g) + e^-(g) \rightarrow X{:}^-(g) \qquad \Delta E_3 = -A_X. \qquad (4.13c)$$

$$X{:}^-(g) \rightarrow X{:}^-(s) \qquad \Delta E_4 = -S_{X:} \qquad (4.13d)$$

$$R\cdot(g) \rightarrow R\cdot(s) \qquad \Delta E_5 = -S_R. \qquad (4.13e)$$

$$X{:}^-(s) + R\cdot(s) \rightarrow (R \overset{\bullet}{-} X)^-(s^*) \qquad \Delta E_6 = -D_{R \overset{\bullet}{-} X}(s^*) \qquad (4.13f)$$

$$\overline{(R\text{–}X)(s) + e^-(g) \rightarrow (R \overset{\bullet}{-} X)^-(s^*) \qquad \Delta E = -A_{RX}(s^*) \qquad (4.13g)}$$

from which we may write

$$A_{RX}(s^*) = A_X. + S_{X:} - D_{R\text{-}X} + D_{R \overset{\bullet}{-} X}(s^*) - S_{RX} + S_R. \qquad (4.14)$$

where, as we have already discussed,

$A_X.$ is the gas phase electron affinity of $X\cdot$,
$S_{X:}$ is the desolvation energy of $X{:}^-$,
S_{RX} is the desolvation energy of RX,

and the meaning of $D_{R \overset{\bullet}{-} X}(s^*)$ can be seen by reference to Figure 4.5, which is the counterpart of Figure 4.4 for negative $A_{RX}(s^*)$, when solvation effects have to be

Table 4.4 Calculated and Experimental Vertical Electron Affinities

Entry	R–X[a]	Calculated[b] (eV)	Experimental (eV)
1	CH$_3$–F	−2.5	−6.1,[c] −4.6,[d] −4.6[e]
2	FCl$_2$C–F	−1.5	−3.9[f]
3	CH$_3$–Cl	−1.3	−3.7,[c] −3.45,[f] −2.9,[g] −2.5,[h] −2.3,[i] −3.2[j]
4	ClCH$_2$–Cl	−0.7	−1.23[f]
5	Cl$_2$CH–Cl	−0.35	−0.35[f]
6	Cl$_3$C–Cl	+0.2	>0[f]
7	CH$_3$–Br	−0.9	−2.7,[k] −2.85,[l] −1.5[m]
8	CH$_3$–I	−0.43	−1.2,[k] −0.5,[n] ≃0[o]
9	CH$_3$–OCH$_3$	−2.6	−6,[c] −5.7[p]
10	CH$_3$–SCH$_3$	−2.2	−3.3,[c] −2.7,[q] −2.8,[r] −2.3[s]
11	CH$_3$–SH	−1.8	—
12	CH$_3$–SeH	−1.5	−1.5[t]
13	CH$_3$–CN	−3.0	−6.5[u]
14	CH$_3$–CCH	−3.4	—
15	CH$_3$–NH$_2$	−3.03	−4.8,[v] −5.2[w]
16	CH$_3$–PH$_2$	−2.5	−3.1,[x] −2.7[y]
17	CH$_3$–AsH$_2$	−2.1	−2.7,[c] −2.3[c]
18	CH$_3$–H	−2.9	−8,[z] −5,[za] −7.5[za]

[a]The line refers to the bond whose vertical electron affinity is observed and/or calculated.

[b]The calculations are carried out using equations 4.5 and 4.12. The data source is summarized in the tables of the Appendix.

[c]From references 16 and 17.

[d]Data for CH$_3$COCH$_2$F, from P. R. Olivato, S. A. Guerrero, A. Modelli, G. Granozzi, D. Jones, and G. Distefano. *J. Chem. Soc. Perkin Trans. 2.* 1505 (1984).

[e]Theoretical evaluation from reference 13.

[f]P. D. Burrow, A. Modelli, N. S. Chiu, and K. D. Jordan. *J. Chem. Phys.* **77**, 2699 (1982).

[g]Data for vinyl chloride from reference 17 and J. K. Olthoff, J. A. Tossell, and J. H. Moore. *J. Chem. Phys.* **83**, 5627 (1985). See also footnote f.

[h]Data for chlorobenzene from reference 17.

[j]Data for chloroacetone from footnote d.

[k]Estimated from the substituent effects seen in entry 3: A_{RX} of methyl chloride is lower than that of an aryl chloride by 1.2–1.4 eV. Thus, adjusting the experimental A_{RX} values of aryl bromides and aryl iodides by this amount leads to estimates of A_{RX} for methyl bromide and methyl iodide.

[l]This value is estimated by applying the 0.6 eV difference between A_{RX} of CCl$_4$ and CBr$_4$ (reference 17) to obtain $A_{CH_3Br} = A_{CH_3Cl} + 0.6$ eV.

[m]Data for p-dibromobenzene from reference 17.

[n]Data for vinyl iodide from reference 17.

[o]Data for p-diiodobenzene from refernce 17.

[p]Data for p-dimethoxybenzene from reference 16.

[q]Data for ethylthioacetone from reference d.

[r]Data for 3,3,6,6-tetramethyl-l-thiacycloheptane. See L. Ng, K. D. Jordan, A. Krebs, and W. Rüger. *J. Am. Chem. Soc.* **104**, 7414 (1982).

[s]Data for p-CH$_3$S–C$_6$H$_4$SCH$_3$ from reference 16.

[t]Data taken from selenofuran. See A. Modelli, M. Guerra, D. Jones, G. Distefano, K. J. Irgolic, K. French, and G. C. Pappalardo. *Chem. Phys.* **88**, 455 (1984).

[u]This is based on a theoretical assignment of the σ^*_{cc} state for acetaldehyde (reference 13). For the experimental data see R. A. Dressler, PhD Thesis, Freiberg, 1985; R. A. Dressler and M. Allan. *Chem. Phys. Lett.* **118**, 93 (1985).

Table 4.4 *(Continued)*

Data taken from trimethylamine, reference 16.
"Data taken from *N,N*-dimethylaminoaniline, reference 16.
'Data taken from trimethylphosphine, reference 16.
'Data taken from *p*-$(CH_3)_2PC_6H_4P(CH_3)_2$, reference 16.
"Reference 16.
'"Data are from K. Rohr. *J. Phys. B. Atom. Mol. Phys.* **13**, 4897 (1980); H. Tanaka, M. Kubo, N. Onodera, and A. Suzuki. *J. Phys. B. Atom. Mol. Phys.* **16**, 2861 (1983). According to a theoretical assignment (reference 13), the t_2 and a_1 states possess $A = -5.3$ and -8.3 eV, respectively.

taken into account. The symbols s and $s*$ refer to equilibrium and nonequilibrium solvation, respectively.

The desolvation energies $S_{X:}$ can be calculated from the expression

$$S_{X:} = (I_{X:}(s*) - I_{X:}(g))/(1 + \rho) \tag{4.15}$$

using the data of Delahay,[7] and setting the gas phase ionization potential of $X:^-$ equal to the electron affinity $A_{X:}(g)$. (The solvent reorganization factor ρ has been defined in equation 4.2.)

The numbers that result from equation 4.15 are very close to the accepted Noyes values[22a] of desolvation energies: for water solvent, equation 4.15 gives 77, 69, and 61 kcal/mol, respectively, as the desolvation energies of Cl^-, Br^-, and I^-, compared to the Noyes values of 75, 69, and 59. Other values[22b] are 75, 73, and 62, respectively.

Using equation 4.15, the value for HO^- is 97 kcal/mol, compared to the suggested[22c] 96–101 kcal/mol and experimentally calibrated values of 104–105 kcal/mol.[22d, e] For N_3^-, the calculated value is 71 kcal/mol, compared to Ritchie's value of 74 kcal/mol[22d] and Pearson's value of 72 kcal/mol.[22e]

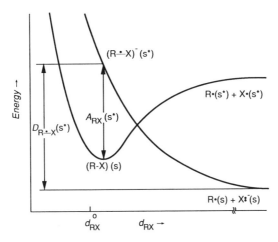

Figure 4.5 Definition of $D_{R \pm X}(s*)$. This figure is the counterpart of Figure 4.4 [for negative $A_{RX}(s*)$] when solvation effects are taken into account.

The data given above refer to water solvent. Values of $S_{X:}$ in DMF solvent can be estimated from those for water together with the free energies of transfer, $\Delta G_t(H_2O \rightarrow DMF)$, of the various anions. Information regarding ΔG_t is available from the work of Abraham[23] and Cox,[24] and can be used to estimate this quantity for other systems of interest. Table 4.5 summarizes the data.

The table includes a quantity, termed $S_{X:}^{\dagger}$, which describes solvation of X:$^-$ with one solvent molecule replaced by R·. In the original literature,[1d,e] this was set equal to the energy required to replace a tightly bound solvent molecule by a nonpolar group R·. If one supposes that there are, on average, four such tightly bound solvent molecules, then the energy needed to replace one of these by R· is approximately one-quarter of the free energy of transfer from a polar solvent to a nonpolar solvent. With benzene as the nonpolar solvent, this leads to

$$S_{X:}^{\dagger} = S_{X:} - \tfrac{1}{4}|\Delta G_t(s \rightarrow \text{benzene})| \qquad (4.16)$$

This is a crude approximation, and it may be better to treat the difference $(S_{X:} - S_{X:}^{\dagger})$ as the result of steric inhibition of solvation in (X:$^-$·R), relative to X:$^-$, although it is not obvious how this might be done in a qualitative manner. Accordingly, it is necessary to compare the estimates given in Table 4.5 against data obtained in a more rigorous manner. In computations by Jorgensen,[25] the free energy of hydration of Cl$^-$ is found to be 5 kcal/mol greater than that of the ion-dipole complex Cl$^-$···CH$_3$Cl. This compares to the 5 kcal/mol estimated in Table 4.5 for the difference in the hydration energies of Cl:$^-$ and Cl:$^-$·CH$_3$. Sim-

Table 4.5 $S_{X:}$, $S_{X:}^{\dagger}$ and ΔG_t (H$_2$O $\rightarrow$ DMF) Data for Anions X:$^-$

		H$_2$O					DMF	
Entry	X$^-$	$S_{X:}$	$(S_{X:}^{\dagger})^a$	Ref.	ΔG_t	Ref.	$S_{X:}$	$(S_{X:}^{\dagger})$
1	F$^-$	104	95	22a,d	13^d	24	91	85
2	Cl$^-$	77^a	72		10	23a	67	64
3	Br$^-$	69^a	65		7	23a	62	60
4	I$^-$	61^a	58		3	23a	58	56
5	HO$^-$	97^a	88		13^e		84	78
6	NC$^-$	73	68	22d	8^e		65	62
7	HS$^-$	79	72	22d	10^e		69	65
8	PhS$^-$	68	64	22d	7^e		61	59
9	CH$_3^-$	63^b	60		3^e		60	57
10	N$_3^-$	71		22d	8	24d	63	
11	HSe$^-$	69^c			7^e		62	

$^a S_{X:}$ values are from equation 4.15. $S_{X:}^{\dagger}$ values are from equation 4.16.
^{b}Calibrated relative to I$^-$ using the Jortner–Noyes equation, with $r(CH_3^-) = 2.1$ Å. See J. Jortner and R.M. Noyes. *J. Phys. Chem.* **70**, 770 (1966).
^{c}Estimated relative to HS$^-$.
^{d}This value is taken from Cox et al.,[24] but is calibrated to fit the ΔG_t values of Abraham.[23]
$^e \Delta G_t$ is matched to a known value for an anion having approximately the same desolvation energy.

ilarly, Jorgensen finds[5] that in DMF the free energy of solvation of Cl^- is 2-3 kcal/mol greater than that of the ion–molecule complex; this compares to the 3 kcal/mol shown in Table 4.5. Finally, we note that the free energies and enthalpies of solvation of HO^- and CH_3O^- in H_2O and DMSO differ by 11-13 kcal/mol.[22d, e, 26] This can be compared to our estimate of 9 kcal/mol for $HO:^-$ versus $HO:^- \cdot CH_3$ (entry 5 of Table 4.5).

Despite its crude nature, it appears that equation 4.16 is useful, and our need for an estimate of $S_{X:}^{\dagger}$ will become clear as we return to our examination of the quantity $D_{R \pm X}(s^*)$. The VB treatment that follows is identical to that given earlier for the gas phase, but with solvation effects now taken into account. Figure 4.6 depicts the changes in energy of $(R \cdot :X^-)$ and $(R:^- \cdot X)$ configurations as R and X fragments are brought from infinity to the equilibrium distance (d_{RX}^o) of the R–X molecule in a solvent. This equilibrium distance is also the R–X distance in the radical anion formed by a vertical electron transfer.

As d_{RX} is varied there is a variation in the solvation about R and X, so that, at d_{RX}^o, solvent molecules occupy the orientations associated with the equilibrated R–X *molecule*. At d_{RX}^o the two configurations are, therefore, in a state of non-equilibrium solvation. Interaction of the configurations causes the lower $(R \cdot :X^-)$ to mix with the upper $(R:^- \cdot X)$ to afford a delocalized radical anion in a state of nonequilibrium solvation. As before, this is represented by ΔE_{int}. We observe that $D_{R \pm X}(s^*)$ can be obtained from the cycle

$$(R \pm X)^-(s^*) \rightarrow (R \cdot :X^-)(s^*) \qquad \Delta E_1 = -\Delta E_{int} \qquad (4.17a)$$

$$(R \cdot :X^-)(s^*) \rightarrow R \cdot (s) + X:^-(s) \qquad \Delta E_2 = -\Delta E_{rep} \qquad (4.17b)$$

$$\overline{(R \pm X)^-(s^*) \rightarrow R \cdot (s) + X:^-(s) \qquad \Delta E = D_{R \pm X}(s^*) \qquad (4.17c)}$$

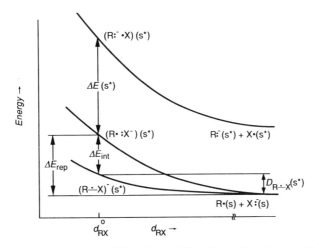

Figure 4.6 The changes in energy of $(R \cdot :X^-)$ and $(R:^- \cdot X)$ configurations as R and X fragments are brought from infinity to the equilibrium distance (d_{RX}^o) of the R–X molecule in a solvent. The lowest curve is a result of mixing of the two configuration curves.

so that

$$D_{R \pm X}(s^*) = -\Delta E_{int} - \Delta E_{rep} \qquad (4.18)$$

The quantity ΔE_{rep} can be obtained from the cycle

$$X:^-(s) + R \cdot (s) \rightarrow X:^-(g) + R \cdot (g) \qquad \Delta E_1 = S_{X:} + S_{R \cdot} \qquad (4.19a)$$

$$X:^-(g) + R \cdot (g) \rightarrow (R \cdot :X^-)(g) \qquad \Delta E_2 = m\sqrt{D_{RR}D_{XX}} \qquad (4.19b)$$

$$(R \cdot :X^-)(g) \rightarrow (R \cdot :X^-)(s) \qquad \Delta E_3 = -S_{X:}^{\dagger} \qquad (4.19c)$$

$$(R \cdot :X^-)(s) \rightarrow (R \cdot :X^-)(s^*) \qquad \Delta E_4 = \rho S_{X:}^{\dagger} \qquad (4.19d)$$

$$\overline{X:^-(s) + R \cdot (s) \rightarrow (R \cdot :X^-)(s^*) \qquad \Delta E = \Delta E_{rep}} \qquad (4.19e)$$

which leads to

$$\Delta E_{rep} = m\sqrt{D_{RR}D_{XX}} + \rho S_{X:}^{\dagger} + S_{X:} - S_{X:}^{\dagger} + S_{R \cdot}. \qquad (4.20)$$

All of the quantities in equation 4.20 have been defined in the preceding discussion.

The configuration mixing term is given by

$$\Delta E_{int} = \tfrac{1}{2} \Delta E(s^*) - \tfrac{1}{2} \sqrt{\Delta E(s^*)^2 + 4h_{RX}^2} \qquad (4.21)$$

where $\Delta E(s^*)$ is the energy difference between the two VB configurations of Figure 4.6, and we assume that solvation does not change the strength of the bond-coupling interactions.

In the gas phase, the energy difference between the two configurations was taken as the difference in electron affinities, $A_X - A_R$. The effect of solvation is shown in Figure 4.7, and is obtained from the cycles

$$(R \cdot :X^-)(g) \rightarrow (R \cdot :X^-)(s) \qquad \Delta E_1 = -S_{X:}^{\dagger} \qquad (4.22a)$$

$$(R \cdot :X^-)(s) \rightarrow (R \cdot :X^-)(s^*) \qquad \Delta E_2 = \rho S_{X:}^{\dagger} \qquad (4.22b)$$

$$\overline{(R \cdot :X^-)(g) \rightarrow (R \cdot :X^-)(s^*) \qquad \Delta E = (\rho - 1)S_{X:}^{\dagger}} \qquad (4.22c)$$

and

$$(R:^- \cdot X)(g) \rightarrow (R:^- \cdot X)(s) \qquad \Delta E_1 = -S_{R:}^{\dagger} \qquad (4.23a)$$

$$(R:^- \cdot X)(s) \rightarrow (R:^- \cdot X)(s^*) \qquad \Delta E_2 = \rho S_{R:}^{\dagger} \qquad (4.23b)$$

$$\overline{(R:^- \cdot X)(g) \rightarrow (R:^- \cdot X)(s^*) \qquad \Delta E = (\rho - 1)S_{R:}^{\dagger}} \qquad (4.23c)$$

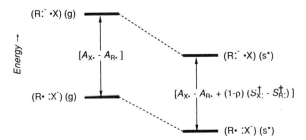

Figure 4.7 Effect of solvation on the energy difference between $(R:^- \cdot X)$ and $(R \cdot :X^-)$ configurations.

From equations 4.18, 4.20, and 4.21, we can write

$$D_{R \pm X}(s^*) = -m\sqrt{D_{RR}D_{XX}} - \rho S_{X:}^\dagger - S_{X:} + S_{X:}^\dagger - S_{R \cdot}$$

$$- \tfrac{1}{2} [(A_{X \cdot} - A_{R \cdot}) + (1 - \rho)(S_{X:}^\dagger - S_{R:}^\dagger)]$$

$$+ \tfrac{1}{2} \sqrt{(A_{X \cdot} - A_{R \cdot} + (1 - \rho)(S_{X:}^\dagger - S_{R:}^\dagger))^2 + 4h_{RX}^2} \quad (4.24)$$

and the final expression for the vertical electron affinity of R–X in a solvent is

$$A_{RX}(s^*) = A_{X \cdot} - D_{RX} - m\sqrt{D_{RR}D_{XX}} - \tfrac{1}{2}(A_{X \cdot} - A_{R \cdot}) - S_{RX}$$

$$- \tfrac{1}{2}(1 - \rho)(S_{X:}^\dagger + S_{R:}^\dagger)$$

$$+ \tfrac{1}{2}\sqrt{(A_{X \cdot} - A_{R \cdot} + (1 - \rho)(S_{X:}^\dagger - S_{R:}^\dagger))^2 + 4h_{RX}^2} \quad (4.25)$$

The data of Table 4.6 are obtained from equation 4.25, with neglect of the S_{RX} term (ca. 1 kcal/mol). The trends in Tables 4.3, 4.4, and 4.6 are much the same:

Table 4.6 A_{RX} Values for CH$_3$X Molecules

Entry	X	$A_{RX}(g)$ (kcal/mol)	$A_{RX}(s^*)$ (kcal/mol)	
			H$_2$O	DMF
1	F	−57	−19.1	−16.5
2	Cl	−30	−0.1	+1.9
3	Br	−21	+7.4	+10.1
4	I	−10	+15.5	+19.1
5	HO	−65	−30.5	−28.2
6	HS	−42	−11.8	−9.7
7	PhS		−7.7	−4.9
8	NC	−68	−39.2	−36.6

$A_{RX}(s^*)$ increases down a column and along a row of the periodic table, as does $A_{RX}(g)$.

4.1.5   Vertical Electron Transfer Energies

For an identity S$_N$2 reaction, for example, $X^- + CH_3X \rightarrow XCH_3 + X^-$, the energy gap of the SCD in the gas phase is given by $(I_{X:} - A_{RX})$. From Tables 4.1, 4.3, and 4.4 we can now construct Table 4.7 for the gas phase. These data and the corresponding values for energy gaps in solution, based on Tables 4.1 and 4.6, will comprise the data base for our detailed examination of the identity process in Chapter 5.

We may, however, note at this stage that the substituents of Table 4.7 have been grouped into "families." Within a so-defined family (entries 1-4, 5-9), there is a progressive decrease in the vertical electron transfer energy gap. There are also equivalences in different families. For example, entries 2 and 6,7 and also entries 3 and 8,9 exhibit approximately equal gaps. The largest energy gap is found for cyanide/methyl cyanide; the smallest is observed with iodide/methyl iodide and also with amide/methylamine.

All of these trends are common to the calculated and experimental data sets. In Chapter 5 our analysis will, therefore, focus upon the calculated trends. This will

Table 4.7   Gas Phase Vertical Electron Transfer Energies for the Identity S$_N$2 Reaction X$^-$ + CH$_3$X → XCH$_3$ + X$^-$

Entry	X	$I_{X:} - A_{RX}$ (eV)[a]	
		Calculated	Experimental
1	F	5.9	8.0(9.6)[b]
2	Cl	4.9	7.1, 7.4
3	Br	4.3	6.1, 5.2
4	I	3.6	4.3,[c] 3.6[d]
5	CCH	6.3	≈9.4
6	OH	4.7	≈7.9
7	OCH$_3$	4.2	7.6
8	SH	4.1	≈5.6
9	SCH$_3$	4.1	5.2
10	CN	6.9	≈10.4
11	NH$_2$	≈3.9	5.6
12	H	3.7	≈5.8[e] (≈8.25)[f]

[a]See Table 4.4 for A_{RX} values.
[b]Data in parentheses are based on $A_{MeF} = -6.1$ eV; others are based on $A_{MeF} = -4.6$ eV.
[c]Using $A_{MeI} = -1.2$ eV.
[d]Using $A_{MeI} = -0.5$ eV, the value for vinyl iodide.
[e]Using $A_{MeH} = -5$ eV.
[f]Using $A_{MeH} = -7.5$ eV.

allow an examination of the S_N2 process, in the gas phase and in solution, using a common data base.

4.2 THE CURVATURE FACTORS (BOND-COUPLING DELAY INDICES)

In the SCD of Figure 4.1, the height of the crossing point, ΔE_c, is seen to be a fraction, f, of the energy gap. Relative to the energy at the reactant extremum of the diagram, we may write

$$\Delta E_c = f(I_{Y:} - A_{RX}) \tag{4.26}$$

As we have pointed out in Section 3.3.2, it is useful to regard f as a bond-coupling delay index. For example, electronic delocalization of the odd electron in $Y \cdot$ or of the odd electron in $(R \overset{\bullet}{\cdot} X)^-$, which together comprise the bond pair of the charge transfer state, will be a significant factor leading to a large f.

If we apply the reasoning of structure **3.14** to this situation, the curve of Figure

4.3

4.1 which correlates the $Y \cdot /(R \overset{\bullet}{\cdot} X)^-$ and $(Y-R)/:X^-$ configurations is seen to couple the two odd electrons (in χ_Y and σ^*_{CX} of **4.3**) along the reaction coordinate. This coupling represents the stabilization that allows the charge transfer state to descend toward the product. Any effect that inhibits or delays this coupling, such as the delocalization just mentioned, should lead to a shallow descent of the charge transfer state (large f).

This idea, which seemed plausible when it was originally proposed,[1] has since been tested using ab initio VB calculations of the SCD for the reaction H^- + CH_3-H.[27] The use of a localized charge transfer state (**4.4a**) leads to $f \approx 0.27$, while **4.4b**, in which the odd electron is delocalized over all of the atoms of $(CH_4)^{\overset{\bullet}{\cdot}}$, leads to $f = 0.42-0.44$.

f = 0.27   f = 0.42 - 0.44

4.4a   **4.4b**

Obviously, delocalization effects might be manifested via $Y\cdot$ and/or via $(R \overset{\pm}{-} X)\cdot^-$. Delocalization via $Y\cdot$ is straightforward: any delocalized nucleophile $(PhCH_2\cdot, N_3\cdot, RCO_2\cdot, \ldots)$ will have a delocalized $Y\cdot$, and shallow descent of the charge transfer state should be expected.

Delocalization in $(R \overset{\pm}{-} X)^-$ requires a more detailed analysis. In particular, we wish to evaluate the contribution of the $(R\cdot:X^-)$ configuration to the VB description of the structure. When this is the dominant configuration, and the odd electron is localized on the union center R, bond coupling with $Y\cdot$ will be most efficient, and a small f should be expected.

For a VB description of the form

$$(R \overset{\pm}{-} X)^- = w_{X:}(R\cdot\ :X^-) + w_{R:}(R:^-\ \cdot X)$$

$$\cdot\ w_{X:} + w_{R:} + 2\sqrt{w_{X:}w_{R:}}\langle R\cdot\ :X^-|R:^-\ \cdot X\rangle = 1 \quad (4.27)$$

if overlap is neglected the weights of the two configurations are $w_{X:}$ and $w_{R:}$. Whenever $w_{R:} \neq 0$, this means that the total interaction of $Y\cdot$ with $(R \overset{\pm}{-} X)^-$ involves a repulsive three-electron contribution (via $Y\cdot :R^-$) together with the stabilizing two electron coupling (via $Y\cdot\ R\cdot$). The effect of delocalization is therefore to place two electrons (rather than one) on the union center R. Thus, an increase in $w_{R:}$ will cause a weakening of the bond coupling with $Y\cdot$ and lead to an increase in f. From equation 4.10, for the interaction of two configurations, the expression for $w_{R:}$ in the gas phase (with neglect of overlap) is

$$w_{R:} = \tfrac{1}{2}[1 - \Delta/\sqrt{\Delta^2 + 4}]; \quad \Delta = |(A_{X\cdot} - A_{R\cdot})/h_{RX}| \quad (4.28)$$

where the difference $A_{X\cdot} - A_{R\cdot}$ is as defined in Figure 4.7. From equation 4.25 and Figure 4.7, the corresponding expression for $w_{R:}$ in a solvent, termed $w_{R:}^*$, is

$$w_{R:}^* = \tfrac{1}{2}[1 - \Delta^*/\sqrt{\Delta^{*2} + 4}]$$

$$\Delta^* = |[(A_{X\cdot} - A_{R\cdot}) + (1 - \rho)(S_{R\cdot X:} - S_{R:X:})]/h_{RX}|$$

$$\approx |[(A_{X\cdot} - A_{R\cdot}) + (1 - \rho)(S_{X:} - S_{R:})]/h_{RX}| \quad (4.29)$$

where we have made the simplifying assumption that $S_{X:}^\dagger - S_{R:}^\dagger \cong S_{X:} - S_{R:}$.

Table 4.8 collects $w_{R:}$ and $w_{R:}^*$ values calculated with equations 4.28 and 4.29, and some qualitative trends may be noted at this stage: the radical anions of the methyl halides possess large $(A_{X\cdot} - A_{R\cdot})$ gaps, and this is reflected in the existence of relatively localized radical anions ($w_{R:} \cong 0.25$). On the other hand, radical anions such as $(CH_3 \overset{\pm}{-} OH)^-$, $(CH_3 \overset{\pm}{-} SH)^-$, and $(CH_3 \overset{\pm}{-} NH_2)^-$ are characterized by smaller $A_{X\cdot} - A_{R\cdot}$ and greater delocalization ($w_{R:} \cong 0.35$–0.45). Radical anions such as $(CH_3 \overset{\pm}{-} CCH)^-$ and $(CH_3 \overset{\pm}{-} CN)^-$ are delocalized because of their strong C–X bonds. Finally, because the differential desolvation term is small, the trends in solution are determined by those in the gas phase.

Table 4.8 $w_{R.}$ and w_R^* Values for $(CH_3 \pm X)^{-a,b}$

| | | | | | w_R^*: | |
Entry	X	$A_{X.} - A_{R.}$	$h_{RX}{}^c$	$w_{R:}$	H_2O	DMF
1	F	76	63.00	0.242	0.206	0.208
2	Cl	82	71.00	0.251	0.239	0.243
3	Br	76	64.60	0.246	0.241	0.242
4	I	69	56.80	0.241	0.243	0.242
5	OCOCH$_3$	76	51.7	0.203		
6	OH	42	70.77	0.357	0.320	0.324
7	OCH$_3$	35	58.00	0.355	—	—
8	SH	52	76.61	0.340	0.328	0.329
9	SCH$_3$	42	76.61	0.370	—	—
10	CN	87	106.2	0.309	0.304	0.305
11	CCH	66	114.89	0.362		
12	NH$_2$	18	78.90	0.450		
13	CH$_2$CH	33	76.78	0.390		
14	CH$_2$Ph	18	71.44	0.436		
15	H	16	95.79	0.453		

$^a A_{X.} - A_{R.}$ and h_{RX} are in kcal/mol.
^{b}The data source is provided in the Appendix.
$^c h_{RX} = m(D_{RR} D_{XX})^{1/2}$.

In general, from the standpoint of the interaction between R and X, delocalization of the radical anion is seen when X has a low gas phase electron affinity and/or forms a strong bond to R. However, if we now recall the calculated values of f associated with **4.4a** and **4.4b**, we realize that we have to take an alternative mode of delocalization into account whenever the substrate possesses two or more identical leaving groups, as in CH_4 or CH_2Cl_2, for example. In the radical anion formed by vertical electron transfer to CH_2Cl_2, the odd electron is delocalized across individual C–Cl bonds in the sense of **4.5a**, already discussed. It is delocalized further *between* the enantiotopic C–Cl bonds in the sense of **4.5b**. The

4.5a   **4.5b**

latter interaction diagram describes the mixing of two bond-localized VB configurations to yield a delocalized radical anion that is stabilized by the quantity DE.[lc,d]
As the number of such VB configurations increases, the overall delocalization of the single electron will also increase, and so will f. This leads us to the
following statement:

Statement 4.1   Any factor that delocalizes the odd electrons in the charge transfer
states $Y \cdot / (R \doteq X)^-$ and $(Y \doteq R)^- / \cdot X$ will cause shallow descent of these states
and f will increase.

The different kinds of delocalization effects are summarized in Table 4.9. Their
effects upon f are shown schematically in Figures 4.8a and b, which are based on
Figure 4.1, and in equations 4.30a–c.

$$f \propto w_{R:} \tag{4.30a}$$

$$f \propto N(X, Y) \tag{4.30b}$$

$$f \propto 1/a_X. \quad \text{and} \quad 1/a_Y. \tag{4.30c}$$

Here $w_{R:}$ has the meaning already discussed, N is the number of leaving groups of
the same type that are attached to the central carbon atom and $a_X.$ and $a_Y.$ are the
coefficients of the odd electrons on the union centers of $X \cdot$ and $Y \cdot$.
    All of these are factors whose effect upon f can to some extent be quantified.
We shall, therefore, focus upon them in our analysis while recognizing that other,
less readily quantifiable, effects may also influence the magnitude of f. For example, *steric effects*, which are the result of overlap repulsion between filled orbitals, will delay bond coupling. This is illustrated in **4.6**, a charge transfer state

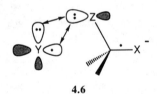

4.6

**Table 4.9   Properties that Lead to Delocalization in the Charge Transfer States
$(Y \cdot R \doteq X)^-$ and $(X \cdot R \doteq Y)^-$**

Entry	Property	Result
1	Stable R:$^-$ (large A_R.)	Delocalized $(R \doteq X)^-$ and $(R \doteq Y)^-$
2	Unstable X:$^-$, Y$^-$ (small A_X., A_Y.)	Delocalized $(R \doteq X)^-$ and $(R \doteq Y)^-$
3	Large h_{RX}, h_{RY}[a]	Delocalized $(R \doteq X)^-$ and $(R \doteq Y)^-$
4	Several identical leaving groups	Delocalized $(RX)^{\doteq}$, for example $(CH_4)^{\doteq}$, $(CH_2X_2)^{\doteq}$
5	π-conjugation in X$\cdot$, Y$\cdot$	Delocalized X$\cdot$, Y$\cdot$

$^a h_{RX} = m(D_{RR}D_{XX})^{1/2}.$

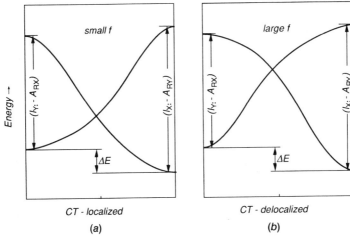

Figure 4.8   (a) Steep descent from the charge transfer state (small f), expected for a localized charge transfer state. (b) Shallow descent from the charge transfer state (large f), an expected property of delocalized radical anions.

in which $Y\cdot$ engages in repulsive three- and four-electron interactions with a substituent Z of the substrate. Such an effect will increase f, but is difficult to quantify.

On the other hand, *electrostatic interactions* in the ground states of the SCD might temper the ascent of the curves and reduce f. This effect is also difficult to formulate and quantify in a generalized form. Finally, secondary configuration mixing into the curves of Figure 4.1 can affect the descent of the charge transfer states.[1] For example, bond coupling in $Y\cdot/(R \overset{\bullet}{-} X)^-$ would be aided by a mixing in of the locally excited configuration $Y:^-(R^+ \ X:^-)$; when this configuration is low in energy a steeper descent of the charge transfer state will occur and f will decrease. Again, to take such an effect into account would require a substantial elaboration of the qualitative VB analysis, with concomitant loss of the insights developed to this point. As we shall find in Chapters 5 and 6, despite the caveats just expressed, our decision to avoid explicit consideration of steric effects, electrostatic interactions and secondary configuration mixing does not have serious consequences.

4.3   THE REACTION ERGICITY

The general effect of the reaction ergicity on the barrier has been treated in Chapter 3 (Section 3.3.4). Here we take note of thermochemical expressions for this quantity.

The quantity appropriate to the SCD of Figure 4.1 is ΔE°. We assume that we can relate this to the reaction enthalpy ΔH° and free energy ΔG°. A simple expression for the ergicity of the gas phase S_N2 reaction $Y^- + R\text{-}X \rightarrow Y\text{-}R + X^-$ is

$$\Delta Q^\circ(g) \approx A_{Y\cdot} - A_{X\cdot} + D_{RX} - D_{RY}; \qquad \Delta Q = \Delta E, \Delta H, \Delta G \quad (4.31)$$

in which A's are electron affinities and D's are bond energies. This equation is appropriate for ΔE°, ΔH°, and ΔG°, provided that the corresponding quantities are used for the electron attachment and bond-breaking energies in equation 4.31. In solution, ergicity can be written as

$$\Delta Q^\circ(s) \approx \Delta Q(g) + S_{Y:} - S_{X:} \tag{4.32}$$

in which S's refer to desolvation energies.

Ergicity is thus determined by the balance between electron affinity differences, bond energy differences, and differential solvation energies. Since these differences will also act upon the other reactivity factors, any discussion of the BEP principle, or any derivation of Brönsted parameters from a rate–equilibrium relationship, must first demonstrate that changes in ΔQ are not accompanied by changes in other barrier factors. As we shall find in Chapter 6, this point has some bearing upon the breakdown of the BEP principle and related concepts.

APPENDIX

Table A4.1 is a compilation of the $A_{X\cdot}$, D_{R-X}, and h_{RX} data that are employed in the calculation of S_N2 reactivity factors in the gas phase. For example, $I_{X:}(g)$ is taken as $A_{X\cdot}$, $A_{RX}(g)$ is given by equation A4.1, and $D_{R\pm X}$ is given by equation

$$A_{RX}(g) = A_{X\cdot} - D_{R-X} + D_{R\pm X} \tag{A4.1}$$

A4.2. In each case, R is methyl and h_{RX} is calculated according to equation A4.3,

$$D_{R\pm X} = -h_{RX} - \tfrac{1}{2}(A_{X\cdot} - A_{R\cdot}) + \tfrac{1}{2}\sqrt{(A_{X\cdot} - A_{R\cdot})^2 + h_{RX}^2} \tag{A4.2}$$

$$h_{RX} = \frac{m\sqrt{D_{RR}D_{XX}}}{2\langle \chi_R | \chi_X \rangle} \tag{A4.3}$$

where the overlap $\langle \chi_R | \chi_X \rangle$ is given a representative value of 0.5.[1b,d]

Table A4.2 collects the corresponding data for second- and third-row elements. Since the primary information required in these cases is not, in general, available, and must be estimated, the A_{RX} values listed here should be regarded as more approximate than in Table A4.1.

Table A4.3 collects A_{RX} data that are derived from $D_{R\pm X}$ values calculated with retention of overlap in the normalization constants. In this case, equations A4.4–A4.6 are the appropriate expressions for the calculation of the overlap repulsion (OR) in $R\cdot:X^-$ and $R:^-\cdot X$.

Table A4.1 $A_{X\cdot}$, D_{R-X}, and h_{RX} Data (kcal/mol) for CH_3X Compounds

X	$A_{X\cdot}$	Source[a]	D_{RX}	Source[b]	h_{RX}	Source[c]
F	78	1	108	1	63.0	1
Cl	83.4	1	84	1	71.0	1
Br	78	1	71	1	64.6	1
I	70.6	2	56	1	56.8	1
OCOCH$_3$	78.2	3	83	2	51.7	2
OH	44 (42.1)	4 (1)	91	1	70.8	2
OCH$_3$	36.7	1	81	1	58.0	2
SH	53.5	1	74	2	76.6	2
SCH$_3$	43.4	1	77	2	76.6	2
SPh	56.9	1	69.4	2	76.6	2
SeH	51	1	—	—	—	—
CN	89 (88)	4 (1)	122	2, 3	106.2	1, 3
CCH	67.8	5	117	3	114.9	1, 3
NH$_2$	19.4 (17.9)	4 (1)	81	1	78.9	1, 2
PH$_2$	29	1	—	—	—	—
H	17.4	2	104	1	95.9	1
CH$_2$Ph	20.4	6	72	1	71.4	3
CH$_2$CN	35	7	73	4	76.8	3
CH$_3$	1.8	8	90	2	—	—
OOH	27.5	9	72	4	51.7	2
CH$_2$COCH$_3$	41.2	1	81	5	—	—

[a]1. B. K. Janousek and J. I. Brauman. In *Gas Phase Ion Chemistry*. M. T. Bowers, Editor, Academic Press, New York, 1979. Volume 2, Chapter 10. 2. E. C. M. Chen and W. E. Wentworth. *J. Chem. Educ.* **52**, 486 (1975). 3. K. Hiraoka, R. Yamdagni, and P. Kebarle. *J. Am. Chem. Soc.* **95**, 6833 (1973). 4. K. Tanaka, G. I. Mackay, J. D. Payzant, and D. K. Bohme. *Can. J. Chem.* **54**, 1643 (1976). 5. B. K. Janousek, J. I. Brauman, and J. Simons. *J. Chem. Phys.* **71**, 2057 (1979). 6. F. K. Meyer, J. M. Jasinski, R. N. Rosenfeld, and J. I. Brauman. *J. Am. Chem. Soc.* **104**, 663 (1982). 7. A. H. Zimmerman and J. I. Brauman. *J. Am. Chem. Soc.* **99**, 3565 (1977). 8. G. B. Elison, P. C. Engelking, and W. C. Lineberger. *J. Am. Chem. Soc.* **100**, 2556 (1978). 9. V. M. Bierbaum, R. J. Schmitt, C. H. DePuy, R. D. Mead, P. A. Schulz, and W. C. Lineberger. *J. Am. Chem. Soc.* **103**, 6262 (1981).

[b]1. R. T. Sanderson. *Chemical Bonds and Bond Energy*. Academic Press, New York, 1976 (Table 10-1). 2. D. F. McMillen and D. M. Golden. *Annu. Rev. Phys. Chem.* **33**, 493 (1982). 3. S. W. Benson, *J. Chem. Educ.* **42**, 502 (1965). 4. Estimated, from substituent effects, e.g., $D_{R-OOH} = D_{R-OMe} - (2/3)\Delta$, where $\Delta = D_{H-OMe} - D_{H-OOH}$. The 2/3 factor represents the ratio of substituent effects upon C–O and H–O, that is, $(D_{Me-OH} - D_{Me-OMe})/(D_{H-OH} - D_{H-OMe})$. 5. Estimated from ΔH_f° data in reference 3 above.

[c]1. J. A. Kerr, M. J. Parsonage, and A. F. Trotman-Dickenson. *Handbook of Chemistry and Physics*. CRC Press, Cleveland, OH, 1976, p. F-204. 2. Estimated values using data from D. F. McMillen and D. M. Golden. *Annu. Rev. Phys. Chem.* **33**, 493 (1982). 3. S. W. Benson. *J. Chem. Educ.* **42**, 502 (1965).

Table A4.2 A_X, D_{R-X}, and h_{RX} Data (kcal/mol) for CH$_3$X Compounds

X	A_X	D_{R-X}	h_{RX}^g	A_{RX}
SeH	51[a]	62[d]	68.9	−31
TeH	≃51[b]	50[d]	64.3	−23
PH$_2$	29[a]	73[e]	68.9	−57
AsH$_2$	≃29[c]	67[f]	60.1	−50
SbH$_2$	≃29[c]	61[f]	53.9	−44

[a]B. K. Janousek and J. I. Brauman. In *Gas Phase Ion Chemistry*. M. T. Bowers, Editor, Academic Press, New York, 1979, Volume 2, Chapter 10.

[b]Estimated to be the same as A_X of SeH.

[c]Estimated to be the same as A_X of PH$_2$.

[d]Estimated as $D_{C-X} = D_{C-S} - \Delta$, where $\Delta = D_{H-S} - D_{H-X}$ (X=Se, Te). The bond energy data are from R. T. Sanderson. *Polar Covalence*. Academic Press, New York, 1983.

[e]Estimated from the data in footnote *f*.

[f]Data from D. F. McMillen and D. M. Golden. *Annu. Rev. Phys. Chem.* **33**, 493 (1982).

[g]Estimated from the D_{X-X} data given by Sanderson (see footnote *d*), and the equation $h_{RX} = m(D_{RR}D_{XX})^{1/2}$ where $m = 1$.

$$N^2[2h_{RX}\langle\chi_R|\chi_X\rangle] = m\sqrt{D_{RR}D_{XX}}$$

$$N^2 = 1/(1 + |\langle\chi_R|\chi_X\rangle|^2) \tag{A4.4}$$

$$OR = -(N')^2[2h_{RX}|\langle\chi_R|\chi_X\rangle|]$$

$$(N')^2 = 1/(1 - |\langle\chi_R|\chi_X\rangle|^2) \tag{A4.5}$$

$$OR = -\frac{1 + |\langle\chi_R|\chi_X\rangle|^2}{1 - |\langle\chi_R|\chi_X\rangle|^2} m\sqrt{D_{RR}D_{XX}} \tag{A4.6}$$

The interaction matrix element is now given by equation A4.7[28] and, using a

$$\langle R\cdot\ :X^-|h_{eff}|R:^-\cdot X\rangle = \frac{1 + |\langle\chi_R|\chi_X\rangle|^2}{1 + |\langle\chi_R\chi_X\rangle|} h_{RX} \tag{A4.7}$$

Table A4.3 Vertical Electron Affinities of CH$_3$–X Compounds Calculated Using Equations A4.1–A4.7

X	A_{RX} (eV)
F	−4.7
Cl	−3.8
Br	−3.0
I	−2.4
OCH$_3$	−4.4
SH	−4.7
CN	−6.7
CCH	−7.5
H	−6.3

representative value of 0.5 for $\langle \chi_R | \chi_X \rangle$,[1b,d] the data of Table A4.3 are obtained. The A_{RX} values that result are more negative than those given in Tables 4.3 and 4.4 of the text but, with the exception of X = SH *versus* OH, the trends in the two sets are the same. The set derived here is closer in magnitude to the experimental ETS values of Table 4.4. The qualitative conclusions of Chapter 5 are not affected by the particular choice of A_{RX}.

REFERENCES

1. (a) S. S. Shaik. *Nouv. J. Chim.* **6**, 159 (1982); (b) S. S. Shaik and A. Pross. *J. Am. Chem. Soc.* **104**, 2708 (1982); (c) S. S. Shaik. *J. Am. Chem. Soc.* **105**, 4359 (1983); (d) S. S. Shaik. *Progr. Phys. Org Chem.* **15**, 197 (1985); (e) S. S. Shaik. *J. Am. Chem. Soc.* **106**, 1227 (1984); (f) S. S. Shaik. *Isr. J. Chem.* **26**, 367 (1985); (g) D. J. Mitchell, H. B. Schlegel, S. S. Shaik, and S. Wolfe. *Can. J. Chem.* **63**, 1642 (1985); (g) S. S. Shaik. *Acta Chem. Scand.* **44**, 205 (1990).

2. (a) M. J. Pellerite and J. I. Brauman. *J. Am. Chem. Soc.* **102**, 5993 (1980); (b) M. J. Pellerite and J. I. Brauman. *J. Am. Chem. Soc.* **105**, 2672 (1983); (c) J. A. Dodd and J. I. Brauman. *J. Am. Chem. Soc.* **106**, 5356 (1984); (d) W. N. Olmstead and J. I. Brauman. *J. Am. Chem. Soc.* **99**, 4219 (1977); (e) C. A. Lieder and J. I. Brauman. *J. Am. Chem. Soc.* **96**, 4028 (1974); (f) J. I. Brauman, W. N. Olmstead, and C. A. Lieder. *J. Am. Chem. Soc.* **96**, 4030 (1974); (g) C. H. DePuy, J. J. Grabowski, and V. M. Bierbaum. *Science* **218**, 955 (1982); (h) D. K. Bohme, G. I. Mackay, and J. D. Payzant. *J. Am. Chem. Soc.* **97**, 4027 (1974); (i) D. K. Bohme and L. B. Young. *J. Am. Chem. Soc.* **92**, 7354 (1970); (j) J. D. Payzant, K. Tanaka, L. D. Betowski, and D. K. Bohme. *J. Am. Chem. Soc.* **98**, 894 (1976); (k) S. E. Barlow, J. M. Van Doren, and V. M. Bierbaum. *J. Am. Chem. Soc.* **110**, 7240 (1988).

3. For mass spectrometric characterization of ion–molecule complexes, see: (a) R. C. Dougherty, J. Dalton, and J. D. Roberts. *Org. Mass Spectrom.* **8**, 77 (1974); (b) R. C. Dougherty and J. D. Roberts. *Org. Mass Spectrom.* **8**, 81 (1974); (c) R. C. Dougherty. *Org. Mass Spectrom.* **8**, 85 (1974).

4. Loose complexes have been observed in S_N2 reactions in acetonitrile solvent. See: (a) J. Hayami, T. Koyanagi, and A. Kaji. *Bull. Chem. Soc. Jpn.* **52**, 1441 (1979); (b) J. Hayami, N. Tanaka, N. Hihara, and A. Kaji. *Tetrahedron Lett.* 385 (1973); (c) J. Hayami, T. Koyanagi, N. Hihara, and A. Kaji. *Bull. Chem. Soc. Jpn.* **51**, 891 (1978).

5. Loose complexes for the Cl^-/CH_3Cl exchange reaction in DMF solvent have been computed ab initio by J. Chandrasekhar and W. L. Jorgensen. *J. Am. Chem. Soc.* **107**, 2974 (1985).

6. Gas phase ionization potentials of anions can be found in: (a) B. K. Janousek and J. I. Brauman. In *Gas Phase Ion Chemistry*. M. T. Bowers, Editor, Academic Press, New York, 1979, Volume 2, Chapter 10; (b) E. C. M. Chen and W. E. Wentworth. *J. Chem. Educ.* **52**, 486 (1975); (c) P. S. Drzaic, J. Marks, and J. I. Brauman. In *Gas Phase Ion Chemistry*. M. T. Bowers, Editor, Academic Press, New York, 1984, Volume 3, Chapter 21. These references provide $A_{Y\cdot}$ rather than vertical $I_{Y\cdot}$ values. For polyatomic nucleophiles vertical $I_{Y\cdot}(g)$ values are expected to be slightly higher than those given here. For example, the vertical $I_{Y\cdot}(g)$ for methyl anion is *ca* 7 kcal/mol. Other differences may be even smaller. See, for example: P. A. Schulz, R. D. Mead, P. L. Jones, and W. C. Lineberger. *J. Chem. Phys.* **77**, 1153 (1982); G. B. Ellison,

P. C. Engelking, and W. C. Lineberger. *J. Phys. Chem.* **86**, 4873 (1982); P. C. Engelking, G. B. Ellison, and W. C. Lineberger. *J. Chem. Phys.* **69**, 1826 (1978); (d) S. G. Lias, J. E. Bartmess, J. F. Liebman, J. L. Holmes, R. D. Levin, and W. G. Mallard. *Gas Phase Ion and Neutral Thermochemistry.* J. Phys. Chem. Ref. Data. **17**, Suppl. 1, 1988.

7. P. Delahay. *Acc. Chem. Res.* **15**, 40 (1982).

8. M. J. Blandamer and M. F. Fox. *Chem. Revs.* **70**, 59 (1970). With $I_{Y:}(s^*) = h\nu_{CTTS} + C$, C is an electron solvation term that varies with the solvent; $C(H_2O) \cong 38$–39 kcal/mol; $C(CH_3CN) \cong 34$ kcal/mol.

9. (a) S. S. Shaik. *J. Org. Chem.* **52**, 1563 (1987); (b) E. Buncel, S. S. Shaik, I. Um, and S. Wolfe. *J. Am. Chem. Soc.* **110**, 1275 (1988).

10. R. A. Marcus. *Annu. Rev. Phys. Chem.* **15**, 155 (1964).

11. See, for example: (a) K. D. Jordan and P. D. Burrows. *Chem. Revs.* **87**, 557 (1987); (b) J. L. Franklin and P. W. Harland. *Annu. Rev. Phys. Chem.* **25**, 485 (1974); (c) G. L. Gutsev and A. I. Boldyrev. *Adv. Chem. Phys.* **61**, 169 (1985); (d) G. Briegleb. *Angew. Chem., Intl. Ed. Engl.* **3**, 617 (1964); (e) J. P. Lowe, *J. Am. Chem. Soc.* **99**, 5557 (1977); (f) P. Kebarle and S. Chowdhury. *Chem. Revs.* **87**, 513 (1987); (g) E. C. M. Chen and W. E. Wentworth. *J. Chem. Phys.* **63**, 3183 (1975); (h) J. Baker, R. H. Nobes, and L. Radom. *J. Comp. Chem.* **7**, 349 (1986); (i) J. A. Pople, P. von R. Schleyer, J. Kaneti, and G. W. Spitznagel. *Chem. Phys. Lett.* **145**, 359 (1988); (j) A. A. Christodoulides, D. L. McCorkle, and L. G. Christophorou. *Electron Molecule Interactions and their Applications.* Academic Press, New York, 1984, Vol. 2, pp. 423–641.

12. M. B. Robin. *Can. J. Chem.* **63**, 2032 (1985).

13. E. Lindholm and J. Li. *J. Phys. Chem.* **92**, 1731 (1988).

14. (a) M. J. W. Boness and G. J. Schulz. *Phys. Rev.* **A9**, 1969 (1974); (b) G. J. Schulz. *Phys. Rev.* **112**, 150 (1958); (c) L. Sanche and G. J. Schulz. *Phys. Rev.* **A5**, 1672 (1972).

15. K. D. Jordan and P. D. Burrow. *Acc. Chem. Res.* **11**, 341 (1978).

16. J. C. Giordan, J. H. Moore, and J. A. Tossell. *Acc. Chem. Res.* **19**, 281 (1986).

17. J. K. Olthoff, *PhD Dissertation.* University of Maryland, USA, 1985.

18. Reference 17, pp. 89–92.

19. R. T. Sanderson. *Chemical Bonds and Bond Energy.* Academic Press, New York, 1976.

20. T. A. Albright, J. K. Burdett, and M. H. Whangbo. *Orbital Interactions in Chemistry.* Wiley-Interscience, New York, 1985, Chapter 2.

21. G. R. Stevenson and R. T. Hashim. *J. Am. Chem. Soc.* **107**, 5794 (1985).

22. (a) R. M. Noyes. *J. Am. Chem. Soc.* **84**, 513 (1962); *ibid.* **86**, 971 (1964); M. H. Abraham. *J. Chem. Soc. Perkin II.* 1893 (1973); D. J. McLennan. *Aust. J. Chem.* **31**, 1897 (1978); (b) S. Goldman and R. Bates. *J. Am. Chem. Soc.* **94**, 1476 (1972); (c) R. Gomer and G. Tryson. *J. Chem. Phys.* **66**, 4413 (1977); (d) C. D. Ritchie. *J. Am. Chem. Soc.* **105**, 7313 (1983); C. D. Ritchie. In *Nucleophilicity.* Advances in Chemistry Series. American Chemical Society, Washington, DC, 1986; (e) R. G. Pearson. *J. Am. Chem. Soc.* **108**, 6109 (1986).

23. (a) M. H. Abraham. *J. Chem. Soc. Perkin Trans. II.* 1375 (1976); (b) M. H. Abraham and J. Liszi. *J. Inorg. Nucl. Chem.* **43**, 143 (1981); (c) M. H. Abraham and J. Liszi. *J. Chem. Soc. Faraday Trans. I.* **74**, 1604 (1978).

24. G. B. Cox, G. R. Hedwig, A. J. Parker, and D. W. Watts. *Aust. J. Chem.* **27,** 477 (1974).

25. J. Chandrasekhar, S. F. Smith, and W. L. Jorgensen. *J. Am. Chem. Soc.* **104,** 3049 (1984); *ibid.* **107,** 154 (1985).

26. (a) E. M. Arnett, D. E. Johnston, and L. E. Small. *J. Am. Chem. Soc.* **97,** 5598 (1975); (b) E. M. Arnett, L. E. Small, R. T. McIver, Jr., and J. S. Miller. *J. Am. Chem. Soc.* **96,** 5638 (1974).

27. G. Sini, S. S. Shaik, J.-M. Lefour, G. Ohanessian, and P. C. Hiberty. *J. Phys. Chem.* **93,** 5661 (1989).

28. S. S. Shaik. In *New Concepts for Understanding Organic Reactions.* NATO ASI Series, Vol. *C267,* J. Bértran and I. G. Csizmadia, Editors, Kluwer Publications, Dordrecht, 1989.

29. D. M. Wetzel, E. A. Solomon, S. Berger, and J. I. Brauman. *J. Am. Chem. Soc.* **111,** 3835 (1989).

5

IDENTITY S_N2 REACTIONS IN THE GAS PHASE AND IN SOLUTION

The principal objective of Chapter 3 was to develop the idea that chemical reactivity can be understood and unified by use of the appropriate state correlation diagram (SCD). In such diagrams, reactivity trends within reaction series are analyzed through the variations of three barrier-controlling factors, namely, the energy gaps of the diagram, and the curvatures and avoided crossing of the intersecting curves.

These factors have been elaborated further and quantified in Chapter 4 for the case of the S_N2 reaction and may now be applied to treat the reactivity patterns found in this most important organic process.

Apart from their inherent historical and practical importance in chemistry, S_N2 reactions include a well-defined identity process ($X^- + R-X$), whose analysis comprises a purely kinetic problem. A detailed consideration of identity S_N2 reactions is thus a necessary starting point for the treatment of more complex S_N2 processes. A recurring theme in this chapter is the deformation that must take place to allow the system to reach the transition state at the crossing point of the SCD. All of the important contributors to reactivity (barriers, transition structures, solvent effects) can be related to this single concept.

5.1 REACTION MECHANISMS IN THE GAS PHASE AND IN SOLUTION

In the gas phase the reaction coordinate of an identity S_N2 reaction (equation 5.1)

$$X:^- + R-X \rightarrow X-R + X:^- \qquad (5.1)$$

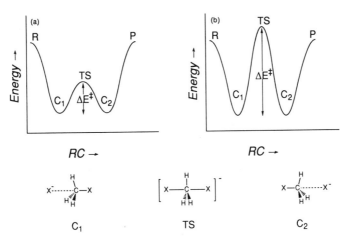

Figure 5.1 Reaction profile for an identity S_N2 reaction: (a) the transition state (TS) has lower energy than the reactants (R) and products (P); (b) the TS has higher energy than R and P.

consists of a double-well potential, as depicted in Figure 5.1. The minima of this figure are ion–dipole complexes, which are separated by a central energy barrier, $\Delta E^{\ddagger}$.[1-10] The overall process therefore consists of a collision step, which generates a reactant ion–dipole complex,[1,3,11] followed by a chemical activation step over a transition state (TS) that contains a pentacoordinated carbon atom. Once this TS has been surmounted, a product ion–dipole complex is formed, whose dissociation leads to the products.[1-10]

Under conditions of collisional deactivation (gas pressure > 1 torr), and at relatively low temperatures, the ion–dipole complexes will lie at or near the bottoms of their wells, and the transformation of one complex to the other will require that most of the central barrier of Figure 5.1 be overcome.[3,12] However, at lower pressures ($\cong 10^{-5}$ torr, as is typical in ICR experiments),[1-3] deactivating collisions are rare and most of the ion–dipole complexes have excess thermal energy. Case a of Figure 5.1 would, therefore, seem to require that the reaction proceed at the collision rate. However, this is not observed experimentally, and the rates of such reactions are often found to be much lower than this.[1-3,6t]

This effect was first explained by Brauman[1,2] in terms of the existence and properties of the central barrier. Since the TS has fewer degrees of freedom than either the reactants or the products, its internal energy levels are less dense (Figure 5.2a). This leads to a free-energy barrier, as is depicted in Figure 5.2b.

Although the double-well potential is now generally accepted, a direct exchange mechanism has also been considered.[2h] Vibrational excitation of the stretching mode of the C–Cl bond in CH_3Cl can bring the system directly to the saddle point and cause the ion–dipole complex to be bypassed. Despite this interesting theoretical result, there is no doubt that, under thermal activation, the mechanism proceeds via the ion–molecule complex.

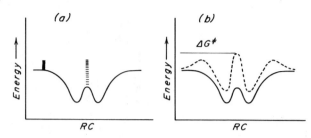

Figure 5.2 (a) Energy profile for identity S$_N$2 reactions. Internal energy levels are indicated schematically by horizontal lines. (b) A free-energy profile (dashed line) versus an enthalpy profile (solid line). The free-energy barrier ($\Delta G^{\ddagger}$) arises from the difference in the density of states. (Reprinted, by permission, from reference 1a. Copyright © 1977 by the American Chemical Society).

Experimental studies reveal that shallow ion–dipole complexes are also observable in acetonitrile solvent.[13] In these complexes the nucleophile is situated behind the C–X bond of RX; addition of methanol causes the complexes to disappear.[13a] Related observations have been made computationally.[9] In DMF an energy well is computed for the ion–dipole complex Cl$^-$ · · · CH$_3$Cl,[9c] but this well is almost nonexistent in water solvent.[9] As is seen in Figure 5.3, the energy curve from the reactants to the ion–dipole complex is almost flat, and the most prominent feature of the reaction coordinate is the existence of the central barrier.

The following sections are concerned with the origins of such barriers, both in the gas phase and in solution, and with the geometries of the transition structures. It will be found that both barriers and structures can be treated as consequences of the deformation of R–X that is associated with the acceptance of a *single electron* from X:$^-$.

5.2 EXPERIMENTAL AND COMPUTATIONAL DATA FOR CENTRAL BARRIERS

There are sufficient data regarding the central barriers of identity S$_N$2 reactions to allow several generalizations concerning the principal reactivity trends. Table 5.1[2] summarizes the barriers that result from the application of an RRKM theoretical treatment and other strategies[2j] to experimental gas-phase rate data.[2] Although the magnitudes of these barriers are sensitive to the assumptions of RRKM theory,[2a, b, j] they are regarded here as experimental barriers. Table 5.2 summarizes the data for barriers obtained in ab initio and semiempirical computations.

A prominent feature of all ab initio calculations is the similarity of the trends in the common data. The effect of electron correlation seemed at one point to be controversial. Thus, although Dedieu and Veillard[6a, b] had concluded that correlation has no effect on the central barrier, Keil and Ahlrichs[6e] found, using the CEPA approximation, a pronounced decrease in energy ($\cong$ 5–7 kcal/mol) in some cases, and an increase in energy in others. Subsequently, the systematic studies of

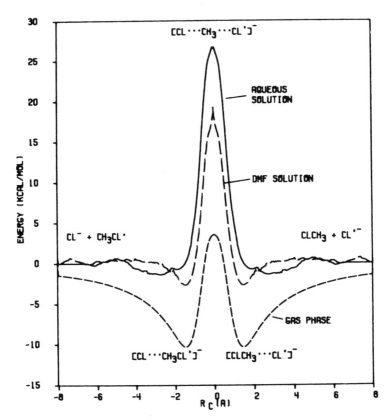

Figure 5.3 Calculated internal energies (kcal/mol) in the gas phase (short dashes) and the potential of mean force in DMF (long dashes) and in aqueous solution (solid curve) for the reaction of Cl⁻ with CH₃Cl, as a function of the reaction coordinate (*RC*), in Ångstroms. (Reprinted, by permission, from reference 9c. Copyright © 1985 by the American Chemical Society).

Cernusak and Urban[6p] suggested that electron correlation uniformly reduces the SCF barrier by as much as 4–12 kcal/mol. However, since correlation also changes the geometry of the transition state by as much as 0.01–0.1 Å for X=F, H,[6n, u, y] and the early examination of correlation effects on the barrier refer to one-point calculations on SCF optimized geometries, the conclusions of Cernusak and Urban may not be final, and do not yet provide uniformly accurate trends. Indeed, recent computations, including correlation, by Vetter and Zulicke,[6q] Truhlar,[6w] Wolfe and Kim,[6x] and Shi and Boyd[6y] indicate that while electron correlation lowers the barrier for X=F, it *raises* the barrier for X=Cl. This leads to an inversion of the relative barriers of the two reactions relative to the SCF level and the conclusions of Pellerite and Brauman.[2a, b] Since none of the other trends of Table 5.2 is inverted by correlation corrections,[6x] it appears that at the present time neither experiment nor theory has provided a definitive conclusion regarding the relative barriers for F and Cl, apart from the view that they are similar.[2j]

Table 5.1   RRKM Derived Central Barriers for Identity S_N2 Reactions

Entry	X	$\Delta E^{\ddagger}$ (kcal/mol)[a]
1	F	26.2
2	Cl	10.2
3	Br	11.2
4	I	$(5.5)^{b}$
5	CH_3CO_2	17.0, 13.2[c]
6	CH_3O	26.6–29.6[d]
7	t-BuO	28.8–30.6[d]
8	CH_3S	24.2, 24.8[c]
9	HCC	41.2, 40.3[c,d]
10	NC	35 (C-attack)
11	CN	≈ 24 (N-attack)
12	$NCCH_2$	≈ 34
13	$PhCH_2$	42; 34[e]
14	$CH_3C(O)CH_2$	≈ 29 (C-attack)
15	$CH_3C(O)CH_2$	≈ 15 (O-attack)
16	PhN	≈ 42
17	F, Cl	13.2[f]

[a]From references 2a and b unless specified.
[b]Estimated from equation 5.12.
[c]From reference 2f.
[d]Different barriers are estimated from the RRKM treatment of CH_3O^-/CH_3Cl and CH_3O^-/CH_3Br. See references 2c and f.
[e]The lower barrier is based on the data of reference 2c.
[f]From references 2i and j.

For the purposes of our analysis, and to illustrate the manner in which *qualitative* reasoning is performed with the SCD model, we have decided to adopt the experimental result, that is, F $\geq$ Cl. Since this is consistent with 4-31G calculations,[4a,b] which comprise the most extensive data set and, at the same time, lead to generally reliable trends,[6p] the discussion which follows employs the 4-31G results. We would have preferred to have theoretical central barriers for all of the halides at a common, reliable, computational level, but this was not yet feasible when this book was written.

In general, the trends in the ab initio and semiempirical calculations[4c,5] are similar, but with some noteworthy exceptions. While all ab initio calculations predict a high barrier for the reaction of H$^-$ with CH_4, MNDO predicts CH_5^- to be a stable intermediate. The INDO(λ) method[5d] rectifies this apparent failure of MNDO, but incorrectly predicts a central barrier of only 0.4 kcal/mol for the reaction of HO$^-$ with CH_3OH. It is therefore necessary to treat semiempirical results with caution[10] and, wherever possible, to calibrate such results against ab initio data.

A comparison of the data of Tables 5.1 and 5.2 reveals that the central barrier is large when X$^-$ is a poor leaving group, for example, H$^-$, H$_2$N$^-$, PhCH$_2^-$,

Table 5.2 Computed Central Barriers for Identity S_N2 Reactions

Entry	X	Ab Initio			Semiempirical		
		$3\text{-}21G^a$	$4\text{-}31G^b$	Large Basis Sets and CI	$MNDO^o$	$AM1^q$	$INDO(\lambda)^s$
1	F	12.2	11.7	$17,^e\ 18\text{-}20,^f\ \approx21,^g\ \approx17,^g$ $17.1,^h\ 20,^h\ 12.9^i$	45.0	r	21.9
2	Cl	5	5.5	$13.9,^e\ \approx12,^g\ \approx15.7,^h$ $13.9,^i\ 12.7,^j\ 6.3,^k$ 17.3^l	10.6	9.05	21.9
3	Br			$12.7,^h\ 11.6^h$	8.0	5.5	
4	I					2.4	
5	HOO		18.5				
6	FO		18.8				
7	HO		21.2		48.2^p		0.4
8	CH_3O		23.5		53.2^p		
9	HS		15.6				
10	HCC		50.4				
11	NC		43.8		26.9^p		
12	CN		28.5				
13	NH_2		≈38				
14	H		52	$61\text{-}66.5,^{f,l}\ 55^m$	-1.9^p		51.7
15	NH_3		$18.7^{c,d}$				
16	H_2O			$10^{d,n}$			

[a] From references 7b and a, respectively.

[b] From references 4a and b unless otherwise specified.

[c] From reference 8a, 4-31G result.

[d] The reaction is X: + CH_3-X^+ → $^+$X-CH_3 + :X.

[e] 6-31G*//6-31G* result (from references 6l and 9a, respectively).

[f] Triple zeta + diffuse functions; data from references 6a–e and 6i.

[g] Double zeta + diffuse functions + MP4/SDTQ4CI corrections from reference 6j. The larger value is the SCF result, the smaller is the MP4 result.

[h] Double zeta + diffuse + polarization functions + MRD/CI corrections from reference 6q. The lower value for F^- and Br^- is the MRD/CI result, the higher value for F^- and Br^- is the SCF result. For Cl^- both values are about 16 kcal/mol.

[i] 6-31+G* + MP4//4-31G calculations from reference 6k.

[j] 6-31G* + MP4//4-31G calculations from reference 6k.

[k] 4-31G + MP4//4-31G calculations from reference 6k.

[l] The highest value is from reference 6f.

[m] 6-31++G** + MP2 corrections, from reference 6o.

[n] From reference 8b: 6-31G** + MP3//3-21G results including ZPE correction.

[o] From references 5a and b, unless otherwise specified.

[p] From reference 5c.

[q] From reference 4c.

[r] AM1 does not afford acceptable results for X = F.

[s] From reference 5d. INDO(λ) takes AO nonorthogonality effects explicitly into consideration through the integral evaluation (see reference 10 for details).

[t] MP2/6-31+G* optimizations. From reference 6x.

Table 5.3 Experimental, Marcus-Derived, and Computed Central Free-Energy Barriers (ΔG°_{298}) for Identity S_N2 Reactions in Solution

Entry	X	$\Delta G^{\ddagger}$ (Expt) and Marcus Equation (Derived)						$\Delta G^{\ddagger}$ (Calc)[j]	
		H_2O[a]	MeOH[a]	Acetone[a]	DMF[a]	Sulfolane[g]	Others	H_2O	DMF
1	F	31.8						26.3	
2	Cl	26.6	28.2	21.8	22.7			31[k]	22
3	Br	23.7	22.9	17.0	18.4				
4	I	22[b]	21.0	16.3	16.0				
5	HO	41.8				17.0			
6	PhO	≈32[c]							
7	HS								
8	PhS	≈35[d]	34[f]						
9	PhSe					≈20 ± 3.5			
10	NC	50.9							
11	NO_3	26.5							
12	CH_3SO_3	34.7	(≈30[e])			24.4			
13	$PhSO_3$	35.6	(≈30[e])			23.6			
14	$TsSO_3$	34.7	(≈30[e])						
15	CF_3SO_3					18.64			
16	FSO_3					≈18	13[h]		
17	SO_4	35.9	(≈30[e])						
18	ClO_4	37.1	(≈31[e])						
19	Me_2O					18.64			
20	H_2O	35.1	(<30[e])						
21	Ph_2S					27.3			
22	C_5H_5N					≈40[i]			
23	$4\text{-}CN\text{-}C_5H_5N$					≈38[i]			

[a]From reference 15 unless otherwise specified.
[b]Recommended value in reference 14.
[c]Derived in this work using the Marcus equation.
[d]Estimated from reference 16c.
[e]Recommended values.
[f]From reference 16c.

[g]From references 16a, b, d, and f.
[h]Reference 17 for the reaction in CH_3CN.
[i]Reference 15b for the reaction in CH_3CN.
[j]From references 9a–c unless otherwise specified.
[k]From reference 9d.

HO^-, CH_3S^-, NC^-, and so on. In contrast, for $X^- = Cl^-$, Br^-, or $X = H_2O$, NH_3, which are good leaving groups, the barrier is small. This trend seems significant, because S_N2 reactivity is normally expressed in terms of the interplay between "nucleophilicity" and "leaving group ability." It is therefore necessary to explain why the barriers of identity reactions should appear to be dominated by "leaving group ability."

Table 5.3 collects the barriers of identity reactions in solution.[15-17] Some of these data are accessible directly;[14-16] the remainder have been derived[15, 17] using the Marcus equation. These latter data are considered to be reliable, because of the consistency of the two kinds of determinations. Nevertheless, some caution is necessary in the evaluation of the data for X = sulfonate or perchlorate (entries 12–18). Based on studies in sulfolane[16a, b] and acetonitrile,[17] the free-energy barriers in water seem to be too high. Recommended values in aqueous solution are indicated in parentheses. Similarly, the barrier for the exchange reaction H_2O + $CH_3OH_2^+$ in water seems too high, in view of the results for $(CH_3)_2O$ + $(CH_3)O(CH_3)_2^+$ in sulfolane.[16a]

The solution trends seen in Table 5.3 resemble the gas-phase trends; more importantly, the trends in the common data from different solvents are invariant. Thus, the trend $F^- > Cl^- > Br^- > I^-$ is the same in the gas phase and in a variety of solvents. Solvation therefore leads to higher central barriers, but does not alter the trends.

5.3 PREDICTIONS OF THE SCD MODEL: THE CENTRAL BARRIER IN THE GAS PHASE

An understanding of these trends requires an understanding of the activation process. For this purpose we derive the barrier by application of the SCD model, and then proceed to rationalize the trends in the barriers and in the *TS* geometries in terms of two equations that contain only thermochemical information concerning bond energies and electron affinities.

5.3.1 The Origin of the Barrier: The Deformation Energy

The SCD appropriate to a gas-phase identity reaction is Figure 5.4, whose anchor states are the ground states and the vertical charge transfer states of the ion–dipole complexes.[18] As explained in Chapters 3 and 4, the transition state occurs at the crossing point, at which the energies of the ground and charge transfer states are the same. To reach this point requires reactant distortions and interactions between the reactants[18g] that combine to stabilize the charge transfer state and concomitantly destabilize the ground state until the two are in resonance at the crossing point. In the S_N2 process, the reactant interactions in the ground state are comprised of electrostatic and exchange contributions which approximately balance each other; destabilization is dominated by the molecular deformations,[18h] and we rely only on the latter to make the following predictions:

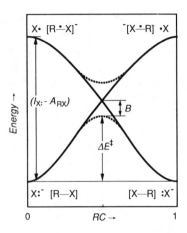

Figure 5.4 State correlation diagram for the identity S$_N$2 reaction. The minima of the curves along the *RC* refer to ion–molecule complexes, and the maxima refer to charge-transfer states. The avoided crossing is given by *B*, $\Delta E^{\ddagger}$ is the central barrier, and $I_{X:} - A_{RX}$ is the electron transfer energy gap. The *RC* is given as the difference in bond orders of the RX and XR linkages.

1. The barrier should result from the deformation required to overcome the vertical electron transfer energy gap and reach the transition state.
2. A correlation should exist between barriers and deformation energies, and these two quantities should correlate with the vertical electron transfer properties of the reactants.

The deformation energy of R–X can be defined as the energy difference between R–X at its geometry in the *TS*, and at its geometry in the ion–dipole complex. This is shown in **5.1**, where $d°$, $\theta°$, and $d^{\ddagger}$, $\theta^{\ddagger}$ are the bond lengths and bond angles in the two structures.

5.1

Figure 5.5 is a plot of calculated deformation energies versus the calculated barriers of identity reactions.[18c] There is a linear relationship between $\Delta E^{\ddagger}$ and ΔE_{def} ($r = 0.979$), which is given by equation 5.2. This relationship states that

$$\Delta E^{\ddagger} = \Delta E_{\text{def}} - 25.1 \pm 3.3 \text{ kcal/mol} \qquad (5.2)$$

differences in intrinsic barriers are mainly the result of the differences in ΔE_{def}; the interaction of X$^-$ with the deformed R–X in the transition state is stabilizing and approximately constant.

This is an important result. It demonstrates that the *barrier can be traced to the deformation energy required to bring CH$_3$X from its geometry in the ion–dipole*

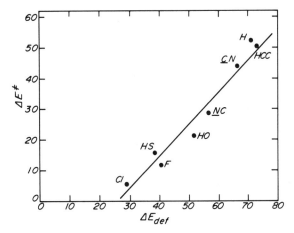

Figure 5.5 A plot of barriers ($\Delta E^{\ddagger}$) versus deformation energies (ΔE_{def}) for eight gas-phase identity reactions. Data are in kcal/mol, from 4-31G calculations. (Reprinted, by permission, from reference 18c).

complex to its geometry in the TS. The SCD provides a physical basis for this observation because of its focus upon the deformation required to achieve resonance at the crossing point of Figure 5.4.

5.3.2 Trends in the Central Barrier and Deformation Energies

For the SCD of Figure 5.4 the barrier is given by equation 5.3. As explained in

$$\Delta E^{\ddagger} = f(I_{\mathrm{X:}} - A_{\mathrm{RX}}) - B \qquad (5.3)$$

Chapter 4, $I_{\mathrm{X:}} - A_{\mathrm{RX}}$ is the vertical electron transfer energy gap, B is the avoided crossing, and f is the curvature index that determines the fraction of the gap which contributes to the activation barrier. The quantity $f(I_{\mathrm{X:}} - A_{\mathrm{RX}})$ is the energy of the crossing point, ΔE_{C}, which is *approximately*[18h] equal to the deformation energy, ΔE_{def}, and B is the interaction energy associated with the delocalization of the charge at the transition state. We may therefore write

$$\Delta E_{\mathrm{def}} \approx \Delta E_{\mathrm{C}} = f(I_{\mathrm{X:}} - A_{\mathrm{RX}}) \qquad (5.4)$$

and

$$\Delta E^{\ddagger} \approx \Delta E_{\mathrm{def}} - B \qquad (5.5)$$

Equation 5.5 anticipates the existence of the linear relationship between $\Delta E^{\ddagger}$ and ΔE_{def} that was seen in Figure 5.5. Such a relationship should be expected to exist whenever the avoided crossing B is either constant or varies over a narrow range. A comparison of equations 5.2 and 5.5 suggests that, at the 4-31G level of theory, B does indeed vary over the narrow range 25.1 ± 3.3 kcal/mol. Equations

5.4 and 5.5 therefore state that the barrier is a consequence of the deformation required to achieve the *TS*. Variations in the barrier will then be caused by variations in $I_{X:} - A_{RX}$ and f. Of course, since B is not strictly a constant, this kind of reasoning may lead to error whenever the different reactions possess very similar primary variables (gap and f).

As discussed in Chapter 4 (Section 4.2), the curvature index f is proportional to the degree of delocalization of the radical anion, $(R\overset{\bullet}{\cdot}X)^-$ in equation 5.6,[6v]

$$f \propto w'_{R:} \tag{5.6}$$

where

$$w'_{R:} = w_{R:}/N \tag{5.7}$$

Here $w_{R:}$ is the weight of the $(R:^- \cdot X)$ configuration in the radical anion $(R\overset{\bullet}{\cdot}X)^-$ and N is a normalization constant that takes into account the number of identical linkages that participate in the delocalization.

A large $w_{R:}$ denotes a delocalized odd electron in $(R\overset{\bullet}{\cdot}X)^-$, which will result in a weak bond-coupling interaction between $X\cdot$ and $(R\overset{\bullet}{\cdot}X)^-$ and lead to shallow stabilization (descent) of the charge transfer state toward the intersection point. This corresponds to a large f. Accordingly, in Chapter 4, we have termed f the bond-coupling delay index. As this index increases, a greater deformation will be required to overcome a given $I_{X:} - A_{RX}$ energy gap. Changes in barriers and in deformation energies are then the result of changes in $w_{R:}$ and in $I_{X:} - A_{RX}$ (Statement 5.1).

Statement 5.1   For a constant $w_{R:}$, the deformation energies and the barriers increase as the vertical electron transfer energy gap $(I_{X:} - A_{RX})$ increases. For a constant energy gap, the deformation energies and barriers increase as the bond-coupling delay index $w_{R:}$ increases.

Statement 5.1 is exemplified by the data of Table 5.4. The deformation energy is large and the barrier is large when either the electron transfer energy gap is large, or when the bond-coupling delay index is large. For example, in entries 3–5 $w_{R:}$ is approximately constant and $\Delta E^{\ddagger}$ and ΔE_{def} increase as the vertical electron transfer energy gap increases. On the other hand, comparison of entries 1 and 4, in which $(I_{X:} - A_{RX})$ is approximately constant, shows that HO$^-$ + CH$_3$OH, the reaction having the larger $w_{R:}$, has the larger $\Delta E^{\ddagger}$ and the larger ΔE_{def}. Similarly, a comparison of entries 3 and 7 shows that H$^-$ + CH$_4$, the reaction having the larger $w_{R:}$, also has the larger barrier and the larger deformation energy.

In general, the trends in the barriers and in the deformation energies are determined by the interplay between $I_{X:} - A_{RX}$ and w_R. The following statement emphasizes this point:

Statement 5.2   The activation process in identity S$_N$2 reactions is a result of the deformation that is required for the movement of a single electron from $X:^-$ to CH$_3$X, coupled to an interchange of bonds.

Table 5.4 Reactivity Indices ΔE_{def} and $\Delta E^{\ddagger}$ for Identity Methyl Transfer Reactions[a,b]

Entry	HCC	$(I_{X:} - A_{RX})$	$w_{R:}$	ΔE_{def}	$w_{R:}(I_{X:} - A_{RX})$	$\Delta E^{\ddagger}$
1	Cl	113	0.251	28.9	28.4	5.5
2	F	135	0.242	40.8	32.7	11.7
3	HS	95	0.340	38.8	32.3	15.6
4	HO	109	0.357	52.1	38.9	21.2
5	HCC	145	0.362	73.5	52.5	50.4
6	NC[c]	159	0.309	67.0	49.1	43.8
7	H	84	0.720	71.7	60.5	52.0
		$(104)^{d}$	$(0.453)^{d}$		(47.1)	

[a]In kcal/mol.
[b]Reactivity factors are calculated from the data in the tables and appendices to Chapter 4. Barriers and deformation energies are based on 4-31G calculations (reference 18c).
[c]C-attack.
[d]These values refer to a bond-localized $(H_3C \doteq H)^{-}$.

Here and elsewhere we use the word "movement" in place of "single electron transfer," to avoid confusion between the S_N2 activation process and well-defined stepwise single electron transfer reactions.

We may comment here on the observation (Section 5.2) that trends in the barriers of identity S_N2 reactions are dominated by the leaving group ability of X. From the relationships $\Delta E_{def} \cong f(I_{X:} - A_{RX})$ (equation 5.4), and $A_{RX} = A_{X:} - D_{R-X} + D_{R:X}$ (equation 4.5), it follows that two properties of X determine the barrier: the C–X bond strength (D_{C-X}) and the electron affinity of X ($A_{X:}$). The barrier will increase as D_{C-X} increases and as $A_{X:}$ decreases.[19] Both of these properties are characteristic of "poor leaving groups."

The existence of a simple thermochemical dependence of the barrier allows variations in identity barriers to be analyzed as X is varied along a row or down a column of the periodic table. The expected trends are summarized in **5.2**. As we

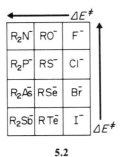

5.2

move down a column, D_{C-X} decreases and $A_{X:}$ either remains constant or increases slightly. Therefore, down each column of the periodic table the barriers will decrease. As we move along a row from left to right, the bond strength D_{C-X} increases slowly, but $A_{X:}$ increases rapidly (ca. 60 kcal/mol from X=NH$_2$ to X=F). Therefore, movement from left to right along a row again leads to a decrease in the barriers. With the possible exception of F versus Cl, all known trends[1-18] are consistent with this analysis, which includes predictions not yet tested experimentally.

5.3.3 Additional Delocalization Factors and Their Effect on the Central Barrier

Whenever X$^-$ is a complex polyatomic group such as benzyl or acetate, additional delocalization effects exist, because of the π-conjugative delocalization in the X· component of the charge transfer state X·$/$(R$\pm$X)$^-$ (see Table 4.9). For example, in PhCH$_2$·, the odd electron is delocalized as in **5.3**. The molecular orbital that

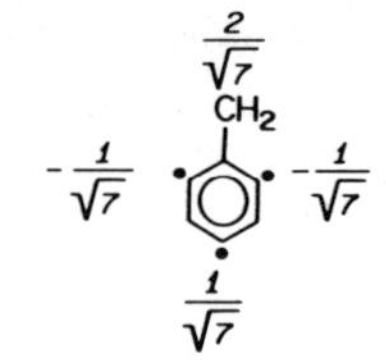

5.3

contains the unpaired electron has a coefficient of $2/\sqrt{7}$ on the CH$_2$ center, which is smaller than the 1.0 coefficient at the oxygen center of HO· (**5.4**). It is clear

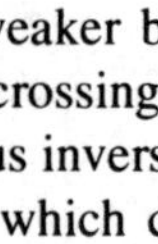

5.4

that a delocalized X· will exhibit weaker bond coupling with (R$\pm$X)$^-$; shallow descent of X·$/$(R$\pm$X)$^-$ toward the crossing point will occur. The magnitude of the bond-coupling delay index f is thus inversely proportional to a_X, the coefficient at the reaction center in the orbital which contains the unpaired electron of X· (equation 5.8).

$$f \propto 1/a_{X:} \tag{5.8}$$

The data of Table 5.5 illustrate this point. Although $(I_{X:} - A_{RX})$ and $w_{R:}$ are smaller in the reaction CH$_3$CO$_2$$^-$ + CH$_3$OCOCH$_3$ (entry 2) than in the reaction Cl$^-$ + CH$_3$Cl (entry 1), acetate has the higher barrier. Similar effects are observed when PhCH$_2$$^-$ + CH$_3$CH$_2$Ph (entry 4), NCCH$_2$$^-$ + CH$_3$CH$_2$CN (entry 5), and CH$_3$COCH$_2$$^-$ + CH$_3$CH$_2$COCH$_3$ (entry 6) are compared to CH$_3$S$^-$ + CH$_3$SCH$_3$ (entry 3). In each case, a higher barrier is associated with delocalization of the odd electron of X· ($1/a_X > 1$ in Table 5.5).

Table 5.5 Reactivity Factors and Barriers for Identity Methyl Transfer Reactions[a, b]

Entry	X	$(I_{X:} - A_{RX})$	$w_{R:}$	$1/a_X^c$	$\Delta E^{\ddagger}$
1	Cl	113	0.251	1	10.2
2	CH_3CO_2	108	0.203	$\sqrt{2}$	17.0, 13.2
3	CH_3S	95	0.370	1	24.2
4	$PhCH_2$	80	0.436	$\sqrt{7}/2$	34–42
5	$NCCH_2$	88	0.390	>1	≈ 34
6	$CH_3C(O)CH_2$	$\approx 88^c$	≈ 0.390	>1	≈ 29 (C-attack)
7	$CH_3C(O)CH_2$	$\approx 105^d$	$\approx 0.360^f$	1^f	≈ 15 (O-attack)

[a] $I_{X:} - A_{RX}$ and $\Delta E^{\ddagger}$ are in kcal/mol.
[b] Barriers are from reference 2a.
[c] Estimated as equal to the value in entry 5.
[d] Calculated for an orbital perpendicular to the π-plane and localized on the oxygen atom of the enolate anion.
[e] a_X is the Hückel HOMO coefficient at the reaction center of X:$^-$
[f] Corresponds to a localized orbital (see footnote d).

5.3.4 On The Ambident Behavior of the Enolate Anion

The ambident nucleophile acetone enolate anion may undergo alkylation by O-attack or by C-attack. According to 4-31G and 3-21G calculations,[20] O-alkylation is preferred and proceeds in the plane perpendicular to the π-system of the anion, as shown in **5.5**.[20a] This geometry avoids the delocalization problem that

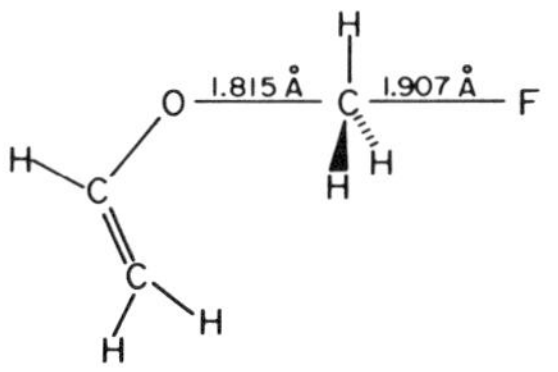

5.5

is associated with π-attack. On the other hand, C-alkylation is constrained to proceed by π-attack and cannot avoid the delocalization effect. This is illustrated in **5.6,** which depicts the orbitals that lead to O-alkylation and C-alkylation in the gas phase.

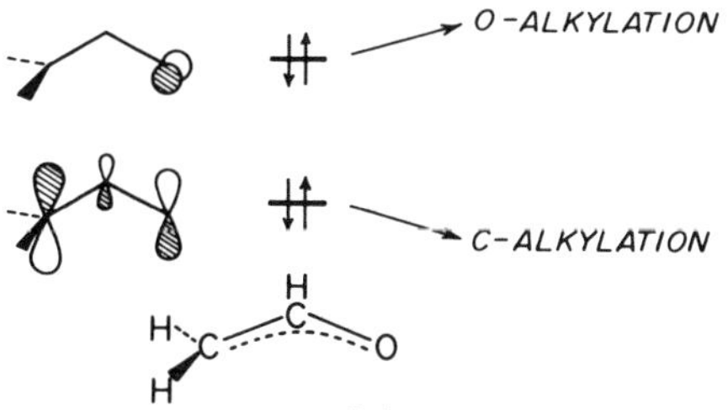

5.6

This effect is seen in entries 6 and 7 of Table 5.5. The O-alkylation pathway is observed to possess a lower barrier compared to C-alkylation.[2a,21] The result is dominated by the delocalization effect which is expressed by an increase in f for the C-alkylation $(1/a_X > 1)$. The same effect is expected to operate in other ambident nucleophiles in which one site of attack makes use of a localized orbital, and a second site of attack involves a delocalized orbital.

5.3.5 Evaluation of Barriers Using the SCD Model

As we have seen, the qualitative behavior of equation 5.3 accounts for all of the trends in the barriers of identity S$_N$2 reactions. To improve the usefulness of this equation, it is necessary to assign values to f and to B.

According to Section 4.2, f is a parameter that describes the steepness of the intersecting curves. For example, for two intersecting parabolas, $f = 0.25$; for two lines, $f - 0.5$; for curves steeper than parabolas, $f < 0.25$, and so on. A chemical property that dominates the behavior of f is the total delocalization index, that is,[18]

$$\overline{w}_{R:} = \frac{w_{R:}}{N \, a_{X:}} \tag{5.9}$$

The magnitude of f for the various cases can be obtained from equation 5.4, namely

$$f = \frac{\Delta E_{def}}{I_{X:} - A_{RX}} \tag{5.10}$$

Using the 4-31G ΔE_{def} data and the $I_{X:} - A_{RX}$ data of Table 5.4 leads to the values of f collected in Table 5.6. As predicted by equation 5.9, f and $w_{R:}$ exhibit similar trends, and we find that $f = (1.25 \pm 0.1)w_{R:}$. Since the two quantities have the same order of magnitude, it is convenient to rewrite equation 5.9 as

$$f = \overline{w}_{R:} \tag{5.11}$$

Table 5.6   Calibrated f Values and $w_{R:}$ Quantities for Identity Methyl Transfer Reactions

Entry	X	f^a	$w_{R:}{}^b$
1	Cl	0.256	0.251
2	F	0.302	0.242
3	HS	0.408	0.340
4	HO	0.478	0.357
5	HCC	0.507	0.362
6	NC	0.421	0.309
7	H	0.854	0.720

$^a f = \Delta E_{def}^{4-31G}/(I_{X:} - A_{RX})$. The data are taken from Table 5.4.
$^b w_{R:}$ values are taken from Table 5.4.

This leads to a set of f values whose magnitudes depend upon an understandable chemical property. We realize that equation 5.11 represents an oversimplification of a complex physical situation, but note that the alternative approach to f would require explicit calculation of the VB curves of each reaction, a formidable project.

The barrier equation is now equation 5.12a, and only the avoided crossing B

$$\Delta E^{\ddagger} = \overline{w}_{R:}(I_{X:} - A_{RX}) - B \qquad (5.12a)$$

remains to be determined. We will assume that B can be determined from equation 5.11 if $\Delta E^{\ddagger}$ and f are known for one reaction.

A well-studied identity methyl transfer reaction is that for X=Cl, for which the experimental barriers are 10–13.2 kcal/mol,[2a,b,i,j] and for which computational results with high-level basis sets including correlation converge to a slightly higher value of 13–16 kcal/mol. An average value $\Delta E^{\ddagger} = 14.5$ has been adopted here. This leads to B ~ 14 kcal/mol, close to the VB ab initio value for the reaction H^{-}/CH_{4}.[6v] This value for B will be used hereafter, and the barrier equation becomes equation 5.12b. This equation contains the two essential features of

$$\Delta E^{\ddagger} = \overline{w}_{R:}(I_{X:} - A_{RX}) - 14 \text{ kcal/mol} \qquad (5.12b)$$

the S_N2 process: the $I_{X:} - A_{RX}$ index, which reflects the movement of a single electron, and the $w_{R:}$ index, which reflects the bond coupling and reorganization associated with the movement of the electron.

Table 5.7 collects reactivity indices and barriers calculated using equation 5.12b, together with 4-31G barriers, experimental barriers, and semiempirical barriers.

Figure 5.6 is a plot of $\Delta E^{\ddagger}$ (equation 5.12b) versus the 4-31G ab initio $\Delta E^{\ddagger}$ values. The correlation is linear ($r = 0.971$). The correlation of $\Delta E^{\ddagger}$ (equation 5.12b) with $\Delta E^{\ddagger}$ (RRKM) is not as good ($r = 0.885$), and is shown in Figure 5.7.

As a further test of the predictive capability of the model, the set of $I_{X:} - A_{RX}$ values based on the A_{RX} of Table A.4.3 and a constant $B = 14$ kcal/mol, are found to reproduce the trends in the barriers with $f = 0.66 w_{R:}$ for the complete set of identity reactions (the factor 0.66 reproduces the 14.5 kcal/mol barrier of Cl^{-} + CH_3Cl). Likewise, with $f = 0.692 w_{R:}$, $B = 14$ kcal/mol, and the experimental $I_{X:} - A_{RX}$ data of Table 4.7, the trends are again the same. Variations in $I_{X:} - A_{RX}$ and $w_{R:}$ are sufficient to conceptualize the trends in the barriers of the identity reactions.

It follows that the notion of an interplay between $(I_{X:} - A_{RX})$ and $w_{R:}$ has captured the essence of the problem, namely that the S_N2 reaction comprises the movement of a single electron synchronized to bond interchange and the inversion mode, as shown in **5.7**.

5.7

Table 5.7 Reactivity Factors, SCD-Calculated Barriers (equation 5.12b), RRKM, and Computational Barriers for Identity Methyl Transfer Reactions[a]

Entry	X	$(I_{X:} - A_{RX})$	$w_{R:}$	$\Delta E^{\ddagger}_{eq5.12}$	$\Delta E^{\ddagger}_{4\text{-}31G}$[b]	$\Delta E^{\ddagger}_{RRKM}$[c]	$\Delta E^{\ddagger}_{MNDO}$[g]	$\Delta E^{\ddagger}_{AM1}$[g]
1	F	135	0.242	18.7	11.7(17.0)	26.2(19)[d]	45	9.05
2	Cl	113	0.251	14.4	5.5(13.9)	10.2; 13.2±2[c]	10.6	5.5
3	Br	99	0.246	10.4		11.2	8.0	2.4
4	I	81	0.241	5.5				
5	NC	159	0.309	35.1	43.8	≈35		
6	CH$_3$CO$_2$	108	0.287	17.0		≈17		
7	HCC	146	0.362	38.9	50.4	41.2		
8	HO	109	0.357	24.9	21.2			
9	CH$_3$O	98	0.355	20.8	23.5	26.6		
10	CH$_3$S	97	0.370	21.2	21.2	24.2		
11	HS	95	0.340	18.3	15.6			
12	HOO	83	0.380	17.4	18.5			
13	H	83	0.720	45.3	52.0			
14	PhCH$_2$	81	0.577	32.5		34[f]		

[a] Reactivity factors are calculated from the data of Chapter 4; $w_{R:}$ is defined by equation 5.9.
[b] From references 4a and b; 6-31G* results are in parentheses.
[c] From reference 2.
[d] Recommended value in parentheses.
[e] From reference 2i.
[f] From reference 2c.
[g] MNDO results from references 5a and b; AM1 results from reference 4c.

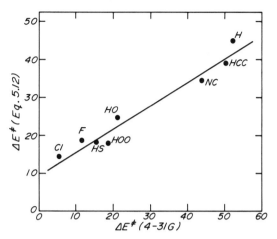

Figure 5.6 A plot of $\Delta E^{\ddagger}$ (equation 5.12) versus $\Delta E^{\ddagger}$ (4-31G). Units are kcal/mol. (Reprinted, by permission, from reference 18c).

5.3.6 Structure–Reactivity Relationships

Equation 5.12b predicts that the ensemble of identity reactions will consist of reaction families, as depicted in Figure 5.8. Each of the lines in this figure refers to reactions in which $w_{R:}$, that is, f, is approximately constant. Within each such series, reactivity is determined by the electron transfer energy gap. Likewise, the difference in reactivity between families having the same gap is determined by $w_{R:}$, that is, a lower barrier is associated with a smaller $w_{R:}$.

The data (Table 5.7) can indeed be organized into groups having approximately constant $w_{R:}$, that is, entries 1–4, 5, 6, 7–12, 13, 14. Within each of these groups the relative barrier heights vary according to $I_{X:} - A_{RX}$, reflecting that aspect of

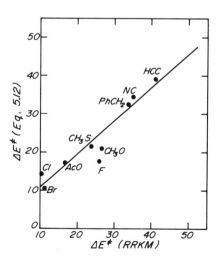

Figure 5.7 A plot of $\Delta E^{\ddagger}$ (eq 5.12) versus $\Delta E^{\ddagger}$ (RRKM).

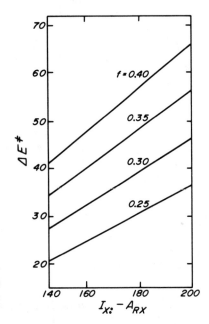

Figure 5.8 A general structure–reactivity plot for identity reactions using $\Delta E^{\ddagger} = f(I_{X:} - A_{RX}) - B$; B is assumed to be constant. Each line represents a family having a constant $w_{R:}$ (f).

the reaction which is determined by the movement of a single electron. If one now compares different groups of reactions, it is seen that relative reactivities are dominated by the magnitudes of $w_{R:}$, with a lower $w_{R:}$ leading to a lower barrier. This latter trend reflects that part of the reaction determined by the bond-coupling effect.

5.4 TRANSITION STRUCTURES: PREDICTIONS OF THE SCD MODEL

An important geometric property of a transition structure is its tightness or looseness. Whether this property is significant in an S$_N$2 reaction requires information concerning the contribution of the C–X stretch to the overall deformation that achieves the transition state. This aspect of transition structures can be examined in two ways: by direct calculation of stretching deformation energies; and by analysis of transition vectors.

The overall deformation energy required for RX to reach its transition state can be partitioned into a component ΔE_r, associated with the bond stretch, and a component ΔE_{θ}, associated with the angular deformation,[18c] as depicted in **5.8**. A plot of ΔE_r versus $\Delta E^{\ddagger}$ is shown in Figure 5.9. Figure 5.10 is the complementary plot of ΔE_{θ} versus $\Delta E^{\ddagger}$. The relationship between $\Delta E^{\ddagger}$ and the stretching deformation ΔE_r is linear (r = 0.954), but ΔE_{θ} is poorly correlated with $\Delta E^{\ddagger}$. In addition, as can be seen in Table 5.8, ΔE_r is larger than ΔE_{θ}, and makes the dominant contribution to the total deformation energy.

This conclusion is consistent with that of an analysis based on examination of transition vectors. Near the top of the barrier the reaction coordinate is described

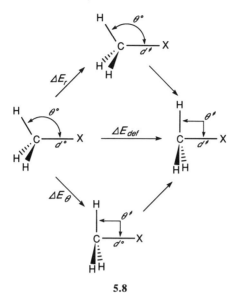

5.8

by a normal mode that has a negative force constant and, hence, an imaginary frequency. This transition vector is the only uniquely defined portion of the reaction coordinate (Chapter 2), and its composition reveals the nature of the deformations that link the *TS* to the reactants and products.

The transition vectors for eight identity reactions are collected in Table 5.9 and illustrated in Figure 5.11. To facilitate the comparisons, in Figure 5.11 the vectors have been shifted by a translation to allow the central carbon to remain stationary.

All of the transition vectors are very similar, despite the great differences in the

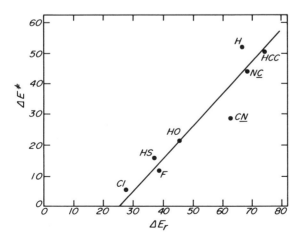

Figure 5.9 A plot of $\Delta E^{\ddagger}$ versus ΔE_r for eight gas-phase identity reactions. Data are in kcal/mol, from 4-31G calculations. (Reprinted, by permission, from reference 18c).

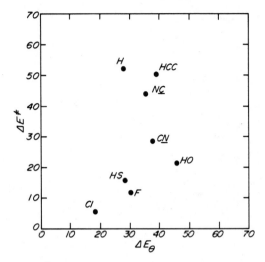

Figure 5.10 A plot of $\Delta E^{\ddagger}$ versus ΔE_{θ} for eight gas-phase identity reactions. Data are in kcal/mol, from 4-31G calculations. (Reprinted, by permission, from reference 18c).

barriers. The modes are seen to involve a component of bend, but in all cases are dominated energetically by a C–X stretch.

Thus, *the C–X stretch is the distinguished deformation leading to the transition state.* Indeed, as discussed in Section 5.1, a theoretical study[2h] of the dynamics of the reaction of chloride ion with methyl chloride shows that vibrational excitation of the C–Cl mode of CH$_3$Cl by four quanta can lead to a direct reaction, while a similar excitation of the CH$_3$ mode does not result in a successful reaction. These findings have several important consequences. First, it is valid to proceed to an examination of the factors that cause loosening of the C–X bond in the transition state. Second, since ΔE_r and $\Delta E^{\ddagger}$ correlate linearly, it should be possible to link

Table 5.8   Calculated Deformation Energies and The Components of These Deformation Energies[a] for Identity Methyl Transfer Reactions

Entry	X	ΔE_r	ΔE_{θ}	ΔE_{def}	$\Delta E^{\ddagger}$
1	H	67.4	28.6	71.7	52.0
2	HCC	74.7	39.5	73.5	50.4
3	NC[b]	68.7	36.0	67.0	43.8
4	CN[c]	63.1	38.3	57.0	28.5
5	HO	45.7	45.9	52.1	21.2
6	HS	37.2	28.7	38.8	15.6
7	F	38.8	30.4	40.8	11.7
8	Cl	27.6	18.5	28.9	5.5

[a] In kcal/mol; 4-31G data are from reference 18c.
[b] C-attack.
[c] N-attack.

Table 5.9 X–CH$_3$–X Transition Vectors in Internal Coordinates

X	$\Delta R(C-X)^{c,d}$	$\Delta\theta(H-C-X)^{d,e}$
H	0.797	0.206
HCC	1.012	0.506
NC^a	1.115	0.481
CN^b	1.125	0.461
HO	1.014	0.480
HS	1.122	0.492
F	1.120	0.531
Cl	1.128	0.439

^{a}C-attack.
^{b}N-attack.
^{c}In atomic units.
^{d}Data from reference 18c.
^{e}In radians.

the looseness of a transition structure to the magnitude of the barrier to achieve this structure. Third, since the *barrier* can be rationalized in terms of the vertical electron transfer energy gap and the bond-coupling delay index, it should be possible to analyze *TS* loosening in the same terms.[18b-d]

5.4.1 Tightness and Looseness of Transition Structure

The 4-31G transition structures for nine identity reactions are shown in Figure 5.12[18c] In such reactions the distortion, which distinguishes one X from another, is $\Delta d^{\ddagger} = d^{\ddagger} - d^{\circ}$, that is, the difference in the C–X bond lengths of the *TS* and the ion–molecule complex. To create a uniform scale of relative C–X stretching for different X's, it is convenient to define the percentage of C–X stretching in the *TS* (%CX‡) as in equation 5.13a. The "looseness" (%L‡) of the *TS* is then defined

$$\%CX^{\ddagger} = 100\ \Delta d^{\ddagger}/d^{\circ} \tag{5.13a}$$

as the sum of the percentages for the two C–X bonds, as shown in equation 5.13b. A large %L‡ is characteristic of a "looser" or more "exploded" transition structure.

$$\%L^{\ddagger} = 2\ (\%CX^{\ddagger}) \tag{5.13b}$$

Geometric data, %CX‡, and %L‡ indices are collected in Table 5.10, along with the corresponding barriers. The %CX‡ vary from 21 (X=Cl) to 58 (X=H). Thus [(H$\cdots$CH$_3\cdots$H)$^-$]‡ is the loosest *TS*, and [(Cl$\cdots$CH$_3\cdots$Cl)$^-$]‡ is the tightest *TS* within this series. In addition, there is a good correlation between $\Delta E^{\ddagger}$ and %L‡. An increase in $\Delta E^{\ddagger}$ is associated with a larger %L‡.

In terms of the SCD model, the factors that control %L‡ are $I_{X:} - A_{RX}$ and $w_{R:}$, and Statement 5.3 summarizes the predicted trends.

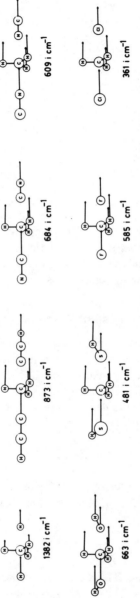

Figure 5.11 Transition vectors for eight identity S_N2 transition structures. (Reprinted, by permission, from reference 18c).

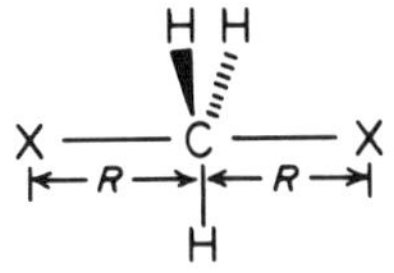

X =	H	1.730	(1.089)
	HCC	2.124	(1.465)
	NC	2.112	(1.459)
	CN	2.013	(1.441)
	HO	1.908	(1.463)
	HS	2.459	(1.914)
	F	1.828	(1.462)
	Cl	2.382	(1.965)

Figure 5.12 C–X bond lengths in S_N2 transition structures from 4-31G calculations. Data in parentheses refer to the C–X bond lengths of CH_3X in the ion–molecule complexes.

Statement 5.3 For a series of identity reactions in which $w_{R:}$ remains constant, $\%L^{\ddagger}$ will increase as the vertical electron transfer energy gap increases. For a series of identity reactions in which the energy gap remains constant, $\%L^{\ddagger}$ will increase as $w_{R:}$ increases.

The data of Table 5.11 demonstrate these points. Entries 1 and 2 have the same $w_{R:}$, and the looser TS (larger $\%L^{\ddagger}$) is found for $X = F^-$, which has the larger electron transfer energy gap. Analogous behavior is seen in entries 3–5. On the other hand, entries 1 and 4 show that, for a constant gap, the TS becomes looser as $w_{R:}$ increases. Evidently delocalization of the radical anion (larger $w_{R:}$) leads to loosening of the transition structure. Consequently, the loosest transition structure is found for X = H, because the radical anion associated with this reaction is the most delocalized.

Table 5.10 C–X Bond Lengths, Percentages of Bond Cleavage in the Transition State, Looseness Indices, and Barriers for Identity Methyl Transfer Reactions[a]

X	d^{ob}	$d^{\ddagger}$	$(\%CX^{\ddagger})^c$	$\%L^{\ddagger d}$	$\Delta E^{\ddagger e}$
H	1.089	1.730	58.0	116.0	52.0
HCC	1.465	2.124	45.0	90.0	50.4
NC	1.459	2.112	44.8	89.6	43.8
NC	1.441	2.013	40.5	81.0	28.5
HO	1.463	1.908	30.4	60.8	21.2
HS	1.914	2.459	28.0	56.0	15.6
F	1.462	1.828	25.0	50.0	11.7
Cl	1.965	2.382	21.1	42.2	5.5

[a] All values are from reference 18c.
[b] C–X bond lengths of CH_3X in the ion-dipole complexes X^-/CH_3X.
[c] Equation 5.13a.
[d] Equation 5.13b.
[e] In kcal/mol.

Table 5.11   Reactivity Factors, Looseness Indices, Deformation Energies, and Barriers for Identity Methyl Transfer Reactions^a

Entry	X	$I_{X:} - A_{RX}{}^b$	$w_{R:}$	%CX‡ (%L‡)	$\Delta E_{def}{}^b$	$\Delta E^{\ddagger b}$
1	Cl	113	0.251	21.1(42.2)	28.9	5.5
2	F	135	0.242	25.0(50.0)	40.8	11.7
3	HS	95	0.340	28.0(56.0)	38.8	15.6
4	HO	109	0.357	30.4(60.8)	52.1	21.2
5	HCC	146	0.362	45.0(90.0)	73.5	50.4
6	NC	159	0.309	44.8(89.6)	67.0	43.8
7	H	83	0.720	58.0(116.0)	71.7	52.0

^{a}%L‡, ΔE_{def}, and $\Delta E^{\ddagger}$ data from reference 18c.
^{b}In kcal/mol.

Table 5.11 also contains a correlation between $\Delta E^{\ddagger}$, ΔE_{def}, and %L‡. This correlation should now be expected, because the same factors, that is, $I_{X:} - A_{RX}$ and $w_{R:}$, determine the trends in every case. The tightest transition state, $(Cl^-/CH_3Cl)^{\ddagger}$, has the lowest deformation energy and the lowest barrier; and the loosest transition state, $(H^-/CH_4)^{\ddagger}$, has the highest deformation energy and the highest barrier. In general, this trend is also observed in calculations with higher-level basis sets,[6] and when correlation corrections are included, with the exception of a possible inversion for X=F, Cl. Related conclusions, that transition state looseness is related to the magnitude of the barrier, have been reached from experimental[22] and theoretical[6x] studies of secondary H/D isotope effects in the alkyl group.

What has thus emerged is a link between deformation energies, barriers, and transition structures, each of which independently reflects the activation process in an S$_N$2 reaction. This unified insight is created by the assertion that the activation process itself is a result of the distortion required to move a single electron from X:$^-$ to X across the CH$_3$ moiety of CH$_3$X (see **5.7**). Reactivity trends in identity reactions of CH$_3$X compounds can be summarized as Statement 5.4.

Statement 5.4   The activation process in an S$_N$2 reaction is a result of the distortion required to achieve resonance between a ground and a charge transfer state, and thereby move a single electron from X to CH$_3$X with a simultaneous interchange of the bonds. Reactants which possess strong C–X bonds and/or low X electron affinities are characterized by large charge transfer energy gaps and delocalized radical anions; these properties lead to high barriers and loose transition structures.

In view of the rather simple thermochemical dependence of barriers on D_{C-X} and $A_{X:}$, and the correlation of barrier heights with *TS* looseness, we can expand our mini-periodic table of trends in identity reactions to **5.9**. Thus, as X$^-$ moves up a column, or from right to left along a row, the looseness of the *TS* and the barrier required to achieve this *TS* both increase. The available trends are in accord with this prediction.[16, 18b, c, f]

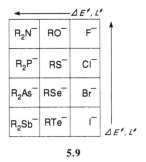

5.9

5.4.2 Relationships Between Transition State Looseness, Barriers, and the Binding Energy of the Transition State[18b, 23]

The existence of a correlation between the height of the barrier and the looseness of the TS merits further discussion. By analogy with the definition of bond energy, the binding energy of the TS (ΔE_b) can be defined as in equation 5.14. The quan-

$$[(XCH_3X)^-]^\ddagger \xrightarrow{\Delta E_b} X:^- + CH_3\cdot + \cdot X \qquad (5.14)$$

tity ΔE_b can be obtained from the thermochemical cycle of equation 5.15, in which $\Delta E^\ddagger$ is the central barrier, D_{id} is the binding

$$[(XCH_3X)^-]^\ddagger \rightarrow X:^- (CH_3X) \qquad \Delta E_1 = -\Delta E^\ddagger \qquad (5.15a)$$

$$X:^-(CH_3X) \rightarrow X:^- + CH_3X \qquad \Delta E_2 = D_{id} \qquad (5.15b)$$

$$CH_3X \rightarrow CH_3X\cdot + \cdot X \qquad \Delta E_3 = D_{C-X} \qquad (5.15c)$$

$$\overline{[(XCH_3X)^-]^\ddagger \rightarrow X:^- + CH_3\cdot + \cdot X \quad \Delta E = \Delta E_b} \qquad (5.15d)$$

energy of the ion–dipole complex, and D_{C-X} is the C–X bond dissociation energy. This leads to equation 5.16.

$$\Delta E_b = D_{C-X} + D_{id} - \Delta E^\ddagger \qquad (5.16)$$

Since all D_{id} values are approximately 10 kcal/mol,[2g, 11] the binding energies of identity transition states will vary as the difference between D_{C-X} and $\Delta E^\ddagger$. When $\Delta E^\ddagger$ is small ($\Delta E^\ddagger \cong D_{id}$), $\Delta E_b \cong D_{C-X}$, and the TS will be bound by an energy that is comparable to the bond energy of CH_3X. Such a transition state will be thermochemically tight.[23] On the other hand, when $\Delta E^\ddagger$ is large, ΔE_b becomes small, and the transition state will approach its dissociation limit ($X:^- + CH_3\cdot + \cdot X$) and be thermochemically loose.[23] If the distinguished deformation which achieves a transition structure is the C–X stretch, thermochemical looseness will coincide with geometric looseness,[23] and a high barrier will be associated with a geometrically loose TS. Since the S_N2 reaction meets this condition,[18c] the correlation between $\Delta E^{\ddagger}$ and $\%L^{\ddagger}$ must follow. However, this need not generally be the case: when bond stretching is not the distinguished distortion, a correlation between $\Delta E^{\ddagger}$ and *geometric* looseness should not be expected, although $\Delta E^{\ddagger}$ will still correlate with the *thermochemical* looseness of the TS.[23,24]

5.4.3   Quantitation of Transition State Looseness With the SCD Model

Statements 5.3 and 5.4 provide qualitative insight into the origins of the trends in transition structure looseness. For quantitative work, this qualitative insight has been reformulated as equation 5.17, which seems to work well.[18d] This equation

$$\frac{I_{X:} - A_{RX}}{G} = \exp\left[\frac{1}{w_{R:}}\left(\frac{\Delta d^{\ddagger}}{d^{\circ}}\right)\right] \qquad (5.17)$$

states that whenever the gap $I_{X:} - A_{RX}$ is equal to some limiting value G, the transition structure will be formed with zero distortion ($\Delta d^{\ddagger} = 0$). As $I_{X:} - A_{RX}$ increases, the amount of bond stretching ($\Delta d^{\ddagger}/d^{\circ}$) will also increase, and in proportion to $I_{X:} - A_{RX}$ and $w_{R:}$. In this way equation 5.17 conveys the qualitative trends of Statements 5.3 and 5.4. As we shall find in Chapter 6, the trends can also be expressed in terms of a linear correlation, $\Delta d^{\ddagger}/d^{\circ} = a[w_{R:}(I_{X:} - A_{RX})] + b$.[25] Other equations may exist that illustrate the same qualitative idea, but equation 5.17 has been found to be convenient.

Rearrangement of equation 5.17 leads to equation 5.18. An upper limit of G

$$\Delta d^{\ddagger}/d^{\circ} = w_{R:} \ln\left[\frac{I_{X:} - A_{RX}}{G}\right] \qquad (5.18)$$

is the gap that leads to a zero barrier. From equation 5.12, this upper limit is $14/w_{R:}$. With this as an estimate for G we obtain equation 5.19.

$$\%CX^{\ddagger} = 100\ \Delta d^{\ddagger}/d^{\circ} = 100\ w_{R:} \ln\left[\frac{w_{R:}(I_{X:} - A_{RX})}{14}\right] \qquad (5.19)$$

Table 5.12 summarizes the predictions of equation 5.19, together with the results of MNDO,[5b] AM1,[4c] and 4-31G[4a,b] calculations. The SCD equation for $\%CX^{\ddagger}$ overestimates the looseness in the case of X=H, but the trends in the data are successfully reproduced.

Equation 5.19 allows transition structures to be grouped into families defined by a constant $w_{R:}$. Within each such family the looseness of the transition structure should vary linearly with $\ln(I_{X:} - A_{RX})$, and the slope of this line will be proportional to $w_{R:}$. The data are plotted in Figure 5.13, and suggest that such linear correlations do exist.[18b,f] The transition structures are seen to fall into families defined by different $w_{R:}$ indices. Extrapolation can then allow other transition structures to be predicted, as is indicated on Figure 5.13 for X=Br, I.

Table 5.12 Calculated (equation 5.19), Ab Initio, and Semiempirical Computed Looseness Indices for Identity Methyl Transfer Reactions[a]

Entry	X	$\%CX^{\ddagger}$ (eq. 5.19)	$\%CX^{\ddagger}$ (MNDO)[c]	$\%CX^{\ddagger}$ (AM1)[d]	$\%CX^{\ddagger}$ (ab initio)[e]
1	F	20.5	21.5		25.0(22.8)
2	Cl	17.7	18.9	20.7	21.1(20.2)
3	Br	13.9	17.6	18.1	
4	I	8.0		12.1	
5	HS	28.4			28.0
6	HO	36.5			30.4
7	HCC	47.8			45.0
8	NC	38.8			44.8
9	H	105.4			58.0
		59.0[b]			

[a] $\%CX^{\ddagger} = 100 \, \Delta d^{\ddagger}/d^{\circ}$.
[b] Calculated using A_{RX} and w_{R} for a bond-localized $(H_3C \overset{\bullet}{-} H)^{-}$.
[c] From references 5a and b.
[d] From reference 4c.
[e] 4-31G data from references 4a and b; data in parentheses are at the 3-21G level from references 7b and 7a respectively.

The plots of transition structures versus $\ln(I_{X:} - A_{RX})$ are reminiscent of the plots of $\Delta E^{\ddagger}$ versus $(I_{X:} - A_{RX})$ (Figure 5.8; cf. Figures 5.13 and 5.15). Consequently, it should be possible to link equations 5.12 and 5.19, to provide a direct relationship between transition structure looseness and the height of the barrier. This is shown as equation 5.20. This equation states that a transition structure

$$\%CX^{\ddagger} = 100 \, \Delta d^{\ddagger}/d^{\circ} = 100 \, w_{R:} \ln \left[\frac{\Delta E^{\ddagger} + 14}{14} \right] \qquad (5.20)$$

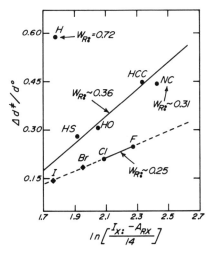

Figure 5.13 A plot of calculated (4-31G) $\Delta d^{\ddagger}/d^{\circ}$ indices against the natural logarithm of the dimensionless quantity $(I_{X:} - A_{RX})/B$, where $B = 14$ kcal/mol. The positions of X=Br, I are obtained by extrapolation.

loosens as the barrier within a reaction series (constant w_R.) increases. The equation is compatible with Statement 5.4, and with **5.9**, and it will be valid whenever the C–X stretch is the distinguished distortion that leads to the transition state.[23] A simpler version of equation 5.20 is $\%CX^{\ddagger} = \alpha\Delta E^{\ddagger} + c$, which will be developed in Chapter 6.[25]

We can conclude that the SCD model leads to very simple, useful expressions for the geometry of the S_N2 *TS*. Other examples exist,[26] in which the position of the *TS* can be expressed in terms of gaps[26b,c] and curvatures.[26a]

5.4.4 Solvation and the Geometries of Identity S_N2 Transition States

Since most of the information concerning S_N2 reactions is based on experimental studies in solution, it is important to determine how, and to what extent, solvation may alter our treatment of the problem. If it could be demonstrated that solvation does not alter transition structures, high-level ab initio computations could then become the most direct method for the exploration of these structures.

In principle, the solvent may cause the transition state to be tighter or looser. The curve shown in **5.10** represents the energy of the X^-/CH_3X transition state

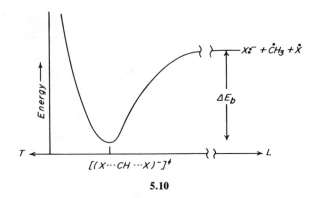

$[(X\cdots CH\cdots X)^-]^{\ddagger}$

5.10

as a function of the tightening (*T*) or loosening (*L*) coordinate. This curve has the appearance of a Morse potential: the increase in energy on the tight side is caused by the repulsive interactions exerted by the decrease in the equilibrium bond lengths of the transition structure. On the loose side, the energy also increases, up to the point at which the transition state fragments into $(X:^- + CH_3\cdot + \cdot X)$.[18b,d] The fragmentation or binding energy is ΔE_b (equation 5.16). When ΔE_b is small (large $\Delta E^{\ddagger}$), solvation might well cause the transition structure to be looser than in the gas phase. Conversely, when ΔE_b is large, the transition structure might experience only a small change in geometry as it is transferred from the gas phase to solution.

Table 5.13 collects the binding energies calculated with equation 5.16, and the bond energies of CH_3X compounds. The binding energies are high and close to

Table 5.13   Ground and Transition State C–X Binding Energies for Identity Methyl Transfer Reactions[a]

Entry	X	D_{C-X}[b]	ΔE_b (gas)[d]	ΔE_b (H_2O)[e]	ΔE_b (DMF)[e]
			Transition State Binding Energies		
1	F	108	92	76	
2	Cl	84	84	58	62
3	Br	71	70	48	53
4	I	56	60	34	40
5	HO	91	80	50	
6	HS	74	60	≈ 40[f]	
7	NC	122	97	70	
8	H	104	62		
9	HCC	117	86		

[a]In kcal/mol.
[b]See the Appendix of Chapter 4 for the data source.
[c]$\Delta E_b = D_{C-X} - \Delta E^{\ddagger} + D_{id}$ (equation 5.16).
[d]Calculated using $D_{id} = 10$ kcal/mol and $\Delta E^{\ddagger}$ values from reference 2.
[e]Calculated using $\Delta G^{\ddagger}$ data (reference 15) instead of $\Delta E^{\ddagger}$ data.
[f]Using $\Delta G^{\ddagger}$ for PhS$^-$ from reference 16c.

the bond energies. This suggests that the transition structure is a tightly bound (implastic) species, much like CH_3X itself. Solvation should not be expected to affect the geometry of such a structure.

An experimental test of this point is possible, based on measurements of heavy-atom kinetic isotope effects in different solvents. When the charges on the entering and leaving groups are the same, rate constants change by several orders of magnitude, but the heavy-atom kinetic isotope effect does not change significantly, even over a wide range of hydroxylic and dipolar nonhydroxylic solvents.[27] In such cases the transition structure does not depend upon the solvent.

Additional computational evidence exists in support of this view (references 9a–e, g–j). The free-energy barrier of the Cl^-/CH_3Cl reaction has been calculated for water and DMF solvents,[9a–c] and the assumption of a constant transition structure, identical to that in the gas phase, reproduces the barriers. Further refinement of these computations[9e] reveals that the optimized transition structure in water (250 molecules) is ca. 2% looser than in the gas phase. Semiempirical calculations of S_N2 transition structures in the gas phase and in the presence of 10 water molecules show only small differences in geometry.[9g–i] The Cl^-/CH_3Cl[7a] and F^-/CH_3F[7b] transition structures have been examined ab initio with and without the addition of two molecules of water. The results of these calculations are shown in **5.11** and

5.11

5.12. It is evident that two water molecules do not alter the gas phase geometry.

$$\left[F\underset{\underset{H}{\overset{}{|}}}{\overset{\overset{H}{|}}{\underset{\quad}{C}}} \,\xrightarrow{\textit{1.78 Å}}\, F \right]^{-} \qquad \left[(H_2O)F \,\xrightarrow{\textit{1.79 Å}}\, \underset{\underset{H}{\overset{}{|}}}{\overset{\overset{H}{|}}{C}} \,\text{——}\, F(H_2O) \right]^{-}$$

5.12a   **5.12b**

Protonation of a leaving group is a substantial perturbation, and can be considered to mimic extreme solvation of the transition state. The transition states of $[(HO\cdots CH_3\cdots OH)^{-}]^{\ddagger}$ and its diprotonated derivative are shown in **5.13a** and **5.13b**.[4a, 7c] Protonation causes a 0.09 Å lengthening per C–O bond, but the same

$$\left[HO\,\text{——}\, \underset{\underset{H}{\overset{}{|}}}{\overset{\overset{H}{|}}{C}} \,\xrightarrow{\textit{1.91 Å}}\, OH \right]^{-} \qquad \left[H_2O \,\xrightarrow{\textit{2.00 Å}}\, \underset{\underset{H}{\overset{}{|}}}{\overset{\overset{H}{|}}{C}} \,\text{——}\, OH_2 \right]^{+}$$

5.13a   **5.13b**

change in geometry is found upon protonation of CH_3OH (see **5.14a** and **5.14b**).[8b]

5.14a   **5.14b**

Similarly, comparison of $[(H_2N\cdots CH_3\cdots NH_2)^{-}]^{\ddagger}$ (**5.15a**)[4a] with the diprotonated transition state **5.15b**[8b] shows a slight loosening, identical to that observed upon protonation of CH_3NH_2 (**5.16a** and **5.16b**).

$$\left[H_2N\,\text{——}\, \underset{\underset{H}{\overset{}{|}}}{\overset{\overset{H}{|}}{C}} \,\xrightarrow{\textit{2.00 Å}}\, NH_2 \right]^{-} \qquad \left[H_3N \,\xrightarrow{\textit{2.09 Å}}\, \underset{\underset{H}{\overset{}{|}}}{\overset{\overset{H}{|}}{C}} \,\text{——}\, NH_3 \right]^{+}$$

5.15a   **5.15b**

5.16a   **5.16b**

Thus, it seems reasonable to regard the identity S$_N$2 transition state as an implastic species, which possesses strong, covalent C–X bonds, like the covalent bond of CH_3X. Solvation will certainly affect electronic polarization of the *TS*, but can have only a slight effect upon its geometry. Solvation is, therefore, an external perturbation of the electronic reorganization which attends the reaction. So far as the solvent is concerned, the transition state is a rigid, negatively charged anion $(XCH_3X)^-$, which does not change its intimate structure perpendicular to the reaction coordinate. With this view, identity S$_N$2 transition structures calculated ab initio are valid for the analysis of reactions in solution, and a separation of gas phase properties from solvation effects becomes an obvious objective.

5.5 SOLVATION AND THE INTRINSIC BARRIERS OF IDENTITY S$_N$2 REACTIONS

Any conceptualization of chemical reactivity must ultimately deal with the problem of solvent effects.[28] This problem is complex, because of the need to treat, both statically and dynamically, large numbers of molecules at a microscopic level.[9,28-34] Such treatments are still in their infancy at the time of the writing of this book,[9,33,34] so that a detailed microscopic understanding of solvent-induced trends within an extensive series of reactions does not exist. An alternative, more tractable but less rigorous, approach is to treat solvation at a macroscopic level and to attempt to provide a physically realistic set of reactivity factors within such a treatment.[18b,d,29-33,35,36]

It is necessary first to define the nature of the problem. We may, for example, focus upon the desolvation of X^- as the anion approaches RX to form a solvated X^-/RX complex. The free-energy barrier for the Cl^-/CH_3Cl reaction has been calculated in water and in DMF (Figure 5.3).[9a-c] In water solvent, the energy curve is flat in the region of the reaction coordinate between the separated reactants and the ion–dipole complex. However, in DMF,[9c] a 3.3 kcal/mol barrier separates the solvated reactants from the solvated ion–dipole complex. This barrier is much smaller than the 22.0 ± 0.05 kcal/mol central barrier, which therefore remains as the most important factor in the reaction.

Pattern recognition studies[37] provide structural insight into reasons for the virtual nonexistence of barriers to desolvation of anions and formation of ion–dipole complexes. In water, the time-averaged solvation spheres of F^- and Cl^- seem to involve either vacant sites or a number of loosely bound water molecules. The average local water structure about F^- consists of the distorted square pyramid **5.17**. This structure allows ready approach of CH$_3$F to displace an axial water

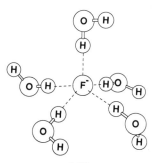

5.17

molecule and form the F$^-$/CH$_3$F complex. In the case of Cl$^-$, the local water structure consists of a distorted facially tricapped trigonal prism, depicted in **5.18**,

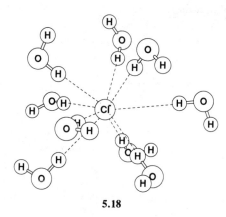

5.18

which contains two loosely bound capping water molecules. These loosely bound molecules undergo ready displacement by CH$_3$Cl to form the solvated Cl$^-$/CH$_3$Cl complex. Further understanding of desolvation will require an understanding of the dynamics of these processes.[38]

5.5.1 State Correlation Diagrams for Identity S$_N$2 Reactions in Solution

As discussed in Section 5.4.4, solvation does not alter the geometry of the identity transition structure. It follows that it is not necessary to consider the electronic wavefunction of the solvent explicitly, because the solvent does not participate in the electronic reorganization that attends the chemical reaction. With this interpretation, the general form of the SCD of an S$_N$2 reaction is the same in the gas phase and in solution.[18d, 35]

Some caution is, however, necessary. Although the solvent does not change the form of the SCD, it may enhance the mixing of the (X:$^-$ R$^+$:X$^-$) VB configuration into each curve and, therefore, into the *TS*. This solvent-induced "polarization" will not be severe for R=CH$_3$, but can be important for R=tBu, ArCH$_2$, and so on. In such cases, the solution SCD should take into account the effect of solvent on the individual VB configurations.

The SCD appropriate for an identity S$_N$2 reaction in solution is Figure 5.14, in which the various states are now solvated,[18b, d, 35] as indicated parenthetically by *s*. Along the reaction coordinate there are changes in the geometry of the reacting system and also changes in the orientations of solvent molecules. At each point there is a specific difference in bond orders, and also a specific orientation of solvent molecules, but the lower and upper curves of the diagram contain the same solvent orientations. This is depicted schematically in **5.19** for the ground state and charge-transfer anchor states of the diagram in water.

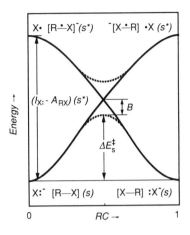

Figure 5.14 State correlation diagram for an identity S$_N$2 reaction in solution. The labels s and s* represent equilibrium and nonequilibrium states of solvation.

This schematic representation is intended to convey the view that the gap of the diagram is the result of a vertical electron transfer. In the ground state, solvent molecules occupy equilibrium positions that are governed by the charge distributions in X$^-$ and RX. In the charge-transfer state, this orientation of solvent molecules constitutes a nonequilibrium arrangement with respect to the charge distributions in X· and (R$\dot{-}$X)$^-$. This nonequilibrium solvation is denoted in Figure 5.14 by s*.

As we move from reactants to products along the reaction coordinate, solvent reorganization will have to occur as the state correlations (equations 5.21a and b) are established, and this reorganization will contribute to the reaction coordinate.

$$[X·/(R\dot{-}X)^-](s*) \rightarrow [(X–R)/:X^-] \quad \text{(equilibrium solvation)} \quad (5.21a)$$

$$[X:^-/(R–X)] \quad \text{(equilibrium solvation)} \rightarrow [(X\dot{-}R)^-/·X](s*) \quad (5.21b)$$

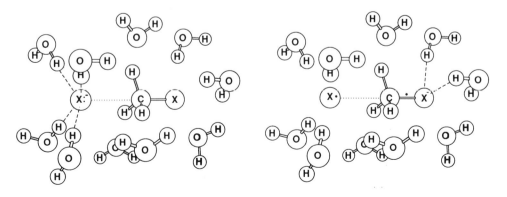

5.19a   **5.19b**

The overall activation process is now the result of a contribution from the molecular distortion associated with the movement of a single electron from X:$^-$ to RX, and a contribution from the *solvent barrier*. This solvent barrier exists because the solvent must reorganize its equilibrium positions to fit the new charge distribution.

This effect is recognized in all treatments of the S_N2 reaction,[18b, d, 31-35] that is, that the solvent barrier is caused by the extensive charge reorganization of the molecular system. In this context the notion of a solvent barrier corresponds to the Marcus concept[39, 40] of a solvent barrier in an electron transfer reaction. In such reactions the "solvent barrier" is calculated from solvent reorganization energies, using the theory of nonequilibrium polarization.[39a-d] This procedure will be used here to separate the *molecular barrier* from the *solvent barrier*. The discussion is based on equation 5.22, which resembles its gas-phase counterpart, equation 5.3. A comparison will be made with some related treatments[31, 32] and a link will then

$$\Delta E_s^{\ddagger} = f(I_{X:} - A_{RX})(s^*) - B \tag{5.22}$$

be drawn between this approach and more traditional structure–solvent relationships.[41]

5.5.2 Reactivity Factors in Solution

In a solvent, the vertical electron-transfer energy gap is the energy associated with the transfer of an electron from X:$^-$ to RX with the solvent molecules frozen (**5.19**). Vertical electron-transfer energies can be estimated by use of thermochemical cycles analogous to those employed in photoemission experiments.[42] For example, equation 5.23 provides the vertical process X:$^-(s) \rightarrow$ X·$(s^*) + e^-(g)$,

$$X:^-(s) \rightarrow X:^-(g) \qquad \Delta E_1 = S_{X:} \ (S > 0) \tag{5.23a}$$

$$X:^-(g) \rightarrow X·(g) + e^-(g) \qquad \Delta E_2 = I_{X:}(g) \tag{5.23b}$$

$$X:^-(g) \rightarrow X·(s) \qquad \Delta E_3 = -S_{X·} \tag{5.23c}$$

$$X:^-(s) \rightarrow X·(s^*) \qquad \Delta E_4 = R[S^*, X:^-] \tag{5.23d}$$

$$\overline{X:^-(s) \rightarrow X·(s^*) + e^-(g) \qquad \Delta E = I_{X:}(s^*)} \tag{5.23e}$$

the ionization of X:$^-(s)$ with the solvent molecules frozen. The vertical ionization potential $I_{X:}(s^*)$ is then

$$I_{X:}(s^*) = I_{X:}(g) + S_{X:} - S_{X·} + R[s^*, X:^-] \tag{5.24}$$

in which $I_{X:}(g)$ is the gas-phase ionization potential of X:$^-$, $S_{X:}$ and $S_{X·}$ are the free energies of desolvation of X:$^-$ and X·, respectively, and the term $R[s^*, X:^-]$ is the energy required to reorganize solvent molecules from their equilibrium positions about X· to their equilibrium positions about X:⁻. This is an energy-demanding process $(R > 0)$.

Reorganization energies associated with X:⁻ anions are known experimentally from photoemission experiments,[42] and their trends resemble those of the corresponding free energies of solvation. Thus the reorganization energies of halide anions decrease in the order $R[s^*, Cl^-] > R[s^*, Br^-] > R[s^*, I^-]$;[31,32] this trend is the same as the trend in the solvation energies. In addition, the reorganization energies, like the solvation energies, vary inversely with the ionic radii of these anions. Apparently the reorganization energy of an anion is proportional to its solvation energy.

The Marcus theory of nonequilibrium polarization leads to equation 5.25 (see

$$R[s^*, X:^-] = (n^{-2} - \epsilon^{-1})/2r \qquad (5.25)$$

Section 4.1.2), in which n is the refractive index of the solvent, and its square equals the optical dielectric constant, ϵ is the static dielectric constant of the solvent, and r is the radius of the cavity required to accommodate an anion in a given solvent.[43] From equations 5.25 and 5.26, (the Born solvation equation), the re-

$$S_{X:} = (1 - \epsilon^{-1})/2r \qquad (5.26)$$

organization energy can be expressed as a fraction, ρ, of the desolvation energy $S_{X:}$ (equation 5.27). This definition of ρ serves two purposes: from desolvation

$$\frac{R[s^*, X:^-]}{S_{X:}} = \rho; \qquad \rho = \frac{\epsilon - n^2}{n^2(\epsilon - 1)} \qquad (5.27)$$

energies of X:⁻, it is possible to estimate reorganization energies in any solvent, and the troublesome description of r has been eliminated.[43]

Equation 5.27 leads to logical trends, in the sense that strong solvation of X:⁻ will be associated with a large solvent reorganization energy. An alternative partitioning of $R[s^*, X:^-]$, into first-shell ("inner") and bulk ("outer") contributions, would be more appropriate theoretically,[42c,d] but not necessarily qualitatively superior. Trends in the hydration energies of anions by small clusters, for example $(H_2O)_4$, parallel trends in the total solvation energies,[44] so that the contribution of "inner-sphere" reorganization will depend upon the total solvation energy.

A potential problem is that equation 5.27 is based upon the continuum model. This model is often dismissed, because it is structureless and cannot account for specific solvation effects such as hydrogen bonding. However, a recent treatment of the hydration problem[43] shows that the Born equation successfully reproduces the experimental hydration enthalpies for salts involving 30 different ions. In addition, equation 5.27 is employed together with the experimental desolvation energy, which includes all of the specific solvent interactions. The reorganization energy that is derived from this equation therefore includes all of the specific interactions. In any event, the specific interactions vary in the same direction as the total desolvation energy: the hydrogen-bonding interaction of H$_2$O with X$^-$ follows the order F$^-$ > Cl$^-$ > Br$^-$ > I$^-$, and is evidently inversely proportional to the size of the ion.[44] The ion–dipole interactions of X$^-$ with DMSO or CH$_3$CN exhibit the same trend.[45] It seems that equation 5.27 should provide physically reasonable trends despite the absence of microscopic details of specific interactions.

The ρ of equation 5.27 is a property of the solvent, which can be termed the "reorganization factor." At 25°C, water has the largest ρ (0.56), other hydroxylic solvents have a somewhat smaller ρ (ca. 0.52), dipolar nonhydroxylic solvents such as DMF and DMSO have ρ values of ca. 0.44, and nonpolar solvents have extremely small ρ values (for benzene, $\rho \cong 0.0093$). These variations suggest that ρ is a parameter that reflects the superstructure of the solvent and its ability to organize itself about a resident ion. Hydroxylic solvents possess highly organized (and organizable) superstructures, and they have large ρ's and large reorganization energies. Nonpolar solvents do not possess an organized superstructure; they are characterized by small ρ and small reorganization energies.

The combination of equations 5.24 and 5.27 leads to equation 5.28 for ver-

$$I_{X:}(s^*) = I_{X:}(g) + (1 + \rho)S_{X:} + S_X. \tag{5.28}$$

tical ionization potentials in solution (see Section 4.1.2).

Vertical electron affinities in solution can be obtained from the thermochemical cycle of equation 4.13, which describes the vertical process in which R–X accepts an electron under frozen geometric and solvent molecular orientations, as in equation 5.29.

$$(R-X)(s) + e^-(g) \rightarrow (R \dot{-} X)^-(s^*) \qquad \Delta E = -A_{RX}(s^*) \tag{5.29}$$

The vertical electron affinity is then equation 5.30, in which $R[s^*, (R \dot{-} X)^-]$ is the reorganization energy required to

$$A_{RX}(s^*) = A_{RX}(g) + S_{R \dot{-} X} - R[s^*, (R \dot{-} X)^-] - S_{RX} \tag{5.30}$$

change the orientations of the solvent molecules about (R–X)$^-$(s) to their positions in (R–X)(s) and, thereby, to create (R$\dot{-}$X)$^-$(s*). From the expression for the reorganization energies, equation 5.27, $A_{RX}(s^*)$ becomes

$$A_{RX}(s^*) = A_{RX}(g) + (1 - \rho) S_{R \dot{-} X} - S_{RX} \tag{5.31}$$

The physical meaning of equation 5.31 is that, in a state of nonequilibrium solvation, the radical anion (R$\dot{-}$X)$^-$ loses part of its solvation energy, and the vertical electron affinity is lowered by the amount $\rho S_{(R \dot{-} X)}$.[18b, 35]

The solvation energies of radical anions are not known, and it is necessary to apply approximate methods to estimate $A_{RX}(s^*)$. These are based on empirical VB calculations of the interaction between the (R$\cdot$:X$^-$) and (R:$^-$$\cdot$X) configurations;[18b, 35] solvation is taken into account in the energies of the configurations. This method has been discussed in Chapter 4 (Section 4.1.2). Here we develop a simpler expression that reproduces the results of the detailed treatment. The desolvation energy of $(R \overset{\cdot}{\cdot} X)^-$ is estimated by equation 5.32 as the weighted sum of the desolvation energies of the individual configurations $(R \cdot : X^-)$ and $(R: {}^- \cdot X)$,

$$S_{R \overset{\cdot}{\cdot} X} \approx w_{X:} S_{R \cdot : X^-} + w_{R:} S_{R: {}^- \cdot X} \tag{5.32}$$

where $w_{X:}$ and $w_{R:}$ are the weights of these configurations in the gas phase (see Appendix). Any effect of solvation upon the electronic distribution in $(R \overset{\cdot}{\cdot} X)^-$ has thus been ignored. The additional approximations, $S_{(R \cdot : X)^-} \cong S_{X:}$ and $S_{(R: \cdot X)^-} \cong S_{R:}$, lead to equation 5.33.

$$S_{R \overset{\cdot}{\cdot} X} \cong w_{X:} S_{X:} + w_{R:} S_{R:} \tag{5.33}$$

Equation 5.31 is now

$$A_{RX}(s^*) \cong A_{RX}(g) + (1 - \rho)[w_{X:} S_{X:} + w_{R:} S_{R:}] - S_{RX} \tag{5.34}$$

With equations 5.28 and 5.34, it is possible to discuss solvent effects upon trends in the vertical electron transfer energy gaps of identity S$_N$2 reactions.

To complete the analysis we need the effect of solvation upon f. The quantity f is proportional to the degree of radical anion delocalization, as given by the weight $w_{R:}^*$ of $(R: {}^- \cdot X)$ under the condition of nonequilibrium solvation of $(R \overset{\cdot}{\cdot} X)^-$ (equation 5.35).

$$(R \overset{\cdot}{\cdot} X)^- (s^*) = w_{X:}^* (R \cdot : X^-) + w_{R:}^* (R: {}^- \cdot X) \tag{5.35}$$

As discussed in Section 4.2, the weight $w_{R:}^*$ depends on the energy difference between the configurations $(R \cdot {}^- : X)$ and $(R: {}^- \cdot X)$, and on their interaction matrix element. This is shown schematically by the VB interaction diagram **5.20**, in which

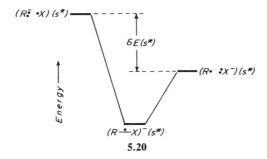

5.20

$\delta E(s^*)$, the energy gap under conditions of nonequilibrium solvation, is given by equation 5.36. In this equation, the differential solvation of the two VB con-

$$\delta E(s^*) = A_{X \cdot}(g) - A_{R \cdot}(g) + (1 - \rho)(S_{X:} - S_{R:}) \tag{5.36}$$

figurations under nonequilibrium conditions is approximated by the differential solvation of the anions X:$^-$ and R:$^-$. The expression for $w_{R:}^*$ is the familiar equation 5.37, which differs from the more accurate equation 4.17 in the substitution of $S_{X:} - S_{R:}$ for $S_{(R:·X^-)} - S_{(R:^-·X)}$. Equation 5.37 can be simplified to 5.38, which can also be reached by a perturbation treatment of **5.20**. Equation 5.38 states that a

$$w_{R:}^* = \frac{1}{2}\left[1 - \frac{\Delta}{\sqrt{\Delta^2 + 4}}\right]$$

$$\Delta = \frac{A_X.(g) - A_R.(g) + (1 - \rho)(S_{X:} - S_{R:})}{|h_{RX}|} \tag{5.37}$$

$$w_{R:}^* \approx \left[\frac{h_{RX}}{A_X.(g) - A_R.(g) + (1 - \rho)(S_{X:} - S_{R:})}\right]^2 \tag{5.38}$$

greater solvation of X:$^-$ than of R:$^-$ will increase the energy difference between the configurations of **5.20** and reduce their mixing.

5.5.3 Trends in the Reactivity Factors

Table 5.14 collects the ionization potentials of X:$^-$ anions in the gas phase and in solution. The following comments can be made, and supplement the discussion of Chapter 4:

1. $I_{X:}(s^*)$ represents the ability of X:$^-$ to donate an electron under the conditions of a frozen solvent orientation. This solvation effect reduces the donor capability of X:$^-$, because of the $(1 + \rho)S_{X:}$ term of equation 5.28, which accounts

Table 5.14   Gas-Phase Ionization Potentials, $I_{X:}$ (g), and Vertical, $I_{X:}$ (s*), and Adiabatic, $I_{X:}$ (s) Ionization Potentials in Solution for X:$^-$ Anions[a]

| Entry | X | $I_{X:}(g)$ | $I_{X:}(s^*)$[b] | | $I_{X:}(s)$[c] | | $\Delta G_t(H_2O \rightarrow DMF)$[d] |
			H$_2$O	DMF	H$_2$O	DMF	
1	F	78	240	213	182	169	13
2	Cl	83	204	183	160	151	10
3	Br	78	186	170	147	140	7
4	I	71	166	156	132	129	3
5	NC	89	203	181	162	154	8
6	HO	44	195	168	141	128	13
7	HS	54	177	156	133	123	10
8	PhS	57	163	147	125	118	7

[a]The data source is in the Appendix of Chapter 4 and in the text of this chapter.
[b]ρ(H$_2$O) = 0.56; ρ(DMF) = 0.48. Values are in kcal/mol and have been rounded off.
[c]$I_{X:}(s) = I_{X:}(g) + S_{X:}$.
[d]$\Delta G_t = S_{X:}$ (H$_2$O) $- S_{X:}$ (DMF).

for the contributions of solvation and solvent reorganization to the ionization process.

2. The effect of solvent reorganization is seen upon comparison to the adiabatic $I_{X:}(s)$ values, which refer to the ionization of $X:^-(s)$ to form an equilibrated $X \cdot (s)$. The $I_{X:}(s)$ values are significantly smaller than the $I_{X:}(s^*)$ values because of the solvent reorganization effect, and the difference is larger in water than in DMF, because the reorganization factor of water is larger than that of DMF. Clearly, as the solvent becomes less organized, the difference $\{I_{X:}(s^*) - I_{X:}(s)\}$ becomes smaller.

3. The anions have smaller $I_{X:}(s^*)$ in DMF than in water, and the difference is always larger than ΔG_t, the free energies of transfer from water to DMF. Once again, this illustrates the effect of the solvent reorganization terms, which amplify ΔG_t.

4. The order of the donor abilities in solution differs from that in the gas phase. Because of their smaller solvation energies, bulkier anions (I^-, HS^-, PhS^-) become better donors in solution.

Table 5.15 summarizes values of A_{RX} and $w_{R:}$ for CH_3X molecules in different media. The two sets of data (I and II) that are presented for each solvent refer, respectively, to the values from the full calculations described in Chapter 4, and to the approximate values from equations 5.34 and 5.37. It can be seen that the approximate treatment reproduces the essential features of the more detailed calculation.

The following trends are found in Table 5.15:

1. Solvation of $(CH_3 \overset{\bullet}{-} X)^-$ makes CH_3X a better electron acceptor. However, nonequilibrium solvation decreases the solvent stabilization of the radical anion by the factor ρ (equation 5.34).

2. As a result of the nonequilibrium solvation, trends in the acceptor properties of CH_3X remain the same as in the gas phase, that is, $A_{MeF} < A_{MeCl} < A_{MeBr} < A_{MeI}$.

3. Solvation causes a polarization of $(CH_3 \overset{\bullet}{-} X)^-$, with a reduction in the contribution of the $(CH_3:^- \cdot X)$ configuration (cf. $w_{R:}$ versus $w_{R:}^*$). This is caused by more effective solvation of the $(CH_3 \cdot :X^-)$ configuration. The different solvation of the two VB configurations increases the energy gap between them and reduces their mixing (equation 5.38). Nevertheless, the overall effect is small, and $w_{R:}^*$ is approximately equal to $w_{R:}$.

5.5.4 Trends in Intrinsic Barriers—Predictions of the SCD Model

The eight reactions examined in Table 5.16 can be grouped into two families. The first, in entries 1–4, possesses smaller $w_{R:}$; the second, in entries 4–8, possesses larger $w_{R:}$. Within each family, the barriers vary in proportion to $(I_{X:} - A_{RX})$. As we have observed previously, this trend is controlled by the donor–acceptor rela-

Table 5.15 A_{RX} and w_R: Properties of CH₃X Molecules in the Gas Phase and in Solution[a,b]

Entry	X	Gas Phase		H₂O[c]				DMF[c]			
				I		II		I		II	
		$A_{RX}(g)$	w_R	$A_{RX}(s^*)$	w_R^*	$A_{RX}(s^*)$	w_R^*	$A_{RX}(s^*)$	w_R^*	$A_{RX}(s^*)$	w_R^*
1	F	-57	0.242	-19	0.206	-16	0.201	-17	0.208	-14	0.204
2	Cl	-30	0.251	-0.1	0.239	+2	0.237	+2	0.243	+4	0.243
3	Br	-21	0.246	+7	0.241	+9	0.240	+10	0.242	+11	0.243
4	I	-10	0.241	+16	0.243	+17	0.243	+19	0.242	+20	0.244
5	NC	-68	0.309	-39	0.304	-37	0.302	-37	0.304	-36	0.305
6	HO	-65	0.357	-30	0.320 (0.309)[d]	-28	0.313	-28	0.324	-26	0.320
7	HS	-42	0.340	-12	0.328	-10	0.321	-10	0.329	-8	0.329
8	PhS	-40	0.331	-8	0.326	-11	0.325	-5	0.328	-9	0.328

[a]Data source and details are in Chapter 4.

[b]A_{RX} in kcal/mol.

[c]Set I is obtained from detailed treatment described in Chapter 4. Set II is obtained from the approximate equations 5.34 and 5.37.

[d]Using S_X: = 106 kcal/mol.

Table 5.16 Reactivity Factors and Barriers for Identity Methyl Transfer Reactions in the Gas Phase, in Water, and in DMF[a]

Entry	X	Gas Phase					H$_2$O			DMF		
		$I_{X:} - A_{RX}$	w_R	$\Delta E^{\ddagger}(I)^b$	$\Delta E^{\ddagger}(II)^c$	$\Delta E^{\ddagger}(III)^d$	$(I_{X:} - A_{RX})(s^*)$	w_R^*	$\Delta E^{\ddagger f}$	$(I_{X:} - A_{RX})(s^*)$	w_R^*	$\Delta E^{\ddagger f}$
1	F	135	0.242	11.7($\approx$19)	45	$\approx$19	259	0.206	31.8	230	0.208	22.7
2	Cl	113	0.251	5.5($\approx$14)	10.6(9.05)	10.2	204	0.239	26.6	181	0.243	18.4
3	Br	99	0.246		8.0(5.5)	11.2	179	0.241	23.7	160	0.242	16.0
4	I	81	0.241		(2.4)	$\approx$6	150	0.243	22.0	137	0.242	
5	NC	159	0.309	43.8		35	242	0.304	50.9	218	0.304	
6	HO	109	0.357	21.2		26.6	225	0.320 (0.309)	41.8	196	0.324	
7	HS	95	0.340	15.6			189	0.328	$\approx$35^g	166	0.329	
8	PhS	97	0.331			24.2^e	171	0.326		152	0.328	

$^a I_{X:} - A_{RX}$ and $\Delta E^{\ddagger}$ values are in kcal/mol.

^{b} 4-31G computed values from references 4a and b. Values in parentheses refer to extended basis sets.

^{c} MNDO computed values from references 5a and b. Values in parentheses are AM1 results from reference 4c.

^{d} RRKM values from references 2a and b. The barrier for F$^-$/CH$_3$F is corrected to the recommended value.

^{e} Value for CH$_3$S$^-$ + CH$_3$–SCH$_3$.

^{f} Values from reference 15.

^{g} Value from reference 16c.

tionship of the reactants, and it reflects that aspect of the reaction governed by the movement of the single electron. Table 5.16 also shows that, for a constant energy gap, the reaction that has the larger $w_{R:}$ has the larger barrier. This latter trend reflects the bond interchange and bond-coupling aspect of the reaction.

Such behavior has already been commented upon for reactions in the gas phase. Since the *trends* are the same in the gas phase and in solution, it follows that *solvation does not alter internal trends in the reactivity factors* but, rather, serves mainly to increase the vertical electron transfer energy gap. This increase makes the activation process more difficult, and leads to an increase in the barrier.

In Figures 5.15 and 5.16, computed[4a, b] gas-phase barriers and experimental[15, 16c] barriers in water solvent have been plotted against the corresponding vertical electron transfer energies. In both cases, the data can be grouped into families which differ in the magnitude of $w_{R:}$. Within each family, the barriers increase as the vertical electron transfer energy increases, and the slope is proportional to $w_{R:}$. These trends follow the general structure–reactivity relationships predicted in Figure 5.8, and reaffirm that the S$_N$2 reaction involves the movement of a single electron. In the gas phase, this is synchronized to the bond rearrangement; in solution, this movement is also coupled to the molecular reorganization of the solvent.

5.5.5 Quantitative Application of the SCD Model to Barriers in Solution

Barriers in solution result from the interplay of $(I_{X:} - A_{RX})(s^*)$ and $w_{R:}^*$. Following equation 5.12, the barrier in solution can be expressed as equation 5.39, and the avoided crossing is assigned the same value, 14 kcal/mol, as in the gas phase.

$$\Delta E_s^{\ddagger} = w_{R:}^* [(I_{X:} - A_{RX})(s^*)] - 14 \text{ kcal/mol} \qquad (5.39)$$

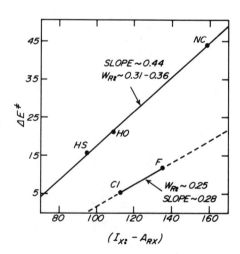

Figure 5.15 A plot of gas-phase barriers calculated at 4-31G versus $I_{X:} - A_{RX}$. Units are kcal/mol.

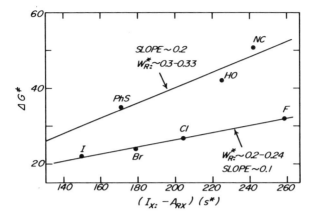

Figure 5.16 A plot of experimental barriers in water solvent versus $I_{X:} - A_{RX}$. Units are kcal/mol.

Table 5.17 compares the barriers calculated with equation 5.39 to experimental barriers.[15, 16c] The calculated barriers are found to be consistently higher than the experimental but, with the exception of the barrier for the reaction HO^-/CH_3OH, for which there is a large uncertainty,[15] a linear correlation ($r = 0.979$) exists between the calculated and the experimental barriers.

It is remarkable that equation 5.39 reproduces all of the trends. Both experiment and theory lead to the same order of intrinsic barriers in solution: $NC^- > HO^- > HS^- > F^- > Cl^- > Br^- > I^-$.

5.5.6 The Molecular Barrier and the "Solvent Barrier"

The combination of equations 5.28 and 5.31 leads to equation 5.40. The sub-

$$(I_{X:} - A_{RX})(s^*) = (I_{X:} - A_{RX})(g) + S_{X^-} - S_{RX} + \rho(S_{X:} - S_{R:X}) \quad (5.40)$$

Table 5.17 Calculated (equation 5.39) and Experimental Free-Energy Barriers for Identity S_N2 Reactions in Solution

Entry	X	H_2O		DMF	
		$\Delta E_s^{\ddagger}$ (eq 5.39)	$\Delta E_s^{\ddagger}$ (expt)[a]	$\Delta E_s^{\ddagger}$ (eq 5.44)	$\Delta E_s^{\ddagger}$ (expt)[a]
1	F	39.4	31.8	33.8	
2	Cl	35.8	26.6	30.0	22.7
3	Br	29.1	23.7	24.7	18.4
4	I	22.5	22.0	19.2	16.0
5	NC	59.6	50.9	52.3	
6	HO	58.0	41.8	49.5	
		$(55.5)^b$		$(46.6)^b$	
7	PhS	41.7	≈ 35	35.9	

[a]Experimental data are from reference 15, except for entry 7 from reference 16c.
[b]Using $w_R^* = 0.309$ (see Table 5.14).

stitution $S_{(R \cdot X)^-} \approx w_{X:}S_{X:} + w_{R:}S_{R:}$ leads to equation 5.41, in which both solva-

$$(I_{X:} - A_{RX})(s^*) \approx (I_{X:} - A_{RX})(g) + 2\rho S_{X:} + w_{R:}(1 - \rho)(S_{X:} - S_{R:}) \quad (5.41)$$

tion and solvent reorganization are seen to increase the vertical electron transfer energy gap.

Since $\Delta E_s^{\ddagger} = w_{R:}^*(I_{X:} - A_{RX}) - B$, we may write equation 5.42. In this equa-

$$\Delta E_s^{\ddagger} \approx w_{R:}^*(I_{X:} - A_{RX})(g) - B + 2\rho w_{R:}^* S_{X:} + w_{R:}^* w_{R:}(1 - \rho)(S_{X:} - S_{R:})$$
$$(5.42)$$

tion, the product $w_{R:}^* w_{R:}$ is small. Since $(S_{X:} - S_{R:})$ is also small, the intrinsic barrier becomes

$$\Delta E_s^{\ddagger} \approx w_{R:}^*(I_{X:} - A_{RX})(g) - B + 2\rho w_{R:}^* S_{X:} \quad (5.43)$$

The first two terms of equation 5.43 are the molecular barrier, designated $\Delta E^{\ddagger}(\text{mol})$. The third term represents the contribution of the solvent to the overall barrier, and it is designated $\Delta E^{\ddagger}(\text{solv})$. As can be seen in Table 5.18, the molecular barrier defined by equation 5.44b differs from the gas phase barrier, $\Delta E_g^{\ddagger}$, because the $w_{R:}^*$ values employed in equation 5.44b are smaller than the $w_{R:}$ values of equation 5.12.

$$\Delta E_s^{\ddagger} = \Delta E^{\ddagger}(\text{mol}) + \Delta E^{\ddagger}(\text{solv}) \quad (5.44a)$$

$$\Delta E^{\ddagger}(\text{mol}) = w_{R:}^*(I_{X:} - A_{RX})(g) - B \quad (5.44b)$$

$$\Delta E^{\ddagger}(\text{solv}) = 2w_{R:}^* \rho S_{X:} \quad (5.44c)$$

The solvent barrier of equation 5.44c refers to the reorganization of the solvent as it attempts to respond to the movement of the single electron. The $w_{R:}$ term of this equation is the delocalization index of $(CH_3 \cdot X)^-(s^*)$, and it is determined primarily by the properties of X and CH_3. At the same time, $w_{R:}$ is the bond-coupling delay index, so that a large $w_{R:}$ is characteristic of a looser X^-/CH_3X transition structure, which will require greater reorganization of the solvent as its equilibrium orientations about the charges of the molecular system are restored.

The calculated solvent barriers in water (Table 5.18) are higher than the molecular barriers, and not very different from the barriers that would have resulted had the process comprised a genuine single electron transfer (SET). For the case of SET, the $w_{R:}$ would be replaced by the quantity 0.25, the Marcus intrinsic barrier for an electron transfer reaction.[40d] Such electron transfer solvent barriers are shown in Table 5.18 as $\Delta E_{et}^{\ddagger}(\text{solv})$.

These considerations project the idea that the S$_N$2 reaction involves two mutually perturbed systems, each of which experiences destruction and restoration of its equilibrium structure. One of these systems is the reactant pair $X:^-/CH_3X$,

Table 5.18 Molecular Barriers, $\Delta E^{\ddagger}$ (mol), Solvent Barriers, $\Delta E^{\ddagger}$ (solv), Gas-Phase Barriers, $\Delta E_g^{\ddagger}$, and "Pure" Electron Transfer Solvent Barriers, $\Delta E_{et}^{\ddagger}$ (s), for Identity Methyl Transfer Reactions[a]

Entry	X	$\Delta E^{\ddagger}$ (mol)[b]	$w_{R:}^*$	$\Delta E_g^{\ddagger c}$	$w_{R:}$	$\Delta E^{\ddagger}$ (solv)[d]	$\Delta E_{et}^{\ddagger}$ (solv)[e]	$(\Delta E^{\ddagger}$ (mol) $+ \Delta E^{\ddagger}$(solv))[f]
1	F	13.9	0.206	18.7	0.242	24.0	29.1	37.9(39.4)
2	Cl	13.0	0.239	14.4	0.251	20.6	21.6	33.6(35.8)
3	Br	9.9	0.241	10.4	0.246	18.6	19.3	28.5(29.1)
4	I	5.7	0.243	5.5	0.241	16.6	17.1	22.1(22.5)
5	NC	34.3	0.304	35.1	0.309	24.9	20.4	59.2(59.6)
6	HO	20.9	0.320	24.9	0.357	34.8	27.2	55.7(58.0)
7	PhS	17.6	0.326	18.1	0.331	24.8	19.0	42.4(41.7)

[a] All barriers are in kcal/mol.
[b] $\Delta E^{\ddagger}$ (mol) $= w_{R:} [I_{X:} - A_{RX}) (g)] - 14$ (equation 5.44b).
[c] $\Delta E_g^{\ddagger} = w_{R:} [I_{X:} - A_{RX}) (g)] - 14$ (equation 5.12).
[d] $\Delta E^{\ddagger}$ (solv) $= 2w_{R:}^* \rho S_{X:}$ (equation 5.44c).
[e] $\Delta E_{et}^{\ddagger}$ (solv) $= 0.25 (2\rho S_{X:})$ (the Marcus model).
[f] The values in parentheses are total barriers $\Delta E_s^{\ddagger}$ obtained from equation 5.39. The comparison of these values to $\Delta E^{\ddagger}$ (mol) $+ \Delta E^{\ddagger}$ (solv) shows the consistency of the partitioning.

which is subjected to the movement of a single electron coupled to bond reorganization. The molecular barrier for this process is only slightly perturbed from that in the gas phase (cf. **5.21a** and **5.21b**). This perturbation arises from enhanced

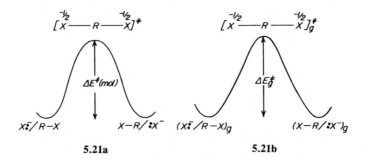

5.21a   **5.21b**

bond coupling due to the solvent polarization ($w_{R:}^* \leq w_{R:}$) of the radical anion.[46]

The second system is the solvent, whose perturbation is coupled to the movement of the charge. For the solvent to restore equilibrium orientations, a reorganization barrier, caused by the deformation in the immediate vicinity of the charges and the adaptation of the bulk solvent to the local events,[30c] must be overcome. These events are illustrated in **5.22**: circles in the initial and final configurations

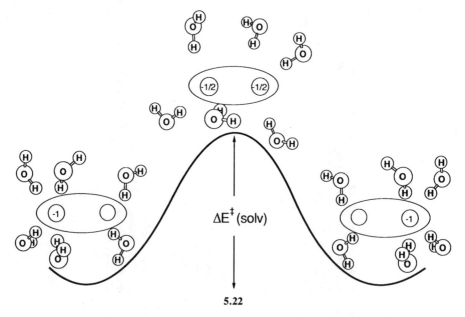

5.22

denote sites within the solvent cavities which will become charged during the process. The solvent barrier is affected by the resident molecular system only to the extent that the molecular system determines the positions of the charges and, therefore, the local events that concern the solvent reorganization. These considerations

Table 5.19   Solvent Barriers for Identity Methyl Transfer Reactions

Entry	X	$\Delta E^{\ddagger}$ (solv)^{a,b,c}				
		H$_2$O	MeOH	DMF	DMSO	Benzene
1	F	24.0	22.7	18.2	17.4	0.3
2	Cl	20.6	19.5	15.6	14.9	0.3
3	Br	18.6	17.5	14.4	13.7	0.2
4	I	16.6	15.8	13.5	12.9	0.2

^{a}All barriers are in kcal/mol. $\Delta E^{\ddagger}$ (solv) $= 2w_{R}^{*}\rho S_{X}$. The aqueous solution values of w_{R}^{*} are used for MeOH. The w_{R}^{*} values of DMF are used for DMSO. The gas-phase w_{R} values are used for benzene.
$^b\rho$(H$_2$O) = 0.56; ρ(MeOH) = 0.55; ρ(DMF) = 0.48; ρ(DMSO) = 0.45; ρ(benzene) = 0.0093.
^{c}Estimated desolvation energies in the different solvents are (in kcal/mol). F$^-$: 104, 100, 91, 93, 68; Cl$^-$: 77, 74, 67, 68, 55; Br$^-$: 69, 66, 62, 63, 51; I$^-$: 61, 59, 58, 59, 47.

are in harmony with the view that the S$_N$2 transition structure is an implastic species whose geometry is not altered by the solvent.[46]

With this interpretation, equation 5.44c can be used to trace the effects of different solvents. This is done in Table 5.19, using the appropriate experimental solvation data.[47-49] In this table the solvents are organized in the order of decreasing ρ, and it is seen that the solvent barrier decreases as ρ decreases. In the case of benzene, the solvent barrier has become negligibly small; in this solvent, the total S$_N$2 barrier will be identical to the molecular barrier. This is not a consequence of weak solvation in benzene, because the solvation energies of X$^-$ anions in benzene are substantial (50–70 kcal/mol).[47b,c] Rather, it is the small reorganization factor of benzene that leads to the disappearance of the solvent barrier. As we have discussed earlier, the ρ factor can be regarded as an index of the superstructure of the solvent. A large ρ reflects a highly organized superstructure, which will require substantial reorganization energy to accommodate the movement of an electron.[50] On the other hand, a small ρ reflects a fluid superstructure, whose reorganization requires little energy, and it leads to a small solvent barrier for the process.

The notion of solvent superstructure that attaches to ρ seems physically realistic, and it can be expected that ρ will be related to other solvent parameters,[28,41a,51] especially solvatochromic parameters, which refer to solvent effects upon electronic transitions. The Z parameter[51e] is defined as the molar transition energy E_T, expressed in kcal/mol, for the charge transfer absorption band of **5.23a** in different

5.23a

5.23b

solvents (equation 5.45), where $\bar{\nu}$ is the wavenumber associated with the excita-

$$Z = E_T(\text{kcal}/\text{mol}) = hc\bar{\nu}N = 2.859 \times 10^{-3} \, \bar{\nu} \, (\text{cm}^{-1}) \qquad (5.45)$$

tion and N is Avogadro's number. The solvatochromic parameter $E_T{}^{51f}$ is based on the transition energy for the longest wavelength absorption band of **5.23b**.

Table 5.20 shows the trends in ρ, Z, and E_T compared to water solvent, for the solvents listed in Table 5.19. Also included in Table 5.20 is the Hildebrand solubility parameter δ_H, which measures the work expended by the solvent as it reorganizes to create a cavity for the solute. Each of these parameters is related to the organization and superstructure of the solvent. It should, therefore, be possible to modulate the sizes of solvent barriers by alteration of the superstructure of the solvent.[52]

It is important to note that while calculations of reorganization energies may be crude, the concept of reorganization energy is rigorous, as are the thermochemical expressions for $I_{X:}(s^*)$ and $A_{RX}(s^*)$ found in equations 5.28 and 5.30. Any improvement in the treatment of $R[s^*, \, X:^-]$ and $R[s^*, \, (R\dot{\cdot}X)^-]$[42d,53] will improve the quantitative capabilities of the approach, but the concepts and trends already encountered will remain. The advent of ab initio Monte Carlo computations of solvent effects[9] is a promising development; and calculations of reaction dynamics in solution are expected eventually to be able to treat the problem at its very heart.[30c,d,32–34]

5.5.7 Solvent Reorganization in the SCD Model and its Relationship to Dynamics Studies and the Equilibrium Solvation Model

Transition state solvation is an important topic that has been the subject of continuous research.[9a–c,29–36,41] One possible treatment of this problem is the adiabatic-equilibrium model, which is based on transition state theory.[9a–c,41] In this model the reaction coordinate is defined by the bonding changes in the molecular system. For example, the reaction coordinate of an identity reaction might be taken as the difference in the lengths of the two bonds of the $(X \cdots C \cdots X)$ moiety. The

Table 5.20 Trends in the Solvent Reorganization Factor ρ and in Other Solvatochromic Parameters Relative to Water

Solvent	ρ/ρ_{H_2O}	$Z/Z_{H_2O}{}^{a,b}$	$E_T/E_{T_{H_2O}}{}^{b}$	$\delta/\delta_{H_2O}{}^{c}$
H$_2$O	1.00	1.00	1.00	1.00
CH$_3$OH	0.98	0.88	0.89	0.62
DMF	0.86	0.72	0.70	0.50
DMSO	0.80	0.75	0.71	0.56
Benzene	0.02	0.57	0.55	

"See equation 5.45.
[b]Data from reference 28b.
[c]Data from reference 51a.

solvent is in thermal equilibrium with the molecular system at each point along this reaction coordinate, as depicted in **5.24**. In this picture, any time the molecular

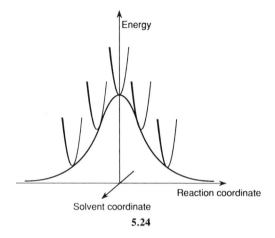

5.24

system achieves the D_{3h} transition structure there will be a successful crossover to product *regardless of the initial solvent configuration*. Therefore, it is sufficient to calculate the gas-phase barrier and then, by Monte Carlo simulations[9a-c] or with a continuum model,[36d] to determine the solvation of the gas-phase barrier.

An alternative model assumes nonadiabatic or nonequilibrium solvation of the *TS*. In this case,[7b, 36] the motion of the solvent is an intimate part of the molecular reaction coordinate, as is depicted in **5.25**, the transition vector of a doubly sol-

$$O-H \leftarrow \!\!\!\!\!\! \rightarrow F \cdots\cdots \overset{\overset{\displaystyle H}{\displaystyle |}}{\underset{\underset{\displaystyle H\ H}{}}{C}} \cdots\cdots F \!\!\rightarrow \ \leftarrow\!\! H - O$$

underneath left: H 0.13 underneath: $H\ H$ -0.13 right: H

5.25

vated F^-/CH_3F *TS*, with emphasis on the motions of the solvent molecules. Motions of the solvent molecules may be sufficient to induce the formation of the transition state. In such a process the solvent is not in equilibrium with the molecular system, and the *TS* is characterized by disequilibrium solvation effects.

This feature is inherent in the Marcus treatment of outer sphere electron transfer,[39] whose reaction coordinate consists exclusively of solvent reorganizational modes.

As we have seen, the SCD model conceptualizes the solvent barrier of an S_N2 reaction as the coupling of solvent reorganization to the bond interchange of the molecular system. This implies that the *TS* possesses both a unique molecular geometry and a unique solvent molecular configuration (equations 5.21a and b).[35b] This is consistent with the nonunity transmission coefficient computed[34] for the molecular D_{3h} structure under the assumption of equilibrium solvation.

Molecular dynamics studies[32-34] have provided additional insights into such nonequilibrium effects. In the work of Warshel et al,[33] and of German and Kuznetsov,[32] curve-crossing diagrams resembling SCD's are used to discuss the molecular dynamics of electron shifts. Hynes et al.[34] have found that the nonequilibrium effects lead to nonunity transmission coefficients because of the frozen solvent motion along the reaction coordinate in the vicinity of the *TS*. This contrasts to the interpretation of Bertrán et al.[36]

The nonequilibrium picture differs somewhat from the traditional view[1a,41] that solvent effects can be understood in terms of differential stabilization of reactants and transition states. With the latter interpretation a higher barrier in solution is the result of more effective solvation of the charge-localized reactant, X$^-$, relative to the charge-delocalized transition state, **5.26**.[1a]

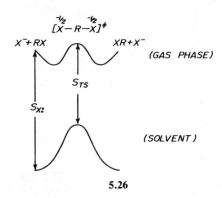

5.26

However, the two approaches are not as different as they may seem. In the transition state approach,[1a] differences between barriers in solution and in the gas phase are the result of the differences in desolvation energies expressed in equation 5.46 (see 5.26). Since $\Delta E_s^{\ddagger} - \Delta E_g^{\ddagger} > 0$ for any solvent, it follows that $S_{X:} - S_{TS} > 0$, and all X$^-$/RX transition states will be less strongly solvated than X$^-$.

$$\Delta E_s^{\ddagger} - \Delta E_g^{\ddagger} = S_{X:} - S_{TS} \tag{5.46}$$

In the SCD model, the barrier $\Delta E_s^{\ddagger} = \Delta E^{\ddagger}(\text{mol}) + \Delta E^{\ddagger}(\text{solv})$. Since the molecular barrier, $\Delta E^{\ddagger}(\text{mol})$, is close to the gas-phase barrier, we may write

$$\Delta E_s^{\ddagger} - \Delta E_g^{\ddagger} \approx \Delta E^{\ddagger}(\text{solv}) \tag{5.47}$$

and, from equation 5.44c, we have

$$\Delta E_s^{\ddagger} - \Delta E_g^{\ddagger} \approx 2w_{R:}^{*}\rho S_{X:} > 0 \tag{5.48}$$

Since the quantity $2w_{R:}\rho S_{X:}$ is always positive, equation 5.48 states that X$^-$/RX transition states are solvated less strongly than X$^-$ in every solvent.

Equating 5.46 and 5.48 gives

$$S_{X:} - S_{TS} \approx 2w_{R:}^{*}\rho S_{X:} \tag{5.49}$$

which rearranges to

$$S_{TS} \approx S_{X:}(1 - 2\rho w_{R:}) \qquad (5.50)$$

Table 5.21 lists transition state desolvation energies calculated with equation 5.50, and the following trends are noted:

1. In each solvent the transition structure is less solvated than is X^-.

2. The difference between solvation of X^- and solvation of the transition structure decreases as one proceeds from hydroxylic solvents to nonhydroxylic to nonpolar solvents. This trend is in harmony with the notion[45] that nonhydroxylic solvents are less sensitive to the ionic radius of X^- than are hydroxylic solvents. For example, benzene barely distinguishes between the large transition structure and the compact X^- ion.

3. Within families (entries 1–4, 5–7), transition structures are more sensitive to solvation when X is small. Thus $[(FCH_3F)^-]^{\ddagger}$ is the most sensitive, within entries 1–4, and $[(HOCH_3OH)^-]^{\ddagger}$ is the most sensitive within entries 5–7.

4. The sensitivity of a transition structure to solvation effects has no geometrical implications. Although $[(FCH_3F)^-]^{\ddagger}$ is looser than $[(ClCH_3Cl)^-]^{\ddagger}$, and is also more sensitive to solvation effects, $[(NCCH_3CN)^-]^{\ddagger}$ is the loosest transition structure of the series, but is almost insensitive to variations in the solvent.

It seems that the equations of the SCD model can be used to probe transition state solvation, with no need to invoke equilibrium solvation. Moreover, the SCD model provides information concerning the structural reorganization of the solvent and its response to local electronic events that are initiated by the chemical reaction. Finally, with appropriate modification, the SCD model can be formulated to discuss dynamic effects of solvent reorganization.[33]

Table 5.21 Calculated Desolvation Energies of Transition States (S_{TS}) and Anions ($S_{X:}$) for Identity Methyl Transfer Reactions[a]

		H_2O		MeOH		DMF		DMSO		Benzene	
Entry	X	S_{TS}	$S_{X:}$	S_{TS}	$S_{X:}$	S_{TS}	$S_{X:}$	S_{TS}	$S_{X:}$	S_{TS}	$S_{X:}$
1	F	80	104	77	100	73	91	76	93	67.7	68
2	Cl	56	77	52	74	51	67	54	68	54.7	55
3	Br	50	69	48	66	48	62	49	63	50.8	51
4	I	44	61	43	59	44	58	46	59	46.8	47
5	NC	48	73	47	70	46	65	48	67		
6	HO	62	97	60	93	58	84	61	86		
7	PhS	43	68	42	65	42	61	44	62		

[a] $S_{TS} = S_{X:} (1 - 2\rho w_{R:}^{*})$. All values are in kcal/mol. $S_{X:}$ values are estimated from references 47–49. See Table 4.5 for the data source.

5.6   OTHER CONCEPTUAL APPROACHES TO THE IDENTITY REACTION

In the preceding sections we have used the SCD model to obtain an understanding and unification of various reactivity trends in identity S$_N$2 reactions. As we noted in Chapter 1, there are other approaches to these problems, and it is appropriate here to reexamine some of these and to search for connections between them and the SCD model. We focus on two main points, namely, (1) the origin and magnitude of the barrier and (2) the transition structure; and we examine how three different models, namely, the Marcus treatment, three-dimensional potential energy surface diagrams (PESD), and frontier molecular orbital (FMO) theory[55] deal with these problems.

5.6.1   The Identity Barrier: Applications of the Marcus Equation, FMO Theory, and PESD

In the Marcus treatment, the identity barrier is a parameter which is used to derive an intrinsic barrier. Once intrinsic barriers are known, the Marcus equation allows a discussion of trends in nonidentity reactions.[4a, b] Clearly, when the Marcus equation is employed in this manner, it neither requires nor provides an understanding of reactivity trends within the identity set.

Nonetheless, the original Marcus treatment of electron transfer contains an avoided crossing diagram,[56] and such diagrams are the basis of the SCD approach. A convergence of the two approaches might then be realized, if the Marcus equation could be rewritten so as to contain an explicit expression for the intrinsic barrier. In its ability to do this, the SCD approach represents a logical conceptual evolution of the Marcus treatment.

Explicit treatments of the identity barrier have been provided by German and Dogonadze,[31a] and by Khostariya,[31b] based on a Marcus-like analysis of electron transfer. Starting from a precursor ion–dipole complex, it is supposed that a thermal fluctuation causes the molecular system to achieve the geometry that is necessary for the electronic resonance. This step is considered to have a low energetic requirement.[32] In a second, rate-determining stage, the solvent rearranges from its initial to its final configuration. This step is treated as a solvent reorganization process, as in the Marcus formalism. There then follow various relaxation steps toward the products. The numerical values of the barriers obtained in these treatments are usually good,[31] perhaps because the solvent reorganization barriers are not very different from the solvent barriers collected in Table 5.18.

The FMO treatment of reactivity focuses upon the stabilizing orbital interaction between the HOMO of a donor molecule and the LUMO of an acceptor molecule. The magnitude of such an interaction is proportional to the square of the overlap between the interacting species, and inversely proportional to the HOMO–LUMO energy gap.[55]

For an S$_N$2 reaction, the LUMO is the σ^*_{CX} MO. The overlap of this MO with the HOMO of X:$^-$ will depend upon the magnitude of the AO coefficient at carbon. Qualitative molecular orbital considerations show[57] that this coefficient increases as the electronegativity difference between C and X increases. Overlap will, therefore, increase from left to right along a row, and it will decrease down a column of the periodic table. If the HOMO–LUMO interaction is dominated by the overlap effect, its magnitude increases from left to right along a row, and decreases from top to bottom down a column. This predicts that barriers should decrease for the series H_2N^-, HO^-, F^-, and also for the series I^-, Br^-, Cl^-, F^-. Only the first of these predicted trends is correct.

The HOMO–LUMO energy gap is also related conceptually to the vertical electron transfer energy gap of the SCD model. Inspection of Table 5.8 suggests that, if the HOMO–LUMO interaction is dominated by the effect of the energy gap, barriers should increase for the series HO^-, F^-, and also for the series I^-, Br^-, Cl^-, F^-. In this case, only the second of these predicted trends is correct.

It follows that, along a row, the barrier is controlled by overlap, and down a column the barrier is controlled by the energy gap. The reasons for this are not clear. It is also not clear what trend would result when the X's belong to different rows and columns. When the qualitative FMO analysis is ambiguous, it becomes necessary to compute the FMO interactions explicitly. However, this requires that the transition structure be known,[58] and the predictive value of FMO theory as a qualitative model is diminished.

The PESD approach employs the diagram shown in **5.27**, in which the lower

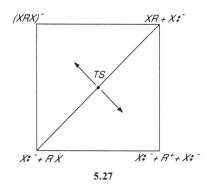

5.27

right-hand corner of the square represents the loose triple ion, and the upper left-hand corner refers to the hypothetical tight pentavalent $CH_3X_2^-$ structure. As we pointed out in Chapter 1, in the gas phase the loose corner is not the triple ion structure but, rather, a covalent structure containing equal contributions from $(X:^- + R\cdot + \cdot X)$ and $(X\cdot + \cdot R + :X^-)$.[18b, c]

Evidently **5.27** cannot be applied to reactions in the gas phase. Moreover, as discussed elsewhere in the literature,[18b] the triple ion character of the loose corner is also uncertain in solution. The nature of this corner depends on the nature of X and, in some cases, solvation does not stabilize the triple ion structure relative to the covalent structures. This is found, in particular, for CH_3X, for which **5.27** is no longer physically realistic.

When **5.27** is meaningful, the PESD will allow discussion of trends in barrier heights, from a consideration of the effects of corner perturbation upon the energy of the transition state. Nevertheless, even in this case the triple ion character of the loose corner arises from a solvent-assisted avoided crossing with $(X:^- + R^+ + :X^-)$ (see Chapter 1). This means that perturbations in the energy of the triple ion corner are not transmitted to the transition state in a straightforward manner. Instead, the contribution of the triple ion structure must be determined from its quantum mechanical mixing into the transition state.[18b,e,59]

5.6.2   The Identity Transition Structure: Applications of the Marcus Equation, FMO Theory, and PESD

The Marcus approach[56] does not deal explicitly with the structure of the identity S$_N$2 transition state. Because a normalized reaction coordinate is used, all identity transition states have $\alpha \doteq 0.5$, by definition.

Since the Marcus equation, like the SCD model, relies upon the concept of an avoided crossing, it is possible that the equation could be modified to allow discussion of transition structure looseness. Such a modification would require that a link be found between intrinsic barriers and looseness.

In the FMO approach, an increase in the HOMO–LUMO interaction leads to an increase in the degree of charge transfer from $X:^-$ to the σ^*_{CX} MO of CH_3X. This increased population of an antibonding orbital will lead to an increase in C–X cleavage. If these remarks apply to the transition state, looser transition structures should be observed whenever $X:^-$ and R–X exhibit a strong HOMO–LUMO interaction. This prediction is not in accord with the findings already examined in this chapter. It follows that the perturbational treatment which forms the basis of FMO theory is not appropriate for the analysis of transition structures.

On the other hand, the PESD approach seems well suited to such an analysis. A change in the relative energies of the loose and tight corners of **5.27** should shift the TS in the direction of the appropriate heavy arrow. However, as we have emphasized, **5.27** is not realistic for gas-phase reactions, because the triple ion structure does not participate in the analysis. In fact, use of a triple ion to make predictions leads to incorrect conclusions.[18c] For example, Cl^-/CH_3Cl, whose triple ion structure has the lowest energy, possesses the tightest transition structure in Table 5.11, whereas H^-/CH_4, whose triple ion structure has the highest energy, represents the loosest transition structure in Table 5.11.

In summary, we find that each of the approaches discussed here provides some insight into reactivity patterns of the identity S$_N$2 reaction, and each has points of intersection with the SCD model. However, the SCD model has been devised specifically to treat reactivity problems and, at the present time, it appears to offer the most satisfying insight into these problems.

5.7   THE IDENTITY REACTION—CONCLUDING REMARKS

The activation process in the identity S$_N$2 reaction has been modeled to account for the molecular distortions and solvent reorganization required to bring ground states and vertical charge transfer states into resonance. In its resonance requirement, this description resembles that given to single electron transfer reactions (SET), but it differs from SET in the nature of the distortions and reorganization that promote the resonance.[35b, 54] In SET reactions, resonance is promoted by motions that do not include bond coupling between the attacking X^- and the molecule. This property of an SET reaction is characterized by a small to zero avoided crossing,[35b, 39, 40, 54d, 56] as depicted in **5.28a.** In contrast, resonance in the S_N2 reaction is promoted by synchronized motions, involving strong bond coupling, which are characterized by a large avoided crossing resonance energy (**5.28b**) ($B \cong 14$ kcal/mol).[18h, i; 60, 61]

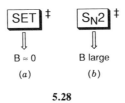

5.28

The factors that determine S_N2 reactivity patterns are the electron transfer energy gaps in the gas phase and in solution, and the radical anion delocalization indices. The quantity $I_{X:} - A_{RX}$ provides information concerning the single electron aspect of the process, the $w_{R:}$ index is associated with bond reorganization (distortion), and the approximately constant B is associated with the strong bond coupling in the TS. A unified description of all reactivity patterns is thereby created: gas phase and molecular barriers originate in the deformation energy required to overcome the electron transfer energy gap; transition structure looseness derives from the stretching distortion associated with the closing of the energy gap; and a solvent barrier results from the reorganization of the equilibrium distribution of the solvent coupled to the charge transfer.

APPENDIX

Approximate Equations Used In Chapter 5

Consider the expression for $(R\dot{\underline{\cdot}}X)^-(s^*)$ (equation A5.1), in which $w_{R:}$ and $w_{X:}$

$$(R\dot{\underline{\cdot}}X)^- = w_{R:}(R:^- \cdot X) + w_{X:}(R\cdot :X^-) \qquad (A5.1)$$

are the weights of the VB configurations. In the gas phase, the energy of $(R\dot{\underline{\cdot}}X)^-(g)$ is given by equation A5.2, where $E_1(g)$ is the gas phase energy of

$$E[(R\dot{\underline{\cdot}}X)^-(g)] = w_{R:}E_1(g) + w_{X:}E_2(g) + 2\sqrt{w_{R:}w_{X:}}\,h_{RX} \qquad (A5.2)$$

$(R:\cdot X)$, and h_{RX} is the interaction matrix element of the two configurations.

Now insert $(R \overset{\cdot}{\cdot} X)^-$ into a medium that solvates the radical anion without changing its intimate geometry. The energy of $(R \overset{\cdot}{\cdot} X)^-(s)$ will be given by equation A5.3, where E_1 and E_2 now refer to the energies of the configurations in a sol-

$$E[(R \overset{\cdot}{\cdot} X)^-(s)] = w_{R:} \, E_1(s) + w_{X:} \, E_2(s) + 2\sqrt{w_{R:} \, w_{X:}} \, h_{RX} \qquad \text{(A5.3)}$$

vent (equations A5.4a and A5.4b). The S terms are the desolvation energies of

$$E_1(s) = E_1(g) - S_{R:}^\dagger \qquad \text{(A5.4a)}$$

$$E_2(s) = E_2(g) - S_{X:}^\dagger \qquad \text{(A5.4b)}$$

the VB configurations, Using equations A5.4, the energy of $(R \overset{\cdot}{\cdot} X)(s)$ is written as equation A5.5. The desolvation energy of $(R \overset{\cdot}{\cdot} X)^-$ can therefore be expressed

$$E[(R \overset{\cdot}{\cdot} X)^- (s)] = E[(R \overset{\cdot}{\cdot} X)^- (g)] - w_{R:} \, S_{R:}^\dagger - w_{X:} \, S_{X:}^\dagger \qquad \text{(A5.5)}$$

as the weighted average of the desolvation energies of the configurations, equation A5.6. To simplify matters, we assume that $S_{R:} \cong S_{R:}^\dagger$, and $S_{X:} \approx S_{X:}^\dagger$. This leads

$$S_{R \overset{\cdot}{\cdot} X} = w_{R:} \, S_{R:}^\dagger + w_{X:} \, S_{X:}^\dagger \qquad \text{(A5.6)}$$

to equation A5.7, used in the text.

$$S_{R \overset{\cdot}{\cdot} X} = w_{R:} \, S_{R:} + w_{X:} \, S_{X:} \qquad \text{(A5.7)}$$

REFERENCES

1. (a) W. N. Olmstead and J. I. Brauman. *J. Am. Chem. Soc.* **99,** 4219 (1977); (b) A. T. Barfknecht, J. A. Dodd, K. E. Salomon, W. E. Tumas, and J. I. Brauman. *Pure Appl. Chem.* **56,** 1809 (1984).

2. (a) M. J. Pellerite and J. I. Brauman. *J. Am. Chem. Soc.* **105,** 2672 (1983); (b) M. J. Pellerite and J. I. Brauman. *J. Am. Chem. Soc.* **102,** 5993 (1980); (c) J. A. Dodd and J. I. Brauman. *J. Am. Chem. Soc.* **106,** 5356 (1984); (d) M. J. Pellerite and J. I. Brauman. *ACS Symp. Ser.* **198,** 81 (1982); (e) C.-C. Han, J. A. Dodd, and J. I. Brauman, *J. Phys. Chem.* **90,** 471 (1986); (f) J. A. Dodd and J. I. Brauman. *J. Phys. Chem.* **90,** 3559 (1986); (g) J. M. Riveros, S. M. José, and K. Takashima. *Adv. Phys. Org. Chem.* **21,** 197 (1985); (h) S. R. Vande Linde and W. L. Hase. *J. Am. Chem. Soc.* **111,** 2349 (1989); (i) S. E. Barlow, J. M. Van Doren, and V. M. Bierbaum. *J. Am. Chem. Soc.* **110,** 7240 (1988); (j) C. H. DePuy, S. Gronert, A. Mullin, and V. M. Bierbaum. *J. Am. Chem. Soc.* **112,** 8650 (1990).

3. G. Caldwell, T. F. Magnera, and P. Kebarle. *J. Am. Chem. Soc.* **106,** 959 (1984).

4. (a) S. Wolfe, D. J. Mitchell, and H. B. Schlegel. *J. Am. Chem. Soc.* **103,** 7694 (1981); (b) D. J. Mitchell. *Theoretical Aspects of S_N2 Reactions.* Ph.D. Thesis, Queen's University, Kingston, Ontario, Canada, 1981; (c) S. Wolfe. Unpublished AM1 computations.

5. (a) F. Carrion and M. J. S. Dewar. *J. Am. Chem. Soc.* **106**, 3531 (1984); (b) M. J. S. Dewar and E. Healy. *Organometallics.* **1**, 1705 (1982); (c) J. W. Viers, J. C. Schug, M. D. Stovall and J. I. Seeman. *J. Comp. Chem.* **5**, 598 (1984); (d) M. V. Bazilevskii, S. G. Koldobskii and V. A. Tikhomirov. *J. Org. Chem. U.S.S.R.* **18**, 795 (1982).

6. (a) A. Dedieu and A. Veillard. *Chem. Phys. Lett.* **5**, 328 (1970); (b) A. Dedieu and A. Veillard. *J. Am. Chem. Soc.* **94**, 6730 (1972); (c) A. J. Duke and R. F. W. Bader. *Chem. Phys. Lett.* **10**, 631 (1971); (d) R. F. W. Bader, A. J. Duke, and R. R. Messer. *J. Am. Chem. Soc.* **95**, 7715 (1973); (e) F. Keil and R. Ahlrichs. *J. Am. Chem. Soc.* **98**, 4787 (1976); (f) C. Leforestier. *J. Chem. Phys.* **68**, 4406 (1978); (g) J. Serre. *Int. J. Quant. Chem.* **26**, 593 (1984); (h) H. B. Schlegel, K. Mislow, F. Bernardi, and A. Bottoni. *Theor. Chim. Acta* **44**, 245 (1977); (i) P. Baybutt. *Mol. Phys.* **29**, 389 (1975); (j) M. Urban, I. Cernusák, and V. Kellö. *Chem. Phys. Lett.* **105**, 625 (1984); (k) S. VandeLinde, W. L. Hase, and H. B. Schlegel. Unpublished results for Cl⁻ + CH₃Cl with different basis sets and MP2, MP3, and MP4 correlation corrections; (l) Carnegie–Mellon Archive. F⁻ + CH₃F calculations with the 6-31G* basis set; (m) C. D. Ritchie and G. A. Chappell. *J. Am. Chem. Soc.* **92**, 445 (1970); (n) Z. Shi and R. J. Boyd. *J. Am. Chem. Soc.* **111**, 1575 (1989); (o) D. Cremer and E. Kraka. *J. Phys. Chem.* **90**, 33 (1986); (p) I. Cernusák and M. Urban. *Coll. Czech. Chem. Commun.* **53**, 2239 (1988); (q) R. Vetter and L. Zülicke. *J. Am. Chem. Soc.* **112**, 5136 (1990); (r) T. Minato and S. Yamabe. *J. Am. Chem. Soc.* **110**, 4586 (1988); (s) T. Minato and S. Yamabe. *J. Am. Chem. Soc.* **107**, 4621 (1985); (t) A. Merkel, Z. Havlas, and R. Zahradnik. *J. Am. Chem. Soc.* **110**, 8355 (1988); (u) Z. Havlas, A. Merkel, J. Kalcher, and R. Janoschek. *Chem. Phys.* **127**, 53 (1988); (v) G. Sini, S. S. Shaik, J. M. Lefour, G. Ohanessian, and P. C. Hiberty. *J. Phys. Chem.* **93**, 5661 (1989); (w) S. C. Tucker and D. G. Truhlar. *J. Phys. Chem.* **93**, 8138 (1989); *J. Am. Chem. Soc.* **112**, 3338 (1990); (x) S. Wolfe and C.-K. Kim. *J. Am. Chem. Soc.* In press; (y) Z. Shi and R. J. Boyd. *J. Am. Chem. Soc.* **112**, 6789 (1990).

7. (a) K. Morokuma. *J. Am. Chem. Soc.* **104**, 3732 (1982); (b) J. Jaume, J. M. Lluch, A. Oliva, and J. Bertrán. *Chem. Phys. Lett.* **106**, 232 (1984); (c) S. Yamabe, E. Yamabe, and T. Minato. *J. Chem. Soc. Perkin Trans. II.* 1881 (1983).

8. (a) I. H. Williams. *J. Am. Chem. Soc.* **106**, 7206 (1984); (b) K. Raghavachari, J. Chandrasekhar, and R. Burnier. *J. Am. Chem. Soc.* **106**, 3124 (1984).

9. (a) J. Chandrasekhar, S. F. Smith, and W. L. Jorgensen. *J. Am. Chem. Soc.* **107**, 154 (1985); (b) J. Chandrasekhar, S. F. Smith, and W. L. Jorgensen. *J. Am. Chem. Soc.* **104**, 3049 (1984); (c) J. Chandrasekhar and W. L. Jorgensen. *J. Am. Chem. Soc.* **107**, 2974 (1985); (d) R. A. Chiles and P. J. Rossky. *J. Am. Chem. Soc.* **106**, 6867 (1984); (e) W. L. Jorgensen and J. K. Buckner. *J. Phys. Chem.* **90**, 4651 (1986); (f) T. Kuzaki, K. Morihashi, and O. Kikuchi. *J. Am. Chem. Soc.* **111**, 1547 (1989); (g) P. A. Bash, M. J. Field, and M. Karplus. *J. Am. Chem. Soc.* **109**, 8092 (1987); (h) K. Ya. Burshtein. *J. Mol. Struct.* (*THEOCHEM*). **153**, 209 (1987); (i) M. V. Bazilevskii and S. G. Koldobskii. *J. Org. Chem. USSR* **20**, 824 (1984); (j) K. Ya. Burshtein and A. N. Isaev. *Bull. Acad. Sci. USSR* **36**, 1858 (1988); (k) Y. S. Kong and M. S. Jhon. *Theor. Chim. Acta* **70**, 123 (1986); (l) I. Černučsák and M. Urban. *Coll. Czech. Chem. Commun.* **49**, 1854 (1984).

10. For a review of computational, theoretical, and experimental studies see M. V. Bazilevskii, S. G. Goldobskii nd V. A. Tikhomirov. *Russ. Chem. Revs.* **55**, 948 (1986).

11. (a) R. C. Dougherty, J. Dalton, and J. D. Roberts. *Org. Mass. Spectrom.* **8**, 77 (1974); (b) R. Yamdagni and P. Kebarle. *J. Am. Chem. Soc.* **93**, 7139 (1971); (c) R. C.

Dougherty and J. D. Roberts. *Org. Mass Spectrom.* **8**, 81 (1974); (d) R. C. Dougherty. *Org. Mass Spectrom.* **8**, 85 (1974).

12. When the TS is lower in energy than the reactants, the reaction rate will increase as the temperature is lowered, because the barrier for the reverse reaction (X:$^-$ $\cdots$ RX) → X:$^-$ + RX is larger than that for the forward reaction (X:$^-$ $\cdots$ RX) → (XR $\cdots$ X:$^-$). The rate constant for traversal of the larger barrier will decrease more quickly as the temperature is lowered, leading to a negative temperature coefficient.

13. (a) J. Hayami, T. Koyanagi, N. Hihara, and A. Kaji. *Bull. Chem. Soc. Jpn.* **51**, 891 (1978); (b) J. Hayami, N. Tanaka, N. Hihara, and A. Kaji. *Tetrahedron Lett.* 385 (1973).

14. For a discussion of the available experimental data, see D. J. McLennan. *Aust. J. Chem.* **31**, 1897 (1978).

15. (a) W. J. Albery and M. M. Kreevoy. *Adv. Phys. Org. Chem.* **16**, 87 (1978); (b) W. J. Albery. *Ann. Rev. Phys. Chem.* **31**, 227 (1980).

16. (a) E. S. Lewis, S. Kukes, and C. D. Slater. *J. Am. Chem. Soc.* **102**, 1619 (1980); (b) E. S. Lewis and D. D. Hu. *J. Am. Chem. Soc.* **106**, 3292 (1984); (c) E. S. Lewis and S. Kukes. *J. Am. Chem. Soc.* **101**, 417 (1979); (d) E. S. Lewis, T. A. Douglas, and M. L. McLaughlin *Isr. J. Chem.* **26**, 331 (1986); (e) E. S. Lewis, M. L. McLaughlin, and T. A. Douglas. In *Nucleophilicity. Adv. Chem. Series.* **215.** Edited by J. M. Harris and S. P. McManus. American Chemical Society, Washington DC, 1987; (f) E. S. Lewis, M. L. McLaughlin, and T. A. Douglas. *J. Am. Chem. Soc.* **107**, 6668 (1985); (g) E. S. Lewis, T. I. Yousaf, and T. A. Douglas. *J. Am. Chem. Soc.* **109**, 2152 (1987).

17. An intrinsic barrier for FSO$_3$$^-$ + CH$_3$OSO$_2$F has been derived by Mitchell[4b] from Arnett's data for the reaction pyridine + CH$_3$OSO$_2$F in acetonitrile. See: E. M. Arnett and R. Reich. *J. Am. Chem. Soc.* **100**, 2930 (1978); *ibid.* **102**, 5892 (1980).

18. (a) S. S. Shaik. *Nouv. J. Chim.* **6**, 159 (1982); (b) S. S. Shaik. *Progr. Phys. Org. Chem.* **15**, 197 (1985); (c) D. J. Mitchell, H. B. Schlegel, S. S. Shaik, and S. Wolfe. *Can. J. Chem.* **63**, 1642 (1985); (d) S. S. Shaik. *Isr. J. Chem.* **26**, 367 (1986); (e) S. S. Shaik and A. Pross. *J. Am. Chem. Soc.* **104**, 2708 (1982); (f) S. S. Shaik. *Can. J. Chem.* **64**, 96 (1986); (g) Note that the gap in Figure 5.4 refers to infinite separation, but is also a good approximation of the geometry of the ion–dipole complex. Because of the overlap of X$\cdot$ with (CH$_3$X)$^-$, the stabilization energy in the ion–dipole ground state is balanced by stabilization in the vertical charge transfer state; (h) Strictly speaking: $\Delta E_c = \Delta E_{def} + \Delta E_{interaction}$. The latter term consists of exchange repulsion, polarization, and electrostatic interactions. These effects partially cancel each other in ion–molecule S$_N$2 reactions.

19. (a) The correlation of the barrier with the methyl cation affinity of X also exhibits this trend;[2a] (b) That barriers should correlate in some manner with D_{C-X} and A_X. has been proposed by R. G. Pearson in *Nucleophilicity* in reference 16e, and by R. F. Hudson. *Chimia* **16**, 173 (1962).

20. (a) D. J. Mitchell and S. Wolfe. Unpublished results found on p. 115 of reference 4b; (b) 3-21G calculations are described in K. N. Houk and M. N. Paddon-Row. *J. Am. Chem. Soc.* **108**, 2659 (1986).

21. M. E. Jones, S. R. Kass, J. Filley, R. M. Barkley, and G. B. Ellison, *J. Am. Chem. Soc.* **107**, 109 (1985).

22. (a) I. Mihel, J. O. Knipe, J. K. Coward, and R. L. Schowen. *J. Am. Chem. Soc.* **101**, 4349 (1979); (b) M. F. Hegazi, R. T. Borchardt, and R. L. Schowen. *J. Am. Chem.*

Soc. **101**, 4359 (1979); (c) J. Rodgers, D. A. Femec, and R. L. Schowen. *J. Am. Chem. Soc.* **104**, 3263 (1982); (d) O.S.-L. Wong and R. L. Schowen. *J. Am. Chem. Soc.* **105**, 1951 (983); (e) C. H. Gray, J. K. Coward, K. B. Schowen, and R. L. Schowen. *J. Am. Chem. Soc.* **101**, 4351 (1979).

23. S. S. Shaik. *J. Am. Chem. Soc.* **110**, 1127 (1988).

24. (a) P. Maitre, P. C. Hiberty, G. Ohanessian, and S. S. Shaik. *J. Phys. Chem.* **94**, 4089 (1990); (b) G. Sini, P. C. Hiberty, and S. S. Shaik. *J. Chem. Soc. Chem. Commun.* 772 (1989).

25. S. S. Shaik, H. B. Schlegel, and S. Wolfe. *J. Chem. Soc. Chem. Commun.* 1322 (1988).

26. (a) A. A. Malinauskas. *J. Org. Chem. USSR* **23**, 643 (1987); (b) M. W. Wong and L. Radom. *J. Am. Chem. Soc.* **110**, 2375 (1988); (c) P. M. W. Gill and L. Radom. *Chem. Phys. Lett.* **132**, 16 (1986).

27. (a) K. C. Westaway. *Can. J. Chem.* **56**, 2691 (1978); (b) K. C. Westaway and Z. Lai. *Can. J. Chem.* **67**, 345 (1989).

28. For recent reviews on the subject of solvent effects see: (a) K. Burger. *Solvation, Ionic and Complex Formation Reactions in Non-Aqueous Solvents.* Elsevier, Amsterdam, 1983; (b) C. Reichardt *Solvent Effects in Organic Chemistry.* Verlag Chimie, Weinheim, 1979; (c) M. H. Abraham. *Progr. Phys. Org. Chem.* **11**, 1 (1974).

29. For analytical models of dynamic solvent effects on polar transformatins see: (a) H. A. Kramers. *Physica (The Hague).* **7**, 284 (1940); (b) G. van der Zwan and J. T. Hynes. *J. Chem. Phys.* **78**, 4174 (1983); **76**, 2993 (1982); (c) J. L. Kurz and L. C. Kurz. *J. Am. Chem. Soc.* **94**, 4451 (1972); *Isr. J. Chem.* **26**, 339 (1985).

30. For static and dynamic treatments of solvent effects at a molecular–semiempirical level see: (a) A. Warshel and S. T. Russell. *Quart. Rev. Biophys.* **17**, 283 (1984); (b) A. Warshel. *Acc. Chem. Res.* **14**, 284 (1981); (c) A. Warshel. *J. Phys. Chem.* **86**, 2218 (1982); (d) J. T. Hynes. *Annu. Rev. Phys. Chem.* **36**, 573 (1985).

31. For semiclassical treatments of solution-phase S_N2 reactions see: (a) E. D. German and R. R. Dogonadze. *Int. J. Chem. Kinet.* **6**, 457, 467 (1974); (b) D. E. Khoshtariya. *Theor. Exp. Chem.* **21**, 397 (1985).

32. For a quantum mechanical model of solution-phase S_N2 reactions, including dynamic effects, see: (a) E. D. German and A. M. Kuznetsov. *Bull. Acad. Sci. USSR* **34**, 1877 (1985); (b) E. D. German and V. A. Tikhomirov. *Bull. Acad. Sci. USSR* **35**, 1733 (1987); (c) E. D. German and A. M. Kuznetsov. *J. Chem. Soc. Faraday Trans. 2* **82**, 1885 (1986).

33. For a reaction dynamics study of identity S_N2 reactions based on VB curve crossing ideas, and with inclusion of solvent, see J.-K. Hwang, G. King, S. Creighton, and A. Warshel. *J. Am. Chem. Soc.* **110**, 5297 (1988).

34. For a molecular dynamics study and analytical modeling of Cl^-/CH_3Cl in water, see: (a) J. P. Bergsma, B. J. Gertner, K. R. Wilson, and J. T. Hynes. *J. Chem. Phys.* **86**, 1356 (1987); (b) B. J. Gertner, J. P. Bergsma, K. R. Wilson, S. Lee, and J. T. Hynes. *J. Chem. Phys.* **86**, 1377 (1987).

35. (a) S. S. Shaik. *J. Am. Chem. Soc.* **106**, 1227 (1984); (b) S. S. Shaik. *Acta Chem. Scand.* **44**, 205 (1990).

36. (a) J. A. Revetllat, J. M. Lluch, A. Oliva, and J. Bertrán. *CR Acad. Sci. Paris.* **T229**, *Série II.* 62 (1984); (b) E. Carbonell, J. L. Andres, A. Lledós, M. Duran, and J. Bertrán. *J. Am. Chem. Soc.* **110**, 996 (1988); (c) J. L. Andres, A. Lledós, M. Duran,

and J. Bertrán. *Chem. Phys. Lett.* **153**, 82 (1988); (d) J. Bertran. In *New Theoretical Concepts for Understanding Organic Reactions*. J. Bertrán and I. G. Csizmadia, Editors, Kluwer Publications. Dordrecht, 1989, p. 231; (e) L. Salem. *Science* **191**, 822 (1976).

37. F. T. Marchese and D. L. Beveridge. *J. Am. Chem. Soc.* **106**, 3713 (1984).

38. For a molecular dynamics study of NaCl → Na$^+$ + Cl$^-$ in water, see O. A. Karim and J. A. McCammon. *J. Am. Chem. Soc.* **108**, 1762 (1986).

39. (a) R. A. Marcus. *J. Chem. Phys.* **24**, 966 (1956); (b) R. A. Marcus. *J. Chem. Phys.* **43**, 679 (1965); (c) R. A. Marcus. *J. Chem. Phys.* **24**, 979 (1956); (d) R. A. Marcus. *J. Chem. Phys.* **38**, 1858 (1963); (e) A summary of Marcus and related treatments by Hush, Levich, Dogonadze, and Chizmadzhev can be found in R. A. Marcus. *Ann. Rev. Phys. Chem.* **15**, 155 (1964).

40. (a) B. S. Brunschwing, J. Logan, M. D. Newton, and N. Sutin. *J. Am. Chem. Soc.* **102**, 5798 (1980); (b) N. Sutin. *Acc. Chem. Res.* **15**, 275 (1982); (c) A. Haim. *Comments Inorg. Chem.* **4**, 113 (1985); (d) L. Eberson. *Electron Transfer Reactions in Organic Chemistry*. Springer-Verlag, Heidelberg, 1987.

41. The treatment of solvent–structure relationships can be found in: (a) reference 28c; M. H. Abraham. *Pure Appl. Chem.* **57**, 1055 (1985); (b) A. J. Parker. *Chem. Revs.* **69**, 1, (1969); (c) P. Haberfield. *J. Am. Chem. Soc.* **93**, 2091 (1971); (d) E. M. Arnett, W. G. Bentrude, J. J. Burke, and P. McC. Duggleby. *J. Am. Chem. Soc.* **87**, 1541 (1965); E. M. Arnett. *J. Chem. Educ.* **62**, 385 (1985); (e) E. Buncel and H. Wilson. *Adv. Phys. Org. Chem.* **14**, 133 (1977); *Acc. Chem. Res.* **12**, 42 (1979); *J. Chem. Educ.* **57**, 629 (1980).

42. (a) P. Delahay. *Acc. Chem. Res.* **15**, 40 (1982); (b) K. von Burg and P. Delahay. *Chem. Phys. Lett.* **86**, 528 (1982); (c) P. Delahay and A. Dziedzic. *Chem. Phys. Lett.* **108**, 169 (1984); (d) P. Delahay and A. Dziedzic. *J. Chem. Phys.* **80**, 5793 (1984).

43. A. A. Rashin and B. Honig. *J. Chem. Phys.* **89**, 5588 (1985).

44. P. Kebarle. *Ann. Rev. Phys. Chem.* **28**, 445 (1977).

45. T. F. Magnera, G. Caldwell, J. Sunner, S. Ikuta, and P. Kebarle. *J. Am. Chem. Soc.* **106**, 6140 (1984).

46. The radical anion polarization will be accompanied by a greater (X:$^-$ CH$_3^+$ X:$^-$) character in the TS, because of improved solvent stabilization of the VB mixing of such a configuration.

47. (a) M. H. Abraham. *J. Chem. Soc. Perkin Trans. II* 1375 (1976); (b) M. H. Abraham and J. Liszi. *J. Inorg. Nucl. Chem.* **43**, 143 (1981); (c) M. H. Abraham and J. Liszi. *J. Chem. Soc. Faraday Trans. I* **74**, 1604 (1978).

48. G. B. Cox, G. R. Hedwig, A. J. Parker, and D. W. Watts. *Aust. J. Chem.* **27**, 477 (1974).

49. Other solvation data appear in: (a) R. M. Noyes. *J. Am. Chem. Soc.* **84**, 513 (1962); *ibid.* **86**, 971 (1964); M. H. Abraham. *J. Chem. Soc. Perkin II.* 1893 (1973); D. J. McLennan. *Aust. J. Chem.* **31**, 1897 (1978); (b) S. Goldman and R. Bates. *J. Am. Chem. Soc.* **94**, 1476 (1972); (c) R. Gomer and G. Tryson. *J. Chem. Phys.* **66**, 4413 (1977); (d) C. D. Ritchie. *J. Am. Chem. Soc.* **105**, 7313 (1983); (e) H. L. Friedman and C. K. Krishnan. *Water, a Comprehensive Treatise*. Vol. 3, Chapter 1. F. Franks, Editor, Plenum Press, London 1973; (f) J. Chandrasekhar, D. C. Spellmeyer, and W. L. Jorgensen. *J. Am. Chem. Soc.* **106**, 903 (1984); (g) J. Jortner and R. M. Noyes. *J. Phys. Chem.* **70**, 770 (1966); (h) E. M. Arnett, L. E. Small, R. T. McIver, Jr., and

J. S. Miller, *J. Am. Chem. Soc.* **96,** 5638 (1974); E. M. Arnett, D. E. Johnston, and L. E. Small. *J. Am. Chem. Soc.* **97,** 5598 (1975).

50. For a related review of solvent properties see G. R. Freeman. *Annu. Rev. Phys. Chem.* **34,** 463 (1983).

51. (a) R. W. Taft, M. H. Abraham, R. M. Doherty, and M. J. Kamlet. *J. Am. Chem. Soc.* **107,** 3105 (1985); (b) M. H. Abraham, M. J. Kamlet, and R. W. Taft. *J. Chem. Soc. Perkin Trans. II.* 923 (1982); (c) R. W. Taft, M. H. Abraham, R. M. Doherty, and M. J. Kamlet. *Nature.* **313,** 384 (1985); (d) M. J. Kamlet, J.-L. M. Abboud, M. H. Abraham, and R. W. Taft. *J. Org. Chem.* **48,** 2877 (1983); (e) E. M. Kosower. *An Introduction to Physical Organic Chemistry.* John Wiley, New York, 1968, pp. 293 ff; (f) K. Dimroth, C. Reichardt, T. Siepmann, and F. Bohlmann. *Ann. Chem.* **661,** 1 (1963).

52. The use of solvent mixtures is a viable strategy. See, for example: (a) L. Menninga and J. B. F. N. Engberts. *Tetrahedron Lett.* 617 (1972); (b) J. R. Haak and J. B. F. N. Engberts. *J. Org. Chem.* **49,** 2387 (1984); (c) W. Karzijn and J. B. F. N. Engberts. *Recl. Trav. Chim.* **96,** 95 (1977); (d) K. Remerie and J. B. F. N. Engberts. *J. Phys. Chem.* **87,** 5449 (1983).

53. E. D. German and Yu. I. Kharkats. *Bull. Acad. Sci. USSR* **34,** 561 (1985).

54. (a) Page 308 of reference 18b; (b) D. Cohen, R. Bar, and S. S. Shaik. *J. Am. Chem. Soc.* **108,** 231 (1986); (c) S. S. Shaik. *J. Org. Chem.* **52,** 1563 (1987); (d) A. Pross. *Acc. Chem. Res.* **18,** 212 (1985).

55. I. Fleming. *Frontier Orbitals and Organic Chemical Reactions.* John Wiley, New York, 1976.

56. R. A. Marcus. *Ann. Rev. Phys. Chem.* **15,** 155 (1964).

57. T. A. Albright, J. K. Burdett, and M. H. Whangbo. *Orbital Interactions in Chemistry.* John Wiley, New York, 1985.

58. S. Wolfe and D. Kost. *Nouv. J. Chim.* **2,** 441 (1978).

59. (a) A. Pross and S. S. Shaik. *Acc. Chem. Res.* **16,** 363 (1983); (b) A. Pross. *Adv. Phys. Org. Chem.* **21,** 99 (1985).

60. See experimental evaluations and a lucid discussion in: (a) T. Lund and H. Lund. *Acta Chem. Scand.* **B42,** 269 (1988); (b) T. Lund and H. Lund. *Acta Chem. Scand.* **B40,** 470 (1986); (c) T. Lund and H. Lund. *Acta Chem. Scand.* **B41,** 93 (1987); (d) T. Lund, S. U. Pedersen, H. Lund, K. M. Cheung, and J. H. P. Utley. *Acta Chem. Scand.* **B41** 285 (1987).

61. For a definition of the S_N2 mechanism as an inner-sphere electron transfer, see M. Chanon. *Bull. Soc. Chim. II.* 197 (1982).

6

NONIDENTITY S_N2 REACTIONS

We have seen that trends in the barriers of identity S_N2 reactions and in the geometries of identity S_N2 transition structures can be predicted, both in the gas phase and in solution. It is, therefore, appropriate to proceed to an analysis of trends in the barriers of nonidentity S_N2 reactions (equation 6.1), and trends in the geome-

$$Y:^- + R\text{–}X \rightarrow Y\text{–}R + {}^-:X \qquad (6.1)$$

tries of nonidentity S_N2 transition structures. We may comment immediately that, regardless of whether a nonidentity reaction is thermoneutral, exoergic, or endoergic, the transition structure will no longer be a bond-symmetric species that lies at the midpoint of the reaction coordinate connecting reactants and products.

The most important distinction between an identity and an exoergic or endoergic nonidentity reaction is the presence of a "thermodynamic driving force" in the nonidentity process. This is defined by ΔE, which denotes the reaction ergonicity in general units (internal energy, enthalpy, free energy). As was discussed in Chapter 1, changes in reactivity that are associated with changes in ΔE are the basis of the rate-equilibrium relationships of physical organic chemistry; indeed, the concept of the thermodynamic driving force evolved from the Bell–Evans–Polanyi principle.[1] This predicts that barriers will decrease within reaction series as the reactions become more exoergic.

For S_N2 reactions, however, even a limited collection of theoretical data (Table 6.1), experimental data in the gas phase (Table 6.2), or experimental data in solution (Table 6.3) is sufficient to illustrate the restricted sovereignty of the BEP principle.[2-4] This is particularly evident in Figure 6.1.

The Leffler–Hammond postulate is the counterpart of the BEP approach in considerations of transition state geometries;[5] it predicts that, within reaction series,

Table 6.1 Computed (4-31G) Reaction Energies (ΔE) and Central Barriers ($\Delta E^{\ddagger}$) for S$_N$2 Reactions[a]

Entry	Reactants	ΔE^b	$\Delta E^{\ddagger}$
1	H$^-$/CH$_3$F	−62.3	7.3
2	HCC$^-$/CH$_3$F	−21.4	19.5
3	HO$^-$/CH$_3$F	−15.4	9.1
4	F$^-$/CH$_3$F	0	11.7
5	H$^-$/CH$_3$CN	−63.1	23.5
6	H$^-$/CH$_3$-CCH	−42.1	34.7
7	F$^-$/CH$_3$-OOH	−6.6	14.1
8	Cl$^-$/CH$_3$Cl	0	5.5

[a]In kcal/mol. Data are from references 2a and b.
[b]Refers to the energy difference between the reactant and product ion–molecule complexes.

Table 6.2 Central Barriers and Reaction Energies for Gas-Phase S$_N$2 Reactions[a]

Entry	Reactants	ΔE^b	$\Delta E^{\ddagger}$ (RRKM)[c]
1	F$^-$/CH$_3$Cl	−28	6.9
2	H$^-$/CH$_3$F	−57	≈16
3	HCC$^-$/CH$_3$F	−24	22.8
4	CH$_3$O$^-$/CH$_3$F	−14	19.9
5	F$^-$/CH$_3$F	0	26.2
6	CN$^-$/CH$_3$Br	−35	8.9
7	Cl$^-$/CH$_3$Cl	0	10.2

[a]In kcal/mol. Data are from reference 3.
[b]These are reaction enthalpies (ΔH).
[c]$\Delta E^{\ddagger}$ (RRKM) values are derived from the kinetic data using RRKM theory.

Table 6.3 Barriers and Reaction Energies for S$_N$2 Reactions in Water Solvent[a]

Entry	Reactants	ΔE	$\Delta E^{\ddagger}$
1	I$^-$/CH$_3$I	0	22.0
2	Br$^-$/CH$_3$I	+2.15	23.9
3	Cl$^-$/CH$_3$I	+0.72	24.6
4	F$^-$/CH$_3$I	+1.2	29.16
5	F$^-$/CH$_3$Br	−0.96	26.53
6	F$^-$/CH$_3$Cl	+0.48	29.64
7	F$^-$/CH$_3$F	0	31.8
8	Cl$^-$/CH$_3$OSO$_2$Ph	−16.5	23.66
9	CN$^-$/CH$_3$OSO$_2$Ph	−52.34	20.76
10	CN$^-$/CH$_3$F	−36.33	25.33

[a]All values are free energies, in kcal/mol. Data are from reference 4a.

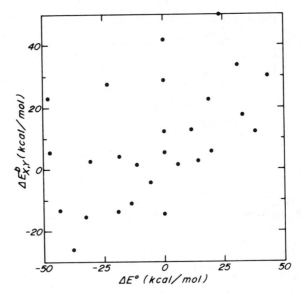

Figure 6.1 A rate-equilibrium plot for the reaction $Y^- + CH_3X \rightarrow YCH_3 + X^-$ in the gas phase.

a transition state will progressively resemble the reactants (products) as ΔE becomes progressively more negative (positive).

The theoretical data collected in Table 6.4 are relevant to these ideas.[2] In this table, the geometry of the transition state is described in terms of the percentage lengthening of the C–X bond in the transition state relative to the C–X bond length in the $Y:^-/CH_3X$ ion–molecule complex (equation 6.2). It can be seen that there

$$\%CX^\ddagger = 100\% \; [(d^\ddagger_{CX} - d^o_{CX})/d^o_{CX}] \qquad (6.2)$$

Table 6.4 Computed (4-31G) Reaction Ergonicities, Percentages of C–X Bond Cleavage, and Central Barriers of S_N2 Reactions[a]

Entry	Y^-/CH_3X	ΔE	$\%CX^{\ddagger b}$	$\Delta E^\ddagger$
1	Cl^-/CH_3Cl	0	21.1	5.5
2	H^-/CH_3F	−62.3	18.7	7.3
3	H^-/CH_3CN	−63.1	31.4	23.5
4	H^-/CH_3CCH	−42.1	36.2	34.7
5	H^-/CH_3OH	−47	26.2	18.5
6	HO^-/CH_3CN	−15.1	35.0	27.0
7	HO^-/CH_3F	−15.4	21.5	9.1
8	HCC^-/CH_3F	−21.4	27.3	19.5
9	F^-/CH_3F	0	25.0	11.7
10	CN^-/CH_3F	−2.7	37.0	26.6

[a]Data are from references 2a and b. Energies are in kcal/mol.
[b]$\%CX^\ddagger$ is defined in equation 6.2.

is no straightforward correlation between $\%CX^{\ddagger}$ and ΔE. Instead, a high degree of C–X cleavage is generally associated with a high barrier,[2c] but neither of these properties correlates with ΔE over the complete data set. It appears that the BEP principle and the Leffler–Hammond postulate are valid only for particular subsets of substituents that form "reaction series." Our problem then is to understand, in general, the reactivity patterns as a whole and, specifically, to determine the factors that define reaction series.

This objective will be pursued in two stages. In the first, by application of the Marcus equation[6] to nonidentity barriers, we recognize the interplay between the intrinsic barrier and the reaction energy. In the second, by application of the SCD model, we find that trends in the barriers, as well as trends in the transition state geometries, are consistent with the notion that the S_N2 reaction involves the movement of a single electron synchronized to the bond interchange.

Our analysis is not based on an exhaustive examination of the vast literature concerning nonidentity reactions. Instead, we begin with an analysis of the ab initio data,[2a,b] continue with experimental gas-phase data,[3,7-9] and turn finally to reactivity trends in solution,[4,10] with special emphasis on the correlations of Albery and Kreevoy.[4a]

6.1 ENERGY BARRIERS AND MARCUS-TYPE RATE–EQUILIBRIUM RELATIONSHIPS

6.1.1 Marcus Treatment of Computational Data for Gas-Phase S_N2 Reactions

Equation 6.3 is a useful form of the Marcus equation; $\Delta E_o^{\ddagger}$ is the intrinsic barrier,[4a]

$$\Delta E^{\ddagger} = \Delta E_o^{\ddagger} + \tfrac{1}{2}\Delta E + \frac{\Delta E^2}{16\Delta E_o^{\ddagger}} \tag{6.3}$$

normally approximated by equation 6.4, the average of the barriers for the identity

$$\Delta E_o^{\ddagger} = \tfrac{1}{2}(\Delta E_{XX}^{\ddagger} + \Delta E_{YY}^{\ddagger}) \tag{6.4}$$

reactions $Y^- + R–Y$ and $X^- + R–X$.

The validity of equation 6.4 can be studied using the SCD model. Following from the discussion of Section 3.3.3, the assumption of the Marcus equation is reasonable, if the avoided crossing interaction does not vary greatly from one identity reaction to another. This seems to be the case, as was found in Chapter 5. The ability of equation 6.4 to pattern the nonidentity data will provide additional support for its use.

There is a complication in the application of equation 6.3 to gas-phase S_N2 reactions, because the Marcus treatment refers to an elementary process.[2a,b,4,6] As seen in Figure 6.2, in the gas phase the reaction coordinate consists of a double-

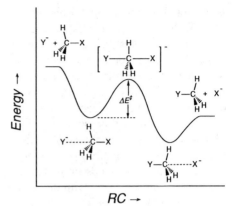

Figure 6.2 The potential energy profile of a gas-phase nonidentity S$_N$2 reaction Y$^-$ + CH$_3$X → YCH$_3$ + X$^-$.

well potential which includes, in addition to the reactants and the products, the ion–dipole complexes Y$^-$/CH$_3$X and X$^-$/CH$_3$Y.[2a, b, 3] Equation 6.3 applies only to the central step of equation 6.5. A modification of equation 6.3 for the treatment

$$Y^-/CH_3X \rightarrow [(Y\text{–}CH_3\text{–}X)^-]^{\ddagger} \rightarrow YCH_3/X^- \qquad (6.5)$$

of the entire reaction coordinate can be found in the original literature.[2a, b]

The ΔE and $\Delta E_o^{\ddagger}$ associated with the central steps of 11 cross reactions are collected in Table 6.5, along with the barriers computed directly at the 4-31G level and the barriers computed with equation 6.3. The Marcus equation is seen to predict the ab initio barriers with good accuracy. A plot of the ab initio barriers versus the barriers predicted by equation 6.3 is linear ($r = 0.99$) with a slope close to

Table 6.5   Marcus (equation 6.3) and Ab Initio (4-31G) Central Barriers for Nonidentity S$_N$2 Reactions Y$^-$ + CH$_3$X → YCH$_3$ + X$^-$[a,b]

Entry	Y, X	ΔE (4-31G)	$\Delta E_o^{\ddagger}$ (eq 6.4)[a]	$\Delta E^{\ddagger}$ (eq 6.3)	$\Delta E^{\ddagger}$ (4-31G)
1	H, NC	−83.1	40.25	9.42	9.7
2	H, CN	−63.4	47.90	21.44	23.5
3	H, F	−62.3	31.85	8.32	7.3
4	H, CCH	−42.4	51.20	32.2	34.7
5	H, OH	−47.1	36.60	16.84	18.5
6	F, SH	−27.0	13.65	3.5	3.5
7	HO, CN	−15.1	32.5	25.4	27.0
8	HO, F	−15.4	16.45	10.6	9.1
9	HCC, F	−21.4	31.05	21.3	19.5
10	F, OOH	−6.6	15.1	12.0	14.1
11	CN, F	−2.7	27.75	26.4	26.6

[a]Data are from references 2a and b. Intrinsic barriers have been calculated using equation 6.4.
[b]The correlation $\Delta E^{\ddagger}$(4-31G) versus $\Delta E^{\ddagger}$(eq 6.3) is linear: $r = 0.990$; slope = 1.07; intercept = −0.712 kcal/mol.

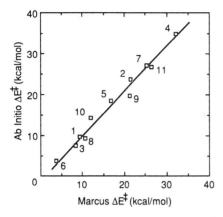

Figure 6.3 A plot of computed (4-31G) $\Delta E^{\ddagger}$ versus $\Delta E^{\ddagger}$ calculated with the Marcus equations 6.3 and 6.4 for the data of Table 6.5. The correlation coefficient $r = 0.990$, and the slope is 1.07.

unity (1.07) (Figure 6.3). The use of equation 6.4 to approximate the intrinsic barrier therefore seems to be reasonable.

It follows that the barriers of nonidentity S_N2 reactions can be understood in terms of the balance between the intrinsic kinetic quantity $\Delta E_o^{\ddagger}$ and the thermodynamic driving force. From the data of Table 6.5 we note that, for a constant ΔE, the barrier $\Delta E^{\ddagger}$ is larger when the kinetic quantity, $\Delta E_o^{\ddagger}$, is larger (entries 2 versus 3, 4 versus 5, 7 versus 8, 6 versus 9). On the other hand, for a constant $\Delta E_o^{\ddagger}$, the barrier decreases as the reaction becomes more exoergic (entries 3 versus 9 versus 7, entries 8 versus 10). This means that, within a reaction series, the BEP principle is obeyed whenever the intrinsic barrier is constant. However, it is not possible at this stage to define the conditions for constancy of $\Delta E_o^{\ddagger}$. These will become clear later, when the SCD model is applied to the problem.

6.1.2 Marcus Treatment of Experimental Gas-Phase Data

The applicability of the Marcus equation to experimental gas-phase data has been discussed by Brauman and his co-workers.[3,7] The quantity that is directly accessible by experiment is the reaction efficiency η, defined in equation 6.6 as the ratio

$$\eta = \frac{k_{\text{obs}}}{k_{\text{coll}}} \tag{6.6}$$

of the observed rate constant (k_{obs}) to the rate of ion–molecule collision (k_{coll}).[8] Experimental reaction efficiencies and reaction enthalpies are summarized in Table 6.6. Although there is no overall correlation between reaction efficiency and reaction energy (ΔE), groups that are associated with large identity barriers (HCC^-, $PhCH_2^-$, CN^-, CH_3S^-) are found to exhibit low efficiencies in their nonidentity reactions.

A more meaningful analysis would involve the correlation of the central barriers of the reactions of Table 6.6 with the appropriate intrinsic barriers and thermodynamic driving forces. To obtain the central barriers of nonidentity S_N2 reactions,

Table 6.6 Efficiencies and Energies of Gas-Phase S_N2 Reactions[a]

Entry	Reactants	Efficiency	ΔE^b
1	CH_3O^-/CH_3Cl	0.3 ± 0.03	-42
2	HCC^-/CH_3Cl	0.024 ± 0.003	-51
3	F^-/CH_3Cl	0.25 ± 0.01	-28
4	CD_3S^-/CH_3Cl	0.03 ± 0.006	-29
5	$PhCH_2^-/CH_3Cl$	0.01	-51
6	Cl^-/CH_3Cl	0.003	0
7	CH_3O^-/CH_3Br	0.40 ± 0.01	-49
8	HCC^-/CH_3Br	0.16 ± 0.02	-59
9	F^-/CH_3Br	0.28	-37
10	$CH_3CO_2^-/CH_3Br$	0.014 ± 0.003	-17
11	CN^-/CH_3Br	0.01	-21
12	CH_3S^-/CH_3Br	0.091	-41
13	$PhCH_2^-/CH_3Br$	0.185	-59
14	Br^-/CH_3Br	>0.008	0
15	H_2P^-/CH_3Cl^c	0.056	
16	H_2P^-/CH_3Br^c	0.47	
17	H_2P^-/CH_3I^c	0.77	

[a]Data are from references 3 and 8a.
[b]These are reaction enthalpies (ΔH), in kcal/mol.
[c]From reference 8d.

Pellerite and Brauman[3] applied an RRKM treatment which led to an approximation of ΔE^b_{YX}, the energy difference between the reactants and the transition states **(6.1)**. If ΔE^b_{YX} is known, or can be approximated, experimental data[9] for the heat

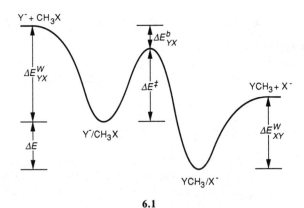

6.1

of formation of the ion–dipole complex, ΔE^w_{YX}, allow the magnitude of the central barrier $\Delta E^{\ddagger}$ of **6.1** to be extracted. Since not all of the required ΔE^w_{YX} data were available experimentally, Pellerite and Brauman assumed[3] that the well depths are a constant 10 kcal/mol for reactions of CH_3Cl, and a constant 11 kcal/mol for reactions of CH_3Br. Although these assumptions cause the $\Delta E^{\ddagger}$ values to be only approximate, the trends that result are of considerable interest.

Application of the Marcus equation, in terms of $\Delta E_0^{\ddagger}$ and ΔE, led Pellerite and Brauman[3] to the intrinsic barriers, $\Delta E_0^{\ddagger}$, collected in Table 6.7. This table also includes the central barriers, $\Delta E^{\ddagger}$, the reaction energies, ΔE, and $\Delta E^{\ddagger}$ values calculated by us using the Marcus equation and the $\Delta E_0^{\ddagger}$ and ΔE data of Pellerite and Brauman.[3]

We observe the same trends as in the computational study. The magnitude of $\Delta E^{\ddagger}$ is determined by the balance between the intrinsic kinetic quantity $\Delta E_0^{\ddagger}$ and the thermodynamic driving force ΔE. For a constant ΔE, the barrier is large when

Table 6.7 Reaction Energies, Intrinsic Barriers, and Central Barriers of Nonidentity S_N2 Reactions[a]

Entry	Y, X	ΔE^b	$\Delta E_0^{\ddagger}$ (RRKM, Marcus)[c]	$\Delta E^{\ddagger}$
1	CH_3O, F	−14	26.4	19.9
2	H, F	−57	38.6	15.4
3	CN, F	−5	30.6	28.2
4	HCC, F	−24	33.8	22.8
5	F, F	0	26.2	26.2
6	CH_3O, Cl	−42	18.4	3.4[e]
7	H, Cl	−86	30.6	2.7
8	CN, Cl	−32	22.6	9.4
9	HCC, Cl	−51	25.8	6.2[e]
10	CH_3S, Cl	−29	17.2	5.8[e]
11	$PhCH_2$, Cl	−51	22.3	4.1
12	F, Cl	−28	18.2	6.9[e]
13	Cl, Cl	0	10.2	10.2[e]
14	CH_3O, Br	−49	20.5(18.9)[d]	3.3[e](2.2)
15	CN, Br	−35	23.1	7.4
16	HCC, Br	−59	26.3	5.0[e]
17	CH_3S, Br	−37	17.7	4.0
18	$PhCH_2$, Br	−59	22.8	2.85[e]
19	F, Br	−37	18.9	4.8
20	CH_3CO_2, Br	−17	14.1	5.2[e](6.9)
21	Br, Br	0	11.2	11.2[e,f]
22	Cl, Br	−8	10.7	≈ 9[g](7.1)

[a] Data are from references 3 and 7. All values are in kcal/mol.

[b] These are reaction enthalpies (ΔH).

[c] $\Delta E_0^{\ddagger}(H^- + CH_3) \approx 51$ kcal/mol, as suggested in reference 3b.

[d] The value in parentheses is obtained as $\Delta E_0^{\ddagger}(CH_3O,Br) = 0.5[\Delta E_0^{\ddagger}(Br,Br) + \Delta E_0^{\ddagger}(CH_3O,CH_3O)]$. The second value is derived from the Marcus equation using the barrier $\Delta E^{\ddagger}$ from the RRKM procedure.

[e] These values are obtained by the RRKM procedure. Others are obtained from intrinsic barrier data using the Marcus equation.

[f] Subject to uncertainty (see reference 3).

[g] Derived using $\Delta E_0^{\ddagger}$ data from reference 3. The value in parentheses is the experimental barrier from reference 8b.

the intrinsic barrier is large (entries 2 versus 16 versus 18, entries 3 versus 22, entries 9 versus 11 versus 14, entries 8 versus 10 versus 12). On the other hand, for a constant intrinsic barrier $\Delta E_o^\ddagger$, $\Delta E^\ddagger$ decreases as the reaction becomes more exoergic (entries 1 versus 5 versus 9 versus 16, entries 6 versus 12 versus 19, entries 3 versus 7). Thus an extremely exoergic reaction (e.g., entry 2) can exhibit a substantial barrier because it has a large intrinsic barrier.

Once again it can be concluded that, within a reaction series, the BEP principle will be obeyed when the intrinsic barrier is constant. Clearly, the Marcus approach is useful for the analysis of the barriers of gas-phase nonidentity S$_N$2 reactions.

6.1.3 Marcus Treatment of Experimental Data in Solution

The applicability of the Marcus equation to S$_N$2 reactions in solution has been discussed by Albery and Kreevoy.[4] It is possible to solve equation 6.3 to obtain values for the intrinsic barriers $\Delta E_o^\ddagger$, from experimental barriers, $\Delta E^\ddagger$, and reaction energies, ΔE. The $\Delta E_o^\ddagger$ calculated in this way can then be compared to those obtained from the approximation of equation 6.4.

The results of Albery and Kreevoy for reactions in water solvent are collected in Table 6.8. There are some notable deviations when these data are compared to the gas-phase reactions. In the gas phase, the relative nucleophilicities are F$^-$ > Cl$^-$ > Br$^-$ (entries 19, 21, 22 of Table 6.7), and CH$_3$O$^-$ (HO$^-$) > CN$^-$ (entries 6, 8 of Table 6.7, and 8, 11 of Table 6.5). Although all of these trends arc reversed in water solvent, Marcus behavior is found to recur in solution. For a constant ΔE, higher barriers are observed for reactions having higher intrinsic barriers (entries 1 versus 6, 9 versus 11 versus 12 versus 17, entries 10 versus 15 versus 24). On the other hand, for a constant $\Delta E_o^\ddagger$, the barrier decreases as the reaction becomes more exoergic (entries 8 versus 9 versus 17, entries 2 versus 3 versus 11).

The reactions of CN$^-$ and HO$^-$ are of special interest. Although these reactions are extremely exoergic compared to others, their barriers are comparable to those of reactions that are almost thermoneutral. This behavior is caused by the large intrinsic barriers of CN$^-$ and HO$^-$ relative to other X's.

Other interesting trends are present in reactions in which X and Y are halide. For each member of this family, the change in ΔE is insignificant, but the barriers vary by up to 8 kcal/mol (entries 11 versus 20). The reactivity trends within this family are again dominated by the variations in the intrinsic barriers.

The internal consistency of the Marcus treatment is seen in Figure 6.4, which compares intrinsic barriers calculated using equation 6.3 and experimental $\Delta E^\ddagger$ and ΔE data, to intrinsic barriers calculated using equation 6.4 and the average of the barriers of identity reactions. Analogous observations have been made[4] for S$_N$2 reactions in methanol, acetone, and DMF, and the conclusions are found to be invariant: reactivity can be interpreted throughout in terms of the interplay between an intrinsic kinetic quantity and a thermodynamic driving force. This statement of the Marcus relationship is consistent with the experimental findings of E. S. Lewis and his co-workers.[10]

The Marcus equation is therefore a generalized rate–equilibrium relationship within which the BEP principle is a special case that pertains to reaction series in

Table 6.8 Reaction Energies, Intrinsic Barriers, and Central Barriers of Nonidentity S_N2 Reactions $Y^- + CH_3X \rightarrow YCH_3 + X^-$ in Water Solvent[a,b]

Entry	Y, X	ΔE	$\Delta E_0^\ddagger$	$\Delta E^\ddagger$
1	Cl, $PhSO_3$	-16.49	31.31	23.66
2	Br, $PhSO_3$	-15.06	30.11	22.94
3	I, $PhSO_3$	-17.21	29.88	21.99
4	CN, $PhSO_3$	-52.34	43.02	20.79
5	HO, $PhSO_3$	-38.48	38.24	21.51
6	NO_3, $PhSO_3$	-17.21	33.22	25.10
7	Cl, NO_3	$+0.72$	27.01	27.49
8	Br, NO_3	$+2.15$	24.86	26.05
9	I, NO_3	0	24.62	24.62
10	HO, NO_3	-21.27	34.66	24.86
11	Cl, F	-0.48	29.40	29.16
12	Br, F	$+0.96$	27.01	27.49
13	I, F	-1.19	28.44	27.96
14	CN, F	-36.33	41.59	25.33
15	HO, F	-22.47	36.33	26.05
16	Br, Cl	$+1.43$	25.33	26.05
17	I, Cl	-0.72	24.14	23.90
18	CN, Cl	-35.85	39.20	22.94
19	HO, Cl	-21.98	34.66	24.62
20	I, Br	-2.15	22.71	21.75
21	CN, Br	-37.28	37.76	21.51
22	HO, Br	-23.42	33.46	22.71
23	CN, I	-35.13	37.52	21.99
24	HO, I	-21.27	32.98	23.18
25	HO, OH	0	41.80	41.80
26	CN, NC	0	50.90	50.90

[a]All values refer to free energies, in kcal/mol.
[b]Data are from reference 4a.

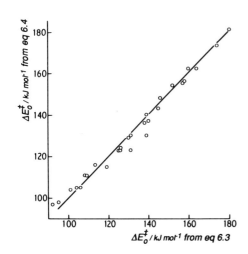

Figure 6.4 A test of the Marcus equation for S_N2 reactions in water solvent, showing a comparison of $\Delta E_0^\ddagger$ calculated from experimental rate data using equation 6.3 with $\Delta E_0^\ddagger$ calculated with equation 6.4 and a known set of symmetrical reactions. The data are from reference 4a.

which only ΔE varies. Such families are rarely observed, and should have no special status in comparison to families that have a constant ΔE and a variable $\Delta E_o^{\ddagger}$ or, in general, families in which both ΔE and $\Delta E_o^{\ddagger}$ vary.

6.2 RATE–EQUILIBRIUM RELATIONSHIPS WITH THE SCD MODEL

The Marcus treatment does not deal with questions such as: (1) why do intrinsic barriers vary in a particular manner? and (2) what determines the existence of a reaction series that possesses a constant $\Delta E_o^{\ddagger}$ and, thereby, obeys the BEP principle? The answers to these questions are found in the SCD model[11] which, like the Marcus treatment, regards the barrier as a composite of an intrinsic quantity and a thermodynamic driving force but, unlike the Marcus treatment, deals *explicitly* with the intrinsic kinetic quantity.

6.2.1 The Intrinsic Barrier in the SCD Model

Figure 6.5 is the state correlation diagram appropriate to a nonidentity S_N2 reaction. The reaction coordinate (RC) is expressed as the difference in the bond orders of C–X and C–Y, and it is normalized to range from zero to unity (see Chapter 3). The initial and final points of the RC correspond to the geometries of the reactant and product ion–dipole complexes, $Y:^-/R-X$ and $X:^-/R-Y$, respectively, in the gas phase and, for convenience, also in solution. The upper states of the diagram are the singlet charge transfer states of the reactants and the products at their encounter geometries. The avoided crossing of the two curves (shown in dashed lines) leads to the reaction profile and to the barrier $\Delta E^{\ddagger}$.

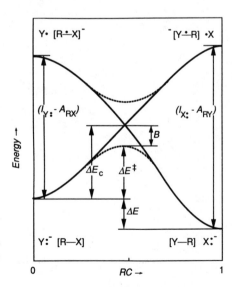

Figure 6.5 State correlation diagram for nonidentity S_N2 reactions $Y^- + CH_3X \rightarrow YCH_3 + X^-$.

The height of the barrier is determined by the deformation required to overcome the vertical electron transfer energy gap and, thereby, achieve resonance and a transition state at the crossing point. The height of the crossing point relative to the reactants depends on four quantities, as explained in **6.2–6.4**.

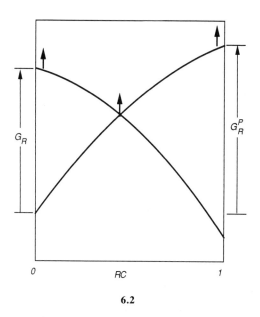

6.2

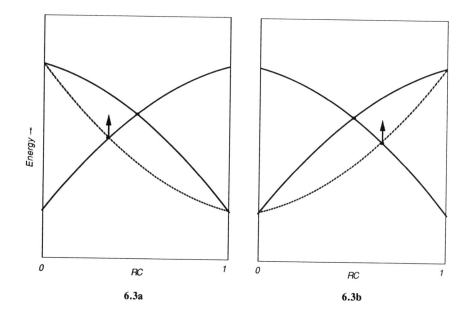

6.3a   **6.3b**

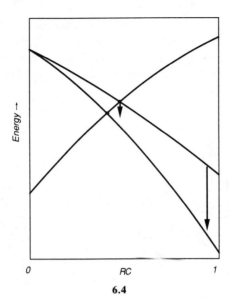

6.4

The first relates to the sizes of the two energy gaps measured from the energy of the reactants. In **6.2** these are denoted by G_R and G_R^P. An increase in the magnitudes of these gaps, with all other factors constant, will raise the energy of the crossing point, as depicted by the heavy arrows. This is presented as Statement 6.1.

Statement 6.1 When all other factors remain constant, an increase in the energies of the charge transfer states Y$\cdot$/(R$\doteq$X)$^-$ and X$\cdot$/(R$\doteq$Y)$^-$, will raise the energy of the crossing point and increase the barrier.

A second quantity is the curvature of the diagrams, illustrated in **6.3a** and **6.3b**. If either of the curves becomes more shallow (dotted line → solid line), the energy of the crossing point increases. This is depicted by the vertical arrows. Since the shallowness increases whenever Y$\cdot$, (R$\doteq$X)$^-$, X$\cdot$, and (R$\doteq$Y)$^-$ become electronically more delocalized, the curvature effect can be described by Statement 6.2.

Statement 6.2 When all other factors remain constant, an increase in the delocalization properties of the odd electrons in the charge transfer species [Y$\cdot$, (R$\doteq$X)$^-$, X$\cdot$, (R$\doteq$Y)$^-$] will raise the energy of the crossing point and increase the barrier.

The third quantity is the reaction energy. As depicted by the arrows in **6.4**, relative stabilization of the product will lower the crossing point. This leads to Statement 6.3.

Statement 6.3 The energy of the crossing point and the barrier will be lowered as ΔE becomes more negative, provided that the substituent effect responsible for this energy change does not cause a concomitant increase in the energies of the charge transfer states or an increase in the delocalization properties of the charge transfer species.

The final barrier factor is B, the avoided crossing interaction of Figure 6.5. As we have noted repeatedly, variations in B are not expected to be dominant. We shall therefore assume for convenience that B remains constant in S_N2 reactions of CH_3X derivatives.

Accordingly, the barrier of a nonidentity S_N2 reaction can be written as equation 6.7, a function of at least five variables. In this equation, G_R and G_R^P are the gaps

$$\Delta E^{\ddagger} = \Delta E^{\ddagger}(G_R, G_R^P, f_R, f_P, \Delta E) \tag{6.7}$$

depicted in **6.2**, and f_R and f_P describe the shallowness of the intersecting curves. In the complete reactivity space defined by the five variables, a series which obeys the BEP principle is a cross section defined by Statement 6.3. However, such series will normally comprise only a limited region of the total space. Many more series should also exist, as will be described.[11a]

The full barrier equation was derived in Chapter 3, equation 3.71, and is restated here as equation 6.8, where $\Delta E^{\ddagger}$ is the barrier and ΔE is the reaction energy

$$\Delta E_f^{\ddagger} \approx \left[f_R + f_P + (1 - f_R - f_P) \frac{\Delta E}{G_R} \right] \frac{G_R G_R^P}{G_R + G_R^P - \Delta E} - B \tag{6.8}$$

in generalized units (enthalpy, free energy). This expression is obtained with the approximation that the two curves of Figure 6.5 are linear in the region of the crossing point.

The relationships among the various electron transfer energy gaps are given in **6.5**: G_R refers to the reactants $Y:^-/R\text{–}X$, and is defined in equation 6.9; G_R^P can

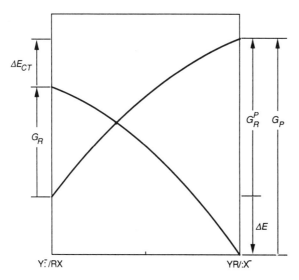

6.5

be defined according to equation 6.10 or 6.11, where ΔE has its usual meaning,

$$G_R = I_{Y:} - A_{RX} \tag{6.9}$$

$$G_R^P = G_P + \Delta E = I_{X:} - A_{RY} + \Delta E; \; (\Delta E < 0 \text{ for exoergic reactions}) \tag{6.10}$$

$$G_R^P = I_{Y:} - A_{RX} + \Delta E_{CT} = G_R + \Delta E_{CT} \tag{6.11}$$

and ΔE_{CT} is the energy difference between the two charge transfer states.

From equations 6.8–6.10, and the condition for the intrinsic barrier ($\Delta E = 0$), we have

$$\Delta E_0^{\ddagger} = (f_R + f_P) \frac{(I_{Y:} - A_{RX})(I_{X:} - A_{RY})}{(I_{Y:} - A_{RX}) + (I_{X:} - A_{RY})} - B \tag{6.12}$$

The curvature indices f_R and f_P are now equated to the total delocalization indices $w_{R:}$ of the charge transfer states, as in the treatment of identity reactions (Chapter 5). This leads to equations 6.13a and b.

$$f_R = \overline{w}_{R:} \; (Y \cdot /(R \dot{-} X)^-) \tag{6.13a}$$

$$f_P = \overline{w}_{R:} \; (X \cdot /(R \dot{-} Y)^-) \tag{6.13b}$$

The avoided crossing B is taken as a constant, and its value is set to 14 kcal/mol, as in the treatment of identity reactions.

With these substitutions, the intrinsic barrier of equation 6.12 now depends on four variables: the electron transfer energy gaps and the delocalization indices of the charge transfer species derived from the reactants and the products. Statement 6.4 follows:

Statement 6.4 For nonidentity S_N2 reactions, $Y^- + R–X \rightarrow Y–R + X^-$, large electron transfer energy gaps ($I_{X:} - A_{RY}$, $I_{Y:} - A_{RX}$) and delocalized charge transfer species ($Y \cdot /(R \dot{-} X)^-$, $X \cdot /(R \dot{-} Y)^-$) will lead to high intrinsic barriers.

The intrinsic barriers of several nonidentity reactions have been calculated with equation 6.12, and are presented in Table 6.9 as $\Delta E_0^{\ddagger}$ (SCD). The data exemplify Statement 6.4. The reactions $H^- + CH_3CCH$ (entry 1) and $CH_3O + CH_3F$ (entry 9) have the same electron transfer energy gaps, but (1) has the larger intrinsic barrier because of the more delocalized nature of the charge transfer species. On the other hand, the reactions $H^- + CH_3CCH$ (entry 1) and $H^- + CH_3OH$ (entry 3) have the same delocalization indices, but (1) has the larger intrinsic barrier because of its larger electron transfer energy gaps. Other comparisons of interest include $H^- + CH_3Cl$ (entry 7) versus $PhCH_2^- + CH_3Cl$ (entry 12), and $HCC^- + CH_3Cl$ (entry 10) versus $CH_3O^- + CH_3F$ (entry 9). It seems clear that the intrinsic barriers of nonidentity reactions and the barriers of identity reactions exhibit similar trends.

Table 6.9 Reactivity Factors[a] and Intrinsic Barriers of Gas-Phase Nonidentity S_N2 Reactions, $Y^- + CH_3X \rightarrow YCH_3 + X^-$

Entry	Y, X	$I_Y: -A_{RX}$	$I_X: -A_{RY}$	f_R	f_P	$\Delta E_o^\ddagger$ (SCD)[b]	$\Delta E_o^\ddagger$ (RRKM)[c]	$\Delta E_o^\ddagger$ (4-31G)[d]
1	H, CCH	96	134	0.362	0.720	46.5	≈46.2	51.2
2	H, CN	86	155	0.309	0.720	42.9	≈43.0	40.3
3	H, OH	83	110	0.357	0.720	36.9	≈38.8	36.6
4	H, F	75	144	0.242	0.720	33.4	≈38.6	31.9
5	CCH, F	125	156	0.242	0.362	27.9	33.8	31.0
6	CN, F	146	146	0.242	0.309	26.2	30.6	27.8
7	H, Cl	48	149	0.251	0.720	21.3	≈30.6	28.8
8	HO, CN	112	154	0.309	0.357	29.2	≈30.8	32.5
9	CH₃O, F	94	139	0.242	0.355	19.5	26.4	16.5
10	CCH, Cl	98	161	0.251	0.362	23.3	25.8	28.0
11	CN, Cl	119	151	0.251	0.309	23.2	22.6	24.7
12	PhCH₂, Cl	51	143	0.251	0.436	11.8	≈22.3	
13	CH₃O, Cl	67	144	0.251	0.355	13.7	18.4	14.5
14	F, SH	120	111	0.340	0.242	19.6	≈25.2	13.7
15	F, Cl	108	140	0.251	0.242	16.1	18.2	8.6
16	CH₃S, Cl	73	135	0.251	0.370	15.4	17.2	≈10.6
17	CH₃CO₂, Br	99	108	0.348	0.203	14.5	14.1	

[a]Except for f_R and f_P, all values are in kcal/mol. The data for the reactivity factors are collected in Chapters 4 and 5.
[b]$\Delta E_o^\ddagger$ (SCD) versus $\Delta E_o^\ddagger$ (RRKM, Marcus) is linear, with $r = 0.95$ and slope 1.03; $\Delta E_o^\ddagger$ (SCD) versus $\Delta E_o^\ddagger$ (4-31G) is linear, with $r = 0.94$ and slope 0.78.
[c]Data taken from references 3 and 7.
[d]Data taken from reference 2b.

Table 6.9 includes, in addition to $\Delta E_o^{\ddagger}$ (SCD), intrinsic barriers from other sources: $\Delta E_o^{\ddagger}$ (RRKM, Marcus) values are taken from Pellerite and Brauman,[3] using equation 6.4 and their suggested value $\Delta E^{\ddagger}$ (H^-/CH_4) $\simeq$ 51 kcal/mol; $\Delta E_o^{\ddagger}$ (4-31G) values are also obtained with equation 6.4 from the ab initio data.[2a, b] There is a linear correlation ($r = 0.95$) between $\Delta E_o^{\ddagger}$ (SCD) and $\Delta E_o^{\ddagger}$ (RRKM, Marcus), and also between $\Delta E_o^{\ddagger}$ (SCD) and $\Delta E_o^{\ddagger}$ (4–31G) ($r = 0.94$).

The existence of such correlations is significant because of the very different origins of the three sets of data. The $\Delta E_o^{\ddagger}$ (SCD) values are obtained from fundamental thermochemical data (Chapter 4); the $\Delta E_o^{\ddagger}$ (RRKM, Marcus) values are obtained from a combined RRKM and Marcus treatment of kinetic data; and $\Delta E_o^{\ddagger}$ (4-31G) values are obtained by ab initio calculations. We can conclude that equation 6.12 captures the physical sense of an intrinsic barrier.

Since the SCD model has thus been linked successfully to the Marcus equation, we may proceed to analyze trends in intrinsic barriers. These barriers will be large whenever the electron transfer energy gaps of the reactants and the products are large, and whenever the charge transfer species are delocalized. The intrinsic barrier thus reflects the *average deformations of reactants and products associated with the movement of a single electron*. These points are summarized, in terms of properties of X and Y, in Statement 6.5, which is reminiscent of the patterns predicted for identity barriers.

Statement 6.5   Strong C–Y and C–X bonds lead to high electron transfer energy gaps ($I_{Y:} - A_{RX}, I_{X:} - A_{RY}$). Low X· and Y· electron affinities lead to delocalized $(R\dot{\cdot}X)^-$ and $(R\dot{\cdot}Y)^-$ radical anions. The combination of strong C–X, C–Y bonds and low X· and Y· electron affinities will cause nonidentity reactions to have high intrinsic barriers.

The trends in Table 6.9 follow Statement 6.5 uniformly. For example, $\Delta E_o^{\ddagger}$ values for reactions that involve F are larger than those that involve Cl, in accord with the relative bond strengths C–F > C–Cl. Likewise, $\Delta E_o^{\ddagger}$ values for reactions that involve CCH and CN are usually high, because of the strong C–CN and C–CCH bonds. On the other hand, H, HO, and CH_3O have higher $\Delta E_o^{\ddagger}$ values than Cl because of the low electron affinities of H·, CH_3O· and HO·, compared to Cl·. The consistency of the conclusions concerning the intrinsic barriers of nonidentity reactions and the barriers of identity reactions should be clear.

6.2.2 The Role of the "Thermodynamic Driving Force"

Qualitative insight concerning the role of the thermodynamic driving force is contained in Statement 6.3, and the quantitative expression of the barrier is found in equation 6.8. Gas-phase barriers calculated with this equation are collected in Table 6.10, along with the reactivity factors and barriers derived from the Marcus equation using gas-phase data and an RRKM analysis.[3,7] Also shown are the barriers calculated with a 4-31G basis set.[2a, b] The correlation between the $\Delta E^{\ddagger}$ (SCD) and the other values is fair but reasonable, in view of the diverse origins of the data and the approximations in each of the methods.

Table 6.10 Reactivity Factors, Reaction Energies, and Central Barriers of Gas-Phase S_N2 Reactions, $Y^- + CH_3X \rightarrow YCH_3 + X^{-a}$

Entry	Y, X	G_R	G_R^p	f_R	f_P	ΔE	$\Delta E^\ddagger$ (SCD)[b]	$\Delta E^\ddagger$ (RRKM)[c]	$\Delta E^\ddagger$ (4-31G)[d]
1	H, F	75	87	0.242	0.720	−57	13.8(14.6)	15.4	7.3
2	H, Cl	48	63	0.251	0.720	−86	0.1(0.8)	2.7	
3	H, Br	39	50	0.246	0.720	−94	−4.6(−3.8)	1.9	19.5
4	HCC, F	125	132	0.242	0.362	−24	17.0(18.7)	22.8	
5	HCC, Cl	98	110	0.251	0.362	−51	3.1(6.8)	6.2	
6	HCC, Br	89	97	0.246	0.362	−59	−1.7(2.6)	5.0	26.6
7	CN, F	146	141	0.242	0.309	−5	23.8(24.0)	28.2	
8	CN, Cl	119	119	0.251	0.309	−32	9.2(10.6)	9.4	
9	CN, Br	110	111	0.246	0.309	−40	4.4(6.6)	7.4	9.1
10	CH$_3$O, F	94	125	0.242	0.355	−14	13.1(14.8)	19.9	
11	CH$_3$O, Cl	67	102	0.251	0.355	−42	−2.4(2.0)	3.4	
12	CH$_3$O, Br	58	90	0.246	0.355	−50	−7.2(−1.9)	2.2(3.3)	
13	CH$_3$S, Cl	74	106	0.251	0.370	−29	3.7(6.7)	5.8	
14	CH$_3$S, Br	65	89	0.246	0.370	−37	−2.0(1.6)	4.0	
15	F, Cl	108	112	0.251	0.242	−28	3.6(4.5)	6.9	
16	F, Br	99	98	0.246	0.242	−37	−1.7(−0.2)	4.8	
17	PhCH$_2$, Cl	51	92	0.332	0.436	−51	−1.0(2.4)	4.1	
18	PhCH$_2$, Br	42	79	0.326	0.436	−59	−6.1(−2.1)	2.9	
19	CH$_3$CO$_2$, Br	99	91	0.348	0.203	−17	6.6(6.7)	6.9(5.2)	
20	H, CN	86	101	0.309	0.720	−54	23.7(23.2)		23.5
21	H, CCH	96	96	0.362	0.720	−38	30.7(29.8)		34.7
22	H, OH	83	71	0.357	0.720	−39	20.0(19.4)		18.5
23	HO, CN	112	140	0.309	0.357	−14	22.8(23.6)		27.0
24	F, SH	120	101	0.340	0.242	−9	15.0(15.2)		3.5

[a]Except for f_R and f_P, all data are in kcal/mol.
[b]The first set of numbers refers to results obtained with equation 6.8; the second set refers to an exact solution of the curve crossing without the assumption of linearity near the crossing point.
[c]Data are from references 3 and 7.
[d]Data are from reference 2.

For small energy gaps, it is conceivable that the avoided crossing can lead to a barrierless energy profile, as shown in **6.6**. Since equation 6.8 calculates the

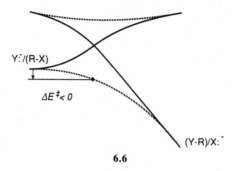

6.6

energy level of the crossing point after the avoided crossing, relative to the energy of the reactant complex, the situation depicted in **6.6** will be manifested by a *negative barrier*. Such cases (e.g., entry 18 of Table 6.10) should be regarded as characteristic of barrierless processes. When this situation is predicted by $\Delta E^{\ddagger}$ (SCD) (entries 3, 12, 16, 18), small but observable barriers are found experimentally. It would be of interest to examine experimentally the reactions $HO^- + CH_3I$ and $PhCH_2^- + CH_3I$ as possible examples of barrierless S_N2 reactions.

6.2.3 BEP-Type Series and Breakdowns of the BEP Principle

Since the SCD model appears to treat the problem quantitatively, we may proceed to an analysis of the relationship between the thermodynamic driving force and other reactivity factors. This will allow the range of sovereignty of the BEP principle to be determined. Table 6.11 summarizes the data for several reaction series.

In the first series (entries 1–3), the vertical electron transfer energy gaps (G_R, G_R^P) and the delocalization indices (f_R, f_P) are approximately constant, and only ΔE varies significantly. In this series the barriers decrease as ΔE becomes more negative; the BEP principle is obeyed, because the principal variable within the series is the thermodynamic driving force.

The second and third series of Table 6.11 (entries 4–6, 7–9) exhibit quite different behavior. In each of these series the delocalization indices are constant, and both ΔE and the vertical electron transfer energy gaps decrease. The effect of the thermodynamic driving force is, therefore, reinforced by the change in the energy gaps, and the barriers ($\Delta E^{\ddagger}$) decrease more rapidly within these series than within the first series (entries 1–3). Each of these latter series also obeys the BEP principle.

Additional insight is afforded by consideration of the relationship between ΔE and G_R. This can be derived, for the reaction $Y^- + R-X \rightarrow Y-R + X^-$, from

Table 6.11   Reaction Families for Gas-Phase S_N2 Reactions, $Y^- + CH_3X \rightarrow YCH_3 + X^-$, from the Reactivity Factors, Reaction Energies, and Central Barriers[a]

Entry	Y, X	G_R	G_R^P	f_R	f_P	ΔE^b	$\Delta E^{\ddagger}$ (SCD)[c]	$\Delta E^{\ddagger}$ (RRKM)[b]
1	Br, Cl	108	112	0.251	0.246	+8	16.5(16.6)	15.1
2	Cl, Cl	113	113	0.251	0.251	0	14.4(−)	10.2
3	F, Cl	108	112	0.251	0.242	−28	3.6(4.5)	6.9
4	H, F	75	87	0.242	0.720	−57	13.8(15.6)	15.4
5	H, Cl	48	63	0.251	0.720	−86	0.1(0.8)	2.7
6	H, Br	39	50	0.246	0.720	−94	−4.6(−3.8)	1.9
7	HCC, F	125	132	0.242	0.362	−24	17.0(18.7)	22.8
8	HCC, Cl	98	110	0.251	0.362	−51	3.1(6.8)	6.2
9	HCC, Br	89	97	0.246	0.362	−59	−1.7(2.6)	5.0
10	HCC, Cl	98	110	0.251	0.362	−51	3.1(6.8)	6.2
11	CH_3S, Cl	74	106	0.251	0.370	−29	3.7(6.7)	5.8
12	F, Cl	108	112	0.251	0.242	−28	3.6(4.5)	6.9
13	H, F	75	87	0.242	0.720	−57	13.8(14.6)	15.4
14	F, Cl	108	112	0.251	0.242	−28	3.6(4.5)	6.9
15	CH_3CO_2, Br	99	91	0.348	0.203	−17	6.6(6.7)	6.9(5.2)
16	H, CN	86	101	0.309	0.720	−54	23.7(23.2)	
17	HO, F	94	125	0.242	0.355	−14	13.1(14.8)	
18	HCC, Br	89	97	0.246	0.362	−59	−1.7(2.6)	5.0
19	CH_3O, Br	58	90	0.246	0.355	−50	−7.2(−1.9)	2.2(3.3)

[a]Except for f values, all data are in kcal/mol.
[b]Data are from references 3 and 7 and refer to enthalpies.
[c]The first of the two numbers is obtained with equation 6.8; the second is calculated from the exact solution of the curve crossing without the assumption of linearity in the region of the crossing point.

the thermochemical cycle of equation 6.14, which leads to equation 6.15, in

$$Y\!:^- + R\text{--}X \rightarrow Y\cdot + (R\dot{\text{--}}X)^- \qquad \Delta E_1 = I_{Y:} - A_{RX} = G_R \qquad (6.14a)$$

$$(R\dot{\text{--}}X)^- \rightarrow R\cdot + :X^- \qquad \Delta E_2 = D_{R\dot{:}X} \qquad (6.14b)$$

$$Y\cdot + R\cdot \rightarrow Y\text{--}R \qquad \Delta E_3 = -D_{R\text{--}Y} \qquad (6.14c)$$

$$\overline{Y\!:^- + R\text{--}X \rightarrow Y\text{--}R + X\!:^- \qquad \Delta E = \Delta E_1 + \Delta E_2 + \Delta E_3} \quad (6.14d)$$

$$\Delta E = (I_{Y:} - A_{RX}) + D_{R\dot{:}X} - D_{R\text{--}Y} = G_R + D_{R\dot{:}X} - D_{R\text{--}Y} \quad (6.15)$$

$$\Delta(\Delta E) \approx \Delta(G_R) \approx \Delta(I_{Y:} - A_{RX}) \qquad (6.16)$$

$$\Delta E(\text{solvent}) \approx \Delta E(g) + S_{Y:} - S_{X:} \qquad (6.17)$$

which the D's are bond energies.

In the first reaction series of Table 6.11 (entries 1–3), R–X (CH$_3$Cl) is constant; $D_{R:X}$ is, therefore, constant. The variables within the series are Y$^-$ = F$^-$, Cl$^-$, Br$^-$. However, since the gas-phase ionization potentials ($I_{Y:}$) of F$^-$, Cl$^-$, and Br$^-$ do not differ greatly, (78.4, 83.4, 78 kcal/mol, respectively), the G_R of equation 6.15 is nearly constant. Accordingly, variations in ΔE are determined by variations in the bond-energy term D_{R-Y}. Since the bond strengths vary in the order $D_{R-F} > D_{R-Cl} > D_{R-Br}$, ΔE becomes more negative down the series, although G_R remains constant.

In each of the second and third reaction series of Table 6.11 (entries 4–6 and 7–9), the nucleophile Y:$^-$ is constant (H$^-$ or HCC$^-$), and the variable is R–X. The electron affinities of these CH$_3$X derivatives vary in the order $A_{CH_3Br} > A_{CH_3Cl} > A_{CH_3F}$; therefore, the gap decreases within each series. According to equation 6.15, a decrease in G_R will lead to a decrease in ΔE, because D_{R-Y} is constant and $D_{R:X}$ is also constant. For example, reference to Table 4.3 shows that for X = F, Cl, Br, $D_{R:X}$ is -27, -29, and -28 kcal/mol, respectively. Consequently, in these cases, variations in ΔE will be determined by variations in G_R. This leads to the relationships of equation 6.16.

$$\Delta(\Delta E) \approx \Delta(G_R) \approx \Delta(I_{Y:} - A_{RX}) \qquad (6.16)$$

In summary, the BEP principle will be obeyed when either of the following conditions is met: (a) the G and f terms remain constant; (b) f remains constant, and $\Delta(\Delta E) \simeq \Delta(G_R)$.[11a]

The next two series of Table 6.11 (entries 10–12, 13–15, 16–17) constitute breakdowns of the BEP principle. In both of these series the most exoergic reaction has either the *highest* barrier or a barrier close to that of a much less exoergic reaction. This trend is associated primarily with an increase in the delocalization properties of the charge transfer species (larger f). This counterbalances the barrier-lowering effect associated with in the increasing exoergicity.

The final series of Table 6.11 (entries 18, 19) also represents a breakdown of the BEP principle. Here the delocalization indices are approximately constant, but ΔE and the electron transfer energy gaps change in opposite directions.

Within a broad set of reactions, a change of X or Y which leads to a decrease in ΔE can thus be accompanied by an increased delocalization of the charge transfer species and/or an increase in the electron transfer energy gaps. Both of these effects will lead to breakdown of the BEP principle. It follows that the BEP principle, and rate–equilibrium relationships in general, are restricted to reaction series for which the variation of ΔE is not counterbalanced by the primary barrier factors.

6.3 REACTIVITY TRENDS IN SOLUTION—THE SCD MODEL

6.3.1 Reactivity Trends in Hydroxylic Solvents

The central barriers of nonidentity S$_N$2 reactions in solution are determined by an interplay of intrinsic gas-phase properties, solvent reorganization effects, and solvation properties. As discussed in Chapter 5, solvation will increase electron transfer energy gaps via the solvent reorganization terms. Furthermore, differential solvation of reactants and products will alter the reaction energy relative to the gas phase. The central barrier will arise from the combination of the molecular deformation and the solvent reorganization that are required to achieve resonance between the ground and charge transfer states.

Table 6.12 lists the reactivity factors for several nonidentity reactions in the gas phase and in water solvent. Solvation is seen to increase the electron transfer energy gaps G_R and G_R^P considerably. As discussed in Chapter 5, this is caused mainly by the increase in the vertical ionization potentials of the anions, $I_{Y:}(s^*)$. On the other hand, the delocalization indices of the radical anions $(R \dot{-} X)^-$ and $(R \dot{-} Y)^-$ (f_R and f_P, respectively) undergo much smaller changes under conditions of nonequilibrium solvation.

The reaction energy ΔE changes significantly. These large changes can be understood in terms of equation 6.17, which states that ΔE(solvent) is modulated

$$\Delta E(\text{solvent}) \approx \Delta E(g) + S_{Y:} - S_{X:} \tag{6.17}$$

relative to $\Delta E(g)$ by the difference in the desolvation energies of the anions $Y:^-$ and $X:^-$. In the case of entry 8 of Table 6.12, this difference is small, because $S_{NC:} \simeq S_{Br:}$. In all other cases ΔE becomes less negative, because $S_{Y:} > S_{X:}$. This is particularly evident in the series of halide exchange reactions, entries 1–7, for which the trend is $S_{F:} \gg S_{Cl:} > S_{Br:} > S_{I:}$. Therefore, in water solvent the reactions of entries 1–7 are almost thermoneutral.

Sets I and II of Table 6.13 refer to the reactivity factors and barriers of halide exchange reactions in water solvent. It is evident that, for such reactions, the effect of ΔE becomes insignificant, and the barriers are determined entirely by the electron transfer energy gaps and the delocalization properties of the charge transfer species. Although the calculated barriers $\Delta E^{\ddagger}(\text{SCD})$ are consistently higher than the experimental barriers, the trends are reproduced by the model, as was found previously for the identity reactions.

The first set (entries 1–4) describes nucleophilic displacement reactions of halides upon CH_3Br. The delocalization indices of this set are almost constant, but the electron transfer energy gaps decrease rapidly down the series. This trend leads to a decrease in the barriers. The orders of ''nucleophilicity'' in water solvent is seen to be $F^- < Cl^- < Br^- < I^-$. This is just the opposite of the trend in the gas phase, namely, $F^- > Cl^- > Br^- > I^-$ (Table 6.7, and entries 1–3 of Table 6.11). *This reversal can be traced to a single property: the vertical ionization of the nucleophile in solution $I_{Y:}(H_2O^*)$.* As seen in the last column of Table 6.13, $I_{Y:}(H_2O^*)$ varies in the order $I^- < Cl^- < Br^- < F^-$, because of the contributions of solvation and solvent reorganization to the ionization process (see Chapter 5).

We can conclude that solvation and solvent reorganization cause I^- to become the best halide ion donor in water solvent. Similar trends will exist whenever the ''nucleophilicity'' of Y^- is followed down a column of the periodic table. For example, in the comparison RO^- vs. RS^-, $I_{HS:}(H_2O^*) = 177$ kcal/mol, and $I_{HO:}(H_2O^*) = 195$ kcal/mol.

Table 6.12 Reactivity Factors for Y^- + CH_3X → YCH_3 + X^- Reactions in the Gas Phase and in Solution

Entry	Y, X	ΔE $(H_2O)^a$	$\Delta E(g)^b$	G_R $(H_2O)^d$	G_R^P $(H_2O)^d$	$G_R(g)$	$G_R^P(g)$	f_R $(H_2O)^d$	f_P $(H_2O)^d$	$f_R(g)$	$f_P(g)$
1	F, Br	-2	-37	233	203	99	98	0.241	0.206	0.246	0.242
2	Cl, Br	0	-8	196	186	104	100	0.241	0.239	0.246	0.251
3	Br, Br	0	0	179	179	99	99	0.241	0.241	0.246	0.246
4	I, Br	0	+8	159	170	92	96	0.241	0.243	0.246	0.241
5	F, Cl	-2	-28	240	221	108	112	0.239	0.206	0.251	0.242
6	F, I	-1	-45	225	184	88	83	0.243	0.206	0.241	0.242
7	Cl, I	0	-16	186	166	93	85	0.243	0.239	0.241	0.251
8	CN, Br	-32(-37)	-35(-40)^c	196	188	110	111	0.241	0.304	0.246	0.309
9	HO, Br	-26	-54	188	190	65	89	0.241	0.320	0.246	0.357
10	HS, Br	-18	-28	169	180	74	92	0.241	0.328	0.246	0.340
11	PhS, Br	-21	-19	156	173	78	99	0.241	0.326	0.246	0.331

$^a\Delta E(H_2O)$ is calculated with equation 6.17, using solvation-free energy data from Chapter 5.

^{b}Enthalpy data, from reference 3.

^{c}Calculated using the formula $\Delta E(g) = A_{Y^-} - A_{X^-} + D_{C-X} + D_{C-Y}$.

^{d}Reactivity factors for reactions in solution are taken from Chapters 4 and 5.

Table 6.13 Reactivity Factors and Barriers (kcal/mol) for Nonidentity S_N2 Reactions $Y^- + CH_3X \rightarrow YCH_3 + X^-$ in Water Solvent

Entry	Y, X	G_R	G_R^P	f_R	f_P	ΔE	$\Delta E^{\ddagger}$ (SCD)[a]	$\Delta E^{\ddagger}$ (expt)[b]	$I_{Y:}$ (H_2O^*) or A_{RX} (H_2O^*)[c]
Series I									
1	F, Br	233	203	0.241	0.206	−2	33.8(34.0)	26.5	240
2	Cl, Br	196	186	0.241	0.239	0	31.8(31.8)	24.6	203
3	Br, Br	179	179	0.241	0.241	0	29.1(−)	23.7	186
4	I, Br	159	170	0.241	0.243	0	25.8(25.8)	21.8	166
Series II									
5	Cl, F	223	242	0.206	0.239	+2	38.4(38.6)	29.6	−19
6	Cl, Cl	203	203	0.239	0.239	0	34.5(−)	26.5	−0.1
7	Cl, Br	196	186	0.241	0.239	0	31.8(31.8)	24.6; 24.7[d]	+7.4
8	Cl, I	189	166	0.243	0.239	0	28.6(28.7)	24.6; 24.9[d]	+16
Series III									
9	CN, Br	196	188	0.241	0.304	−37	26.1(27.7)	21.5	203
10	HO, Br	188	190	0.241	0.320	−26	30.2(31.5)	22.8	195
11	HS, Br	169	180	0.241	0.328	−18	29.4(30.5)	<22.8	177
12	PhS, Br	156	173	0.241	0.326	−21	25.2(26.7)	<22.8	163

[a] See footnote c of Table 6.11.
[b] From reference 4a.
[c] In series I and III, this quantity is $I_{Y:}(H_2O^*)$; in series II, this quantity is $A_{RX}(H_2O^*)$.
[d] From reference 13.

The second set of data in Table 6.13 (entries 5–8) describes nucleophilic attack by Cl$^-$ upon CH$_3$X (X = F, Cl, Br, I). Once again, the only significant variation is the decrease in the electron transfer energy gaps down the series, which leads to a decrease in the barriers. The order of ''leaving group ability'' in water solvent (CH$_3$I > CH$_3$Br > CH$_3$Cl > CH$_3$F) originates in the vertical electron affinities $A_{RX}(H_2O^*)$. These vary in the order $A_{CH_3I} > A_{CH_3Br} > A_{CH_3Cl} > A_{CH_3F}$.

We can conclude that, when ΔE is constant, the trend in the relative reactivities of CH$_3$X molecules originates in their ability to accept a single electron, $A_{RX}(s^*)$. Other trends of this sort can be generated from the $A_{RX}(s^*)$ data of Chapter 5; for example, *under conditions of constant* ΔE, the relative reactivities are predicted to be: CH$_3$I > CH$_3$Br > CH$_3$Cl > CH$_3$F > CH$_3$SH > CH$_3$OH > CH$_3$CN.

In summary, the thermodynamic driving force does not contribute to the trends seen in entries 1–8 of Table 6.13. These trends originate in the single electron transfer properties of the reactants Y$^-$/RX. The BEP principle, and rate–equilibrium relationships generally, would not be applicable to such series.

The third set of Table 6.13 (entries 9–12) refers to reactions in water solvent that exhibit significant variation in ΔE. In entries 9 and 10 the electron transfer energy gap prefers HO$^-$ + CH$_3$Br, but f_P prefers CN$^-$ + CH$_3$Br. The dominant factor is then ΔE, which is more negative for CN$^-$ + CH$_3$Br, and this reaction has the smaller barrier.

It is noteworthy that in the gas phase HO$^-$ + CH$_3$Br is the more exoergic of the two reactions (Table 6.12, entries 8 and 9), and HO$^-$ is a much better nucleophile than CN$^-$.[2a,b,3,8a] The solvent effect arises, in this case, from the strong solvation of HO$^-$ compared to CN$^-$. Thus, once again we find that the solvent can cause reactivity reversals whose origins are predictable by analysis of the interplay between the single electron transfer energy gaps, the f indices, and the reaction ergonicity.

6.3.2 Reactivity Trends in Nonhydroxylic Solvents

Some of the trends just discussed recur in nonhydroxylic solvents such as DMF and DMSO, but others are altered. Table 6.14 lists the reactivity factors and barriers in DMF solvent for nucleophilic attack of Cl$^-$ upon CH$_3$X (X = F, Cl, Br, I). In this solvent, ΔE becomes progressively more negative and the electron transfer energy gaps decrease rapidly down the series. The combination of these effects

Table 6.14 Reactivity Factors and Barriers (kcal/mol) for Nonidentity S$_N$2 Reactions, Y$^-$ + CH$_3$X → YCH$_3$ + X$^-$, in DMF Solvent

Entry	Y, X	ΔE	G_R	G_R^P	f_R	f_P	$\Delta E^\ddagger$ (SCD)	$\Delta E^\ddagger$(expt)[a]	A_{RX} (DMF*)
1	Cl, F	+5	199	216	0.208	0.243	34.0	>22.7	−17
2	Cl, Cl	0	181	181	0.243	0.243	30.0	22.7	+2
3	Cl, Br	−3	173	165	0.242	0.243	25.9	18.4	+10
4	Cl, I	−7	164	147	0.242	0.243	21.1	16.0	+19

[a]From reference 4a.

leads to a gradual decrease in the barriers. The decrease in the electron transfer energy gap originates in the electron affinity of RX, $A_{RX}(DMF^*)$, which becomes more positive down the series. We can conclude that, in DMF, the reactivity trend $CH_3I > CH_3Br > CH_3Cl > CH_3F$ is caused mainly by the more effective single-electron acceptor property $A_{RX}(s^*)$, with a reinforcing contribution from the thermodynamic driving force. The trend in DMF is the same as in the gas phase[8] and in water solvent.[4,12,13] In the gas phase, the order $CH_3I > CH_3Cl > CH_3Br > CH_3F$ is established by $A_{RX}(g)$ and ΔE, while in water solvent only $A_{RX}(H_2O^*)$ is important.

In some nonhydroxylic solvents the trend in the reactivities of Y^- is the same as in the gas phase: $F^- > Cl^- > Br^-\ I^-$.[8] This can be seen in Table 6.15,[12,14] which illustrates the following trends:

1. For halide exchange reactions, the relative reactivities of Y^- are greatly condensed in nonhydroxylic solvents, compared to hydroxylic solvents (entries 1–3).
2. The reactivity order of Y^- in nonhydroxylic solvents is variable. In some cases, the trends $Cl^- > Br^-$ resemble the gas-phase trends; in other cases, the trends $Br^- > Cl^-$ and $I^- > Cl^-$ (entries 3, 6, 8, 9–12) resemble the trends in hydroxylic solvents.

Table 6.15 Effect of Solvent on the Relative Nucleophilicities of Y^- in Nonidentity S_N2 Reactions $Y^- + CH_3X \rightarrow YCH_3 + X^-$

Entry	RX	Relative Reactivity	Solvent	Reference
1	CH_3I	$I > Br > Cl$ ($10^3:40:1$)	MeOH	12
2	CH_3I	$Cl > Br$ ($2:1$)	DMF	12
3	CH_3I	$Br \geq I > Cl$ ($1.9:1.7:1$)	Me_2CO	14j
4	CH_3Br	$I > Cl$ ($159:1$)	MeOH	12
5	CH_3Br	$Br > I > Cl$ ($22:12:1$)	Me_2CO	14j
6	CH_3Br	$I > Cl > Br$	DMF	4a
7	CH_3Cl	$I > Br > Cl$	H_2O, MeOH	4a
8	CH_3Cl	$Br > I > Cl$ ($2.4:1.4:1$)	Me_2CO	14g–i
9	$PhC(O)CH_2Cl$	$Br > Cl$	MeCN	14i
10	$PhC(O)CH_2SCN$	$Br \geq Cl$	MeCN	14i
11	$PhC(O)CH_2SeCN$	$Br > Cl$	MeCN	14i
12	$PhCH_2Cl$	$F \geq Cl$ ($1.3:1$)	MeCN	14f
13	CH_3OTs	$I > Br > Cl$ ($50:16:1$)	MeOH	12
14	CH_3OTs	$Cl > Br > I$ ($2.1:1.2:1$)	Molten salt	14e
15	CH_3OTs	$Cl > Br > I$ ($5:3:1$)	DMF	12
16	$n-C_3H_7OTs$	$F > Cl > Br > I$ ($830:8.3:3.3:1$)	DMSO	14b
17	$n-C_4H_9OBs$	$Cl > Br > I$ ($18:5:1$)	Me_2CO	14a
18	CH_3I	$ArS > ArO$ ($2.51 \times 10^3:1$)	MeOH	12
19	$n-C_4H_9I$	$R^1S > R^2O$ ($7.2 \times 10^4:1$)	DMSO	14k
20	$n-C_4H_9Cl$	$R^1S > R^2O$ ($1.2 \times 10^4:1$)	DMSO	14k
21	$PhCH_2Cl$	$R^1S > R^2O$ ($1.35 \times 10^4:1$)	DMSO	14k

3. In the reactions of halide anions with alkyl sulfonates (entries 14–18), the trend is uniformly the same as in the gas phase (F$^-$ > Cl$^-$ > Br$^-$ > I$^-$), and opposite to the trend in hydroxylic solvents.

4. In DMSO under conditions of constant pK_a,[14k] the trend in the reactivities of thia and oxyanions in RS$^-$ > RO$^-$ (entries 20–22). This is the same as the trend in methanol solvent at constant pK_a, and opposite to the gas-phase trend.

Table 6.16 summarizes the reactivity factors in the gas phase, in water, solvent, and in DMF. Entries 1–4 refer to nucleophilic attach upon CH$_3$Br by Y$^-$ = F$^-$, Cl$^-$ Br$^-$, I$^-$. In gas phase, the reaction energy becomes progressively more positive down the series. In water solvent, the ΔE(H$_2$O) are constant, and the ΔE(DMF) exhibit intermediate behavior. The difference between water and DMF originates in the free energy of transfer of Y$^-$, ΔG_t(Y$^-$)(H$_2$O → DMF).[15] The transfer from water to DMF causes the solvation energy of F$^-$ to decrease by ca. 13 kcal/mol, but that of I$^-$ is decreased by only ca. 3 kcal/mol.

In each of the solvents the vertical electron transfer energy gaps are significantly higher than in the gas phase. The gaps are especially large for reactions that involve F$^-$ which, therefore, becomes a poor electron donor. The differences between the G_R gaps and also between the G_R^P gaps are smaller in DMF than in water. The following conclusions may be drawn from these observations:

1. The effect of ΔE will lead to the reactivity order F$^-$ > Cl$^-$ > Br$^-$ > I$^-$ in DMF.

2. The effect of the electron transfer energy gaps will lead to the opposite reactivity order F$^-$ < Cl$^-$ < Br$^-$ < I$^-$ in DMF, and this effect is attenuated relative to water solvent.

The result of (1) and (2) is a balance between the reactivity factors, and the relative reactivities are more compressed in DMF than in water. This compression is seen more clearly in Table 6.17, which also includes the experimental barriers.

This analysis predicts that a reversal of reactivity may occur when ΔE differs significantly from zero, and F will become the most reactive nucleophile. In Table 6.15 we have already seen such a reversal when RX is an alkyl tosylate or brosylate. The reaction energies for Y$^-$ + CH$_3$OTs (Y$^-$ = F$^-$, Br$^-$, Cl$^-$, I$^-$) are shown in entries 5–8 of Table 6.17. These reactions are exoergic in water, and more so in DMF; and the exoergicity in DMF increases in the order F$^-$ > Cl$^-$ > Br$^-$ I$^-$. In such series the effect of ΔE will dominate, and the gas-phase trend will reappear: F$^-$ > Cl$^-$ > Br$^-$ > I$^-$.

All of these examples illustrate how complex nucleophilicity patterns can be analyzed by a detailed consideration of electron transfer energy gaps and bond coupling/bond interchange factors. It is thus found that nucleophilicity may follow trends in vertical ionization potentials, as in the case of halides in water solvent, or a complex combination of ionization potential, bond strength, and bond-coupling effects, as in the case of halides in DMF solvent, or hydroxide versus cyanide in solution.

Table 6.16 Reactivity Factors for Y⁻ + CH₃X YCH₃ + X⁻ Reactions in the Gas Phase, in Water, and in DMF, and Free Energies of Transfer of Y⁻ Anions from Water to DMF[a]

Entry	Y, X	$\Delta E(g)$[b]	$\Delta E(H_2O)$[c]	$\Delta E(DMF)$[c]	$G_R(g)$	$G_R^P(g)$	$G_R(H_2O)$	G_R^P (H_2O)	$G_R(DMF)$	G_R^P (DMF)	ΔG_t ($H_2O \rightarrow DMF$)[d]
1	F, Br	−37	−2	−8	99	98	233	203	203	178	13
2	Cl, Br	−8	0	−3	104	100	196	186	173	165	10
3	Br, Br	0	0	0	99	99	179	179	160	160	7
4	I, Br	+8	0	+3	92	96	159	170	146	154	3
5	HO, Br	−54	−26	−32	65	89	188	190	158	166	13
6	HS, Br	−28	−18	−21	74	92	169	180	146	159	10
7	PhS, Br	−19	−21	−20	78	99	156	173	137	155	7

[a]All values are in kcal/mol.
[b]These are reaction enthalpies.
[c]Determined using equation 6.17 and the solvation-free energies of Chapter 4.
[d]Values for free energy of transfer of Y⁻ are estimated in reference 15.

Table 6.17 (Set I) Central Barriers for Y⁻ + CH₃X → YCH₃ + X⁻ Reactions in Solution and Reaction Energies for Y⁻ + CH₃OSO₂Ph → YCH₃ + PhSO₃⁻ Reactions in Solution (Set II)[a]

Set I

Entry	Y, X	Water		DMF	
		$\Delta E^{\ddagger}$ (SCD)	$\Delta E^{\ddagger}$ (expt)[b]	$\Delta E^{\ddagger}$ (SCD)	$\Delta E^{\ddagger}$ (expt)[b]
1	F, Br	33.8	26.5	25.9(25.5)[c]	
2	Cl, Br	31.8	24.6	25.9	17.9
3	Br, Br	29.1	23.7	24.7	18.4
4	I, Br	25.8	21.8	23.3	17.2
		$\Delta(\Delta E^{\ddagger}) = 8.0$	$\Delta(\Delta E^{\ddagger}) = 4.7$	$\Delta(\Delta E^{\ddagger}) = 2.7$	$\Delta(\Delta E^{\ddagger}) = 1.2$

Set II

Entry	Y, X	ΔE (H₂O)[d]	ΔE (DMF)[e]
5	F, OSO₂Ph	−16	−29
6	Cl, OSO₂Ph	−17	−27
7	Br, OSO₂Ph	−15	−22
8	I, OSO₂Ph	−17	−20

[a]All values are in kcal/mol.

[b]Data are from reference 4a.

[c]The value in parentheses is calculated using $\Delta E = -9$ kcal/mol and $G_R(\text{DMF}) = 201$ kcal/mol (cf. the corresponding values in Table 6.16, entry 1).

[d]Data are from reference 4a.

[e]$\Delta E(\text{DMF})$ is determined using $\Delta G_t(\text{Y}^-; \text{H}_2\text{O} \rightarrow \text{DMF})$ from reference 15. $\Delta G_t(\text{PhSO}_3^-; \text{H}_2\text{O} \rightarrow \text{DMF})$ is assumed to be zero.

6.4 REACTION SERIES IN PHYSICAL ORGANIC CHEMISTRY

As we have seen, the central barriers of nonidentity S_N2 reactions are determined by an interplay of electron transfer properties and reaction thermodynamics. Within this context we may return to the question of whether there are any reaction series that provide true rate–equilibrium relationships. Such reaction series would be characterized by constant delocalization indices and electron transfer energy gaps, and only the reaction energy would vary. Substituent effects upon rates would then be determined only by substituent effects upon equilibria.

Entries 1–3 of Table 6.11, which refer to gas-phase reactions of F^-, Cl^-, and Br^- with CH_3Cl, comprise the only such series that we have encountered. Its existence must be considered to be fortuitous; other series contain multiple variations of the reactivity factors, and there is no obvious way to design a genuine rate–equilibrium study. It follows that rate–equilibrium studies seldom reflect equilibrium effects alone.[10, 11a]

Most commonly, the reaction series that are used to study rate–equilibrium relationships are Hammett series based on the variation of a remote substituent, as in the variation of Z in **6.7**. To determine whether such a series is a true rate–

6.7

equilibrium series or involves variations in the other reactivity factors as well, it is necessary to determine the substituent dependence of the reactivity factors.

When the nucleophilic center remains constant, as in **6.7**, the delocalization indices are also approximately constant. This is illustrated in Table 6.18 for the reactions of a series of oxygen and sulfur anions with CH_3Cl. Although there is

Table 6.18 Gas-Phase Electron Affinities of Y· Groups and Delocalization Indices for $Y^- + CH_3Cl \rightarrow YCH_3 + Cl^-$ Reactions in the Gas Phase and in Solution

Entry	Y^-	A_Y.[a]	Gas Phase[c]		Water		DMF	
			f_R	f_P	f_R	f_P	f_R	f_P
1	PhO^-	54	0.251	0.292	0.239		0.243	
2	$o\text{-}CH_3OC_6H_4O^-$	54[b]	0.251	≈ 0.292	0.239		0.243	
3	$o\text{-}ClC_6H_4O^-$	59[b]	0.251	≈ 0.276	0.239		0.243	
4	HS^-	54	0.251	0.340	0.239	0.328	0.243	0.329
5	CH_3S^-	44	0.251	0.370	0.239		0.243	
6	PhS^-	57[b]	0.251	0.331	0.239	0.326	0.243	0.328

[a] A_Y. values are in kcal/mol; the data are collected in Chapter 4.

[b] From J. H. Richardson, L. M. Stephenson, and J. I. Brauman. *J. Am. Chem. Soc.* **97**, 2967 (1975).

[c] $f_R = w_{R:}$ of $(CH_3 \pm Cl)^-$; $f_P = w_{R:}$ of $(CH_3 \pm Y)^-$.

significant variation of the substituents attached to the nucleophilic centers, the variation in f_P [$f_P = w_{R:}$ of $(CH_3 \dot{=} Y)$] is not large, because the electron affinity of $Y \cdot$, $A_{Y \cdot}$, does not vary greatly.

In a series such as **6.7** the vertical electron transfer energy gaps will be dominated by the variation in the ionization potential of $Y:^-$ as the ring substituent Z is varied. We can express the variation in G_R and G_R^P as follows:

$$\Delta(G_R) \approx \Delta(G_R^P) \approx \Delta(I_{Y:}^v);$$

$$[I_{Y:}^v = \text{vertical ionization potential; in solution } I^v = I(s*)] \quad (6.18)$$

This states that, as the vertical ionization potential of $Y:^-$ decreases, the electron transfer energy gaps G_R and G_R^P will also decrease.

The reaction energy for the series **6.7** is given in equation 6.19 in terms of

$$\Delta E = I_{Y:}^a - I_{X:}^a + D_{C-X} - D_{C-Y} \quad (6.19)$$

adiabatic ionization potentials of $Y:^-$ and $X:^-$, either in the gas phase or in solution, and the corresponding bond energies D_{C-X} and D_{C-Y}. Since RX in **6.7** is constant, variations in ΔE will be determined by variations in $I_{Y:}^a$ and D_{C-Y}, as expressed in equation 6.20.

$$\Delta(\Delta E) = [\Delta(I_{Y:}^a) - \Delta(D_{C-Y})] \approx [\Delta(I_{Y:}^v) - \Delta(D_{C-Y})];$$

$$(I_{Y:}^v \approx I_{Y:}^a \text{ in the gas phase; } I_{Y:}^v \approx I_{Y:}^a + \rho S_{X:} \text{ in solution}) \quad (6.20)$$

It follows from equations 6.18 and 6.20 that, at best, the reaction series **6.7** will contain variations in both ΔE and the electron transfer energy gaps, and will not be a true rate–equilibrium series.[10] Table 6.19 illustrates this point for reactions of substituted pyridines with CH$_3$I.[16, 17] In this series the variations in ΔE approximately parallel the increase in the gas-phase ionization potentials of the pyridines. The rate variations are, therefore, influenced by changes in both ΔE and the electron transfer energy gaps. Using the language of the Marcus approach, we can state that rate variations are influenced by changes in both ΔE and $\Delta E_o^{\ddagger}$, as Lewis has found in numerous cases.[10]

Table 6.19 Ergonicities and Barriers for Reactions of Substituted Pyridines with Methyl Iodide in Acetonitrile Solvent and Gas-Phase Vertical Ionization Potentials of the Pyridines

Pyridine Substituent	ΔE^a	$I_{Y:}^v(g)^b$	$\Delta E^{\ddagger a}$
None	−11.93	221.4	22.15
3-Chloro	−6.9	226.1	23.45
2-Chloro	−2.72	236.5	25.04

$^a\Delta E$ and $\Delta E^{\ddagger}$ are free energies, in kcal/mol, from reference 16.
^{b}Vertical ionization potentials of the pyridines are from reference 17.

For a reaction series such as **6.7**, when the variation in ΔE is proportional to the variation in the electron transfer energy gaps, we may write

$$\Delta(\Delta E) = k\Delta(G_R) \approx k\Delta(G_R^P); \quad (k, \text{ proportionality factor}) \quad (6.21)$$

Studies of $\Delta E^{\ddagger}$ versus ΔE will then reveal rate–equilibrium relationships, and concomitant studies of $\Delta E^{\ddagger}$ versus $I_{Y:}$ will reveal rate–electron transfer relationships. However, it must now be realized that ΔE *is no longer an independent reactivity variable*; caution must be exercised in the interpretation of the Brönsted coefficients of such rate–equilibrium plots.[10, 11a, 18]

6.4.1 An Overview of Reactivity Trends in Nonidentity Reactions

Since the barriers of nonidentity reactions depend on electron transfer energy gaps, delocalization indices, and reaction thermodynamics, the barrier function can be written as equation 6.22, which is identical to equation 6.7. It is possible to

$$\Delta E^{\ddagger} = \Delta E^{\ddagger}(G_R, G_R^P, f_R, f_P, \Delta E) \quad (6.22)$$

experiment in different ways within the space defined by equation 6.22. With a knowledge of thermochemical relationships one could design reaction series that possess constant ΔE. Reactivity trends that are dominated by the electron transfer properties of the reactants and the products would then be investigated. Alternatively, one could design series that possess constant f_R and f_P and thereby investigate the effects of ΔE and the electron transfer energy gaps. Other possibilities also exist: a change in solvent will change both the electron transfer energy gaps and ΔE. There are many possibilities, and the experimental strategy is made clear with the aid of the SCD model.

6.5 TRANSITION STATE GEOMETRIES IN NONIDENTITY REACTIONS

6.5.1 Bonding Asymmetry, Earliness, Looseness, and Bond Orders of Transition States

In Chapter 5 we found that definition of the percentage of bond cleavage in the transition state relative to the ground state leads to a unified and convenient scale for the discussion of transition structures. The extension of this approach to nonidentity reactions is straightforward.[2c, 11a] For the forward reaction of $Y^- + RX \rightarrow YR + X^-$, the percentage elongation of the C–X bond is given by equation 6.23, where d_{CX}^o is the R–X bond length in the reactant ion–molecule complex,

$$\%CX^{\ddagger} = 100\% \, [(d_{CX}^{\ddagger} - d_{CX}^o)/d_{CX}^o] \quad (6.23)$$

and $d_{CX}^{\ddagger}$ is the bond length in the transition state. The corresponding percentage elongation of the C–Y bond is given by equation 6.24.

$$\%CY^{\ddagger} = 100\% \, [(d_{CY}^{\ddagger} - d_{CY}^{o})/d_{CY}^{o}] \qquad (6.24)$$

The difference between $\%CX^{\ddagger}$ and $\%CY^{\ddagger}$ is the bond-cleavage asymmetry of the transition state, equation 6.25, and the sum of $\%CX^{\ddagger}$ and $\%CY^{\ddagger}$ provides in-

$$\%AS^{\ddagger} = \%CY^{\ddagger} - \%CX^{\ddagger} \qquad (6.25)$$

formation concerning the looseness, $\%L^{\ddagger}$, of the transition state, equation 6.26. As we shall see, the quantities $\%AS^{\ddagger}$ and $\%L^{\ddagger}$ can be understood with the SCD

$$\%L^{\ddagger} = \%CX^{\ddagger} + \%CY^{\ddagger} \qquad (6.26)$$

model, and related to the forward and reverse barriers in an empirically useful manner.

The C–X and C–Y bond orders are also useful geometric indices. They are defined in equation 6.27 by the Pauling relationship, in which a_{CX} and a_{CY} are

$$n_{CX}^{\ddagger} = \exp\,[-(d_{CX}^{\ddagger} - d_{CX}^{o})/a_{CX}] \qquad (6.27a)$$

$$n_{CY}^{\ddagger} = \exp\,[-(d_{CY}^{\ddagger} - d_{CY}^{o})/a_{CY}] \qquad (6.27b)$$

constants characteristic of the C–X and C–Y bonds, respectively. As depicted in **6.8**, it is convenient to scale these constants so that the bond orders in the transition

6.8

states of identity reactions are required to be 0.5.[2a,b] Some a_{CX} values obtained in this way[2a,b] are collected in Table 6.20.

The requirement that $n_{CX}^{\ddagger} = 0.5$ for an identity reaction leads to a convenient definition of transition state asymmetry in a nonidentity reaction. As can be seen in Table 6.21, the quantity $n_{CX}^{\ddagger} + n_{CY}^{\ddagger}$ is then unity for many nonidentity transition states. It follows that, with the condition that $n_{CX}^{\ddagger} = 0.5$ for identity reactions, the total bond order in nonidentity reactions is conserved, and the total bond order "length" of the transition state is thus normalized to unity. This allows the bond-order asymmetry of a transition state to be described in terms of the deviations of the individual bond orders from 0.5, as illustrated in **6.9**. This shows that, since the bond order distance between X and Y remains unity, the transition structure

Table 6.20 Values of the Pauling Proportionality Constant a_{CX} from 4-31G Calculations on CH_3X and $[X \cdots CH_3 \cdots X]^{-a}$

X	a_{CX} (Å)
F	0.600
OCH_3	0.687
OH	0.691
OOH	0.703
OF	0.704
Cl	0.729
NH_2	0.795
SH	0.825
NC^b	0.858
H	0.936
CN^c	0.948
CCH	0.959

[a] From references 2a and b.
[b] Refers to N attack.
[c] Refers to C attack.

Table 6.21 Individual and Total Bond Orders of $[YCH_3X]^-$ Transition Structures[a]

Y	X	$n_{CY}^{\ddagger}$	$n_{CX}^{\ddagger}$	$n_{CY}^{\ddagger} + n_{CX}^{\ddagger}$
F	H	0.593	0.406	0.999
F	OH	0.546	0.455	1.001
F	CN	0.442	0.558	1.000
F	CCH	0.491	0.510	1.001
F	OOH	0.438	0.566	1.004
F	SH	0.342	0.651	0.993
F	NH_2	0.591	0.421	1.012
F	OCH_3	0.509	0.492	1.001
F	OF	0.384	0.634	1.018
F	NC	0.393	0.606	0.999
H	OH	0.445	0.549	0.994
H	CN	0.389	0.610	0.999
H	CCH	0.423	0.570	0.993
H	NC	0.346	0.662	1.008
H	NH_2	0.482	0.513	0.995
HO	CN	0.426	0.583	1.009
HO	CCH	0.463	0.538	1.001
NH_2	CN	0.405	0.604	1.010

[a] Ab initio results from references 2a and b.

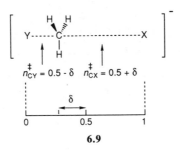

6.9

$[(YRX)^-]^{\ddagger}$ is related to its two identity congeners by a movement, δ, of R along the constant bond-order distance. This allows the bond-order asymmetry of the transition state to be written as equation 6.28.

$$AS^{\ddagger}(n^{\ddagger}) = n_{CX}^{\ddagger} - n_{CY}^{\ddagger} = 2\delta \qquad (6.28)$$

It should be noted that the $AS^{\ddagger}(n^{\ddagger})$ of equation 6.28 is not the same as the $\%AS^{\ddagger}$ of equation 6.25; the latter comprises an absolute definition of asymmetry, while the former refers to asymmetry relative to a hypothetical transition state produced, without bond-order changes, from two identity transition states (**6.9**). We shall find that the SCD model allows both types of asymmetry to be predicted from simple curve-crossing considerations.

6.5.2 Bond-Order Asymmetry in the Transition State and the Position of the Transition State Along the Reaction Coordinate

For a normalized reaction coordinate which varies between zero and unity, the bond-order asymmetry of the transition state coincides with its position along this reaction coordinate. The position of a transition state in such cases has been discussed in Chapter 3, and is given by equation 6.29. This states that transition

$$RC^{\ddagger} = \tfrac{1}{2}(n_{CY}^{\ddagger} - n_{CX}^{\ddagger} + 1) = \tfrac{1}{2}[1 - AS^{\ddagger}(n^{\ddagger})] \qquad (6.29)$$

states for which $AS^{\ddagger}(n^{\ddagger}) = 0$ are located at $RC^{\ddagger} = 0.5$. Such transition states are bond-symmetric and have equal C–X and C–Y bond orders. A nonsymmetric transition state $[AS^{\ddagger}(n^{\ddagger}) \neq 0]$ will be located at $RC^{\ddagger} \neq 0.5$, and have $n_{CX}^{\ddagger} \neq n_{CY}^{\ddagger}$.

The position of the transition state can be rationalized from first principles with the SCD model. According to Chapter 3 (equation 3A.14), $RC^{\ddagger}$, the position of the crossing point, is given by equation 6.30. Since this has the same form as

$$RC^{\ddagger} \approx \frac{1}{2} + \frac{f_R(G_R - \Delta E) - f_P(G_P + \Delta E) + \Delta E}{G_R + G_P} \qquad (6.30)$$

equation 6.29, the condition for existence of a bond order-asymmetric transition state is given by equation 6.31. The role of ΔE is illustrated in **6.10**. The change from an identity reaction to a nonidentity reaction, in which only ΔE varies, will

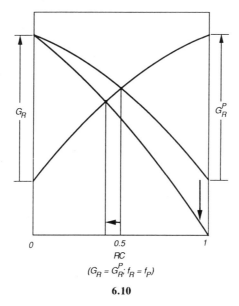

$$(G_R = G_R^P, f_R = f_P)$$

6.10

lead to bond order-asymmetry in the transition state. As indicated by the heavy arrows in **6.10**, if the reaction becomes exoergic, $RC^\ddagger < 0.5$ and $n_{CX}^\ddagger > n_{CY}^\ddagger$.

$$\frac{f_R G_R - f_P G_P + \Delta E(1 - f_R - f_P)}{G_R + G_P} \neq 0 \qquad (6.31)$$

The role of the electron transfer energy gap is illustrated in **6.11**, which shows

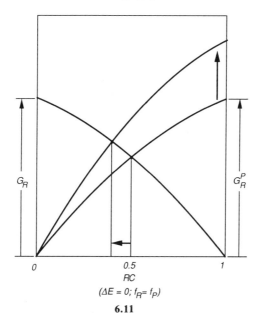

$$(\Delta E = 0; f_R = f_P)$$

6.11

that bond-asymmetry can exist in *thermoneutral reactions*. The heavy arrows of **6.11** show that an increase in one of the gaps, with all other barrier factors unchanged, is equivalent to a change in ΔE. An increase in G_R^P leads to a bond order-asymmetric transition state in which $RC^{\ddagger} < 0.5$ and $n_{CX}^{\ddagger} > n_{CY}^{\ddagger}$.

The role of the delocalization indices is illustrated in **6.12**. In this case, bond

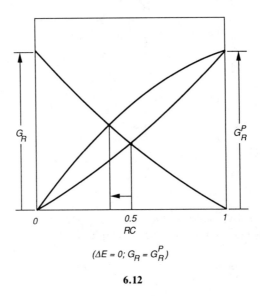

$$(\Delta E = 0;\ G_R = G_R^P)$$

6.12

order-asymmetry is created in a thermoneutral reaction by differences in the delocalization indices of the reactants and the products. The heavy arrows in **6.12** show that an increase in f_P will lead to a bond-asymmetric transition state with $RC^{\ddagger} < 0.5$ and $n_{CX}^{\ddagger} > n_{CY}^{\ddagger}$.

The effect of ΔE should now be expected, but the existence of bond-asymmetry in *thermoneutral* reactions is less obvious. Entries 1–3 of Table 6.22 refer to thermoneutral gas phase reactions; entries 4–6 refer to thermoneutral reaction in water solvent. As seen in the $RC^{\ddagger}$ column, equation 6.30 predicts that each of the nonidentity thermoneutral reactions will possess a bond-asymmetric transition state. The effect can be seen to originate in changes to both f and G. In entry 2, bond asymmetry is caused mainly by $G_R^P > G^R$. In entry 1, bond asymmetry is caused by the combination $G_R^P > G_R$ and $f_P > f_R$.

It is important to recognize that the bond order–asymmetry and the related quantity, the position of the transition state along the reaction coordinate, do not provide explicit information regarding the individual bonds. This point is illustrated in Table 6.23, using ab initio transition structures.[2a, b] Entry 5 exhibits the second earliest transition state, based on $RC^{\ddagger}$, but its $\%CX^{\ddagger}$ is larger than that of entries 1, 3, and 6, which are later transition states. The terms "early" and "late" therefore have meaning only in the sense of bond order–asymmetry.

Table 6.22 Reactivity Factors, Reaction Ergonicities, Positions of the Transition State Along the Reaction Coordinate, Bond Asymmetries, and Transition State Bond Orders for $Y^- + CH_3X \rightarrow YCH_3 + X^-$ Reactions

Entry	Y, X	ΔE (expt)[a]	ΔE (4-31G)	G_R	G_R^P	f_R	f_P	$(RC^{\ddagger})$[b]	$RC^{\ddagger}(\Delta E = 0)$[b]	4-31G %$CX^{\ddagger}$ − %$CY^{\ddagger}$	$n^{\ddagger}_{CX}$	$n^{\ddagger}_{CY}$
1	CH₃S, F	−3.6(~0)[c]	—	100	126	0.242	0.380	0.385	0.395	—	—	—
2	HO, CCH	~0[c]	~0	122	133	0.362	0.357	0.487	0.486	6	0.538	0.463
3	HO, OH	0.0[c]	0.0	109	109	0.357	0.357	0.500	0.500	0	0.500	0.500
4	Br, Br	0.0[d]		179	179	0.241	0.241	0.500	0.500			
5	Cl, Br	0[d]		196	186	0.241	0.239	0.507	0.507			
6	I, Br	0[d]		159	170	0.241	0.243	0.478	0.478			
7	Cl, Br	−8[c]		104	100	0.246	0.251	0.474				
8	I, Br	+8[c]		92	96	0.246	0.241	0.533				

[a]Calculated from bond energy and electron affinity data. See Chapter 4.
[b]Equation 6.30.
[c]Gas phase reactions.
[d]Reactions in water solvent. See Table 6.16.

Table 6.23 Percentages of Bond Breaking, %CX‡ and %CY‡, Percentage of Bond Asymmetry, %AS‡, Bond Orders, $n_{CX}^{\ddagger}$ and $n_{CY}^{\ddagger}$, Bond-Order Asymmetry, AS$^{\ddagger}(n^{\ddagger})$, Reaction Energies, and Position of the Transition State Along the Reaction Coordinate, RC‡, for Nonidentity Reactions Y$^-$ + CH$_3$X → YCH$_3$ + X^{-a}

Entry	Y, X	%CX‡	%CY‡	%AS$^{\ddagger b}$	$n_{CX}^{\ddagger}$	$n_{CY}^{\ddagger}$	AS$^{\ddagger}(n^{\ddagger})^c$	RC$^{\ddagger d}$
1	F, F	25.0	25.0	0.0	0.500	0.500	0.0	0.500
2	F, CN	37.6	32.1	5.5	0.558	0.442	0.116	0.526
3	H, OH	26.2	70.2	44.0	0.549	0.445	0.105	0.259
4	H, CCH	36.2	74.6	38.4	0.570	0.423	0.148	0.221
5	H, CN	31.4	81.9	50.5	0.610	0.389	0.222	0.154
6	H, F	19.0	61.5	32.8	0.593	0.406	0.186	0.100

^{a}Geometric data refer to 4-31G calculations from references 2a and b.
^{b}Equation 6.25.
^{c}Equation 6.28.
^{d}Equation 6.30.

6.5.3 Percentages of Bond Cleavage, Absolute Asymmetry and Looseness of the Transition State, and the Leffler–Hammond Hypothesis

Bond Cleavage in the Transition State and the Magnitude of the Barrier. In Chapter 5 we found that %CX‡, the percentage elongation of the C–X bond in the identity transition state, can be predicted from a knowledge of the vertical electron transfer energy gap and the delocalization index. An increase in either of these reactivity factors leads to a higher percentage of bond elongation, because of molecular deformation dominated by C–X stretching is required to achieve resonance at the crossing point of the SCD. Equation 6.32 restates the linear relationship between calculated[2a,b] %CX‡ and the height of the crossing

$$\%CX^{\ddagger} = 1.1\ w_{R:}(I_{X:} - A_{RX}) - 10.2 \tag{6.32}$$

point. Since this height differs from the height of the identity barrier by B, the avoided crossing interaction, it is not surprising that %CX‡ is also linearly related (equation 6.33) to the identity barrier (Chapter 5). We now demonstrate that

$$\%CX^{\ddagger} = a\Delta E^{\ddagger} + b; \quad a \text{ and } b \text{ are constants} \tag{6.33}$$

similar correlations exist for nonidentity reactions, equation 6.34.

$$Y^- + CH_3X \overset{\Delta E_f^{\ddagger}}{\underset{\Delta E_r^{\ddagger}}{\rightleftarrows}} YCH_3 + X^- \tag{6.34}$$

Figure 6.6 is a plot of calculated %CX‡ versus the central barriers, $\Delta E^{\ddagger}$, of 14 S$_N$2 reactions.[2c] These include thermoneutral, exoergic, endoergic, forward, and reverse barriers which span a range of ca. 80 kcal/mol in activation energy and

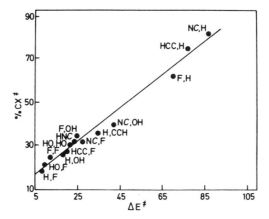

Figure 6.6 A plot of %CX‡ versus the central barriers (kcal/mol) of reactions Y$^-$ + CH$_3$X → YCH$_3$ + X$^-$. Each data point refers to the specific Y,X pair. (Reprinted, by permission, from reference 2c).

ca. 70% in bond stretching. The plot is seen to be linear (r = 0.992; for a larger set of 30 S$_N$2 reactions, r = 0.985^{2c}), so that bond cleavage of the leaving group in the transition state increases as the magnitude of the S$_N$2 barrier increases. This finding can be expressed as equations 6.35 and 6.36.

$$\%CX^‡ = a\Delta E_f^‡ + b \tag{6.35}$$

$$\%CY^‡ = a\Delta E_r^‡ + b \tag{6.36}$$

Table 6.24 reviews the properties that are associated with high barriers and high percentages of bond cleavage in the transition state. With each nucleophile, methyl fluoride has the smallest %CX‡ and methane has the largest %CX‡. In all cases the %CX is determined by the interplay of A_{RX} and the degree of radical anion delocalization ($w_{R:} = f_R$). Accordingly, CH$_3$F, which has the highest A_{RX} and the lowest f_R, exhibits the smallest %CX‡, and CH$_4$, which has the largest f_R, shows

Table 6.24 Calculated Reaction Energies and Percentages of Bond Cleavage in S$_N$2 Transition States Involving CH$_3$X Molecules and the Reactivity Factors of These Processes

Entry	R–X	Range of %CX^{‡a}	A_{RX} (kcal/mol)^b	f_R (= $w_{R:}$)^b
1	CH$_3$–F	18.4 → 25.0	−57	0.242
2	CH$_3$–OH	26.2 → 30.4	−65	0.357
3	CH$_3$–CCH	36.2 → 45.0	−78	0.362
4	CH$_3$–H	− → 58.0	−66	0.720

^{a}4-31G calculations, from references 2a, b.
^{b}Data are from Chapter 5.

the largest $\%CX^{\ddagger}$. In the cases of CH_3-OH and CH_3-CCH, intermediate $\%CX^{\ddagger}$ are found, and the trend $\%CX^{\ddagger}(CH_3-CCH) > \%CX^{\ddagger}(CH_3-OH)$ is determined primarily by the poorer electron acceptor capability of CH_3-CCH (more negative A_{RX}). If we recall the factors that primarily determine A_{RX} and $w_{R:}$, we can understand why X's that form strong bonds will be associated with high barriers, and require a high percentage of bond cleavage to reach the transition state.

Transition State Looseness and the Magnitude of the Intrinsic Barrier.
In equation 6.26 we have defined transition state looseness as the sum of $\%CX^{\ddagger}$ and $\%CY^{\ddagger}$. If we now combine equations 6.35 and 6.36 to obtain equation 6.37, we find a correlation of looseness with the sum of the barriers of the forward and reverse reactions.

$$\%L^{\ddagger} = \%CX^{\ddagger} + \%CY^{\ddagger} = a(\Delta E_f^{\ddagger} + \Delta E_r^{\ddagger}) + 2b \tag{6.37}$$

Figure 6.7 is a plot of $\%L^{\ddagger}$ versus the sum of the central barriers[2c] for the 4-31G data set.[2a,b] This plot includes variations of X and Y across the periodic table, and its linearity ($r = 0.975$) implies that the geometric looseness of the transition state increases as the sum of the forward and reverse barriers increases.

If we combine equation 6.37 with the Marcus equation 6.3, and neglect the quadratic term of the latter, we find (equation 6.38) that the looseness of the

$$\Delta E_f^{\ddagger} + \Delta E_r^{\ddagger} \approx 2\Delta E_o^{\ddagger} \tag{6.38}$$

transition state is predicted to correlate with the magnitude of the intrinsic barrier, $\Delta E_o^{\ddagger}$. This point is seen clearly in Table 6.25, which displays computed intrinsic barriers and transition state looseness indices of nonidentity reactions.

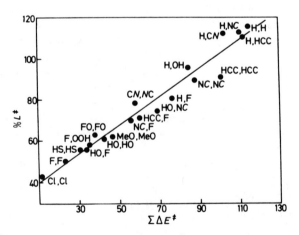

Figure 6.7 A plot of transition state looseness $\%L^{\ddagger}$ versus the sum of the central barriers $\Sigma\Delta E^{\ddagger} = (\Delta E_f^{\ddagger} + \Delta E_r^{\ddagger})$ of nonidentity reactions $Y^- + CH_3X \rightarrow YCH_3 + X^-$. (Reprinted, by permission, from reference 2c).

Table 6.25 Intrinsic Barriers (kcal/mol) and Looseness of Transition States of Nonidentity S_N2 Reactions $Y^- + CH_3X \rightarrow YCH_3 + X^{-a}$

Entry	Y, X	$\Delta E_0^{\ddagger}$	$\%L^{\ddagger} (= \%CX^{\ddagger} + \%CY^{\ddagger})$
1	Cl, Cl	5.5	42.2
2	F, F	11.7	50.0
3	HS, SH	15.6	56.0
4	HO, F	16.5	56.1
5	HO, OH	21.2	60.8
6	CH_3O, OCH_3	23.5	61.8
7	NC, F	27.8	69.7
8	HCC, F	31.0	71.0
9	H, F	31.9	80.2
10	HO, CN	32.5	74.1
11	HCC, OH	35.8	74.2
12	H, OH	36.6	96.4
13	NC, CN	43.8	89.6
14	H, CN	47.9	113.3
15	HCC, CCH	50.4	90.0
16	H, H	52.0	116.0

[a]From 4-31G calculations; references 2a and b.

Statement 6.6 summarizes the trends in the looseness of the nonidentity transition state:

Statement 6.6 X and Y groups that are associated with high vertical electron transfer energy gaps, $I_{X:} - A_{RY}$ and $I_{Y:} - A_{RX}$, and delocalized odd electrons in the charge transfer species $X\cdot$, $(CH_3\overset{\bullet}{\cdot}Y)^-$, $Y\cdot$, and $(CH_3\overset{\bullet}{\cdot}X)^-$, will cause high forward and reverse barriers, high intrinsic barriers, and lead to transition structures having highly stretched bonds.

Absolute Asymmetry of the Transition State and the Leffler–Hammond Hypothesis. According to equation 6.25, the absolute asymmetry of the transition state is given by the difference $\%CY^{\ddagger} - \%CX^{\ddagger}$. From equations 6.35 and 6.36, we may therefore write equation 6.39, which states that the reaction ergonicity determines the absolute asymmetry of the transition state.

$$\%AS^{\ddagger} = \%CY^{\ddagger} - \%CX^{\ddagger} = a(\Delta E_r^{\ddagger} - \Delta E_f^{\ddagger}) = -a\Delta E \qquad (6.39)$$

Equation 6.39 seems to reflect the Leffler–Hammond hypothesis. *However, it must be noted that the equation is valid despite the general breakdown of rate-equilibrium relationships for this series of reactions (Figure 6.1).* Therefore, equation 6.39 comprises a novel structure–reactivity relationship, as summarized in Statement 6.7:

Statement 6.7 The absolute asymmetry of an S_N2 transition state increases as the reaction becomes more exoergic.

Some caution is needed here. Although absolute asymmetry ($\%AS^{\ddagger}$) and bond order–asymmetry [$AS^{\ddagger}(n^{\ddagger})$] are similar concepts, the latter is related to the position of the TS along the reaction coordinate (equation 6.29), and the former is related to the geometry of the TS. Therefore Statement 6.7 does not necessarily imply that a more exoergic reaction possesses an earlier transition state.

6.5.4   Characterization of Transition State Geometries in Solution

The linear relationships encountered in equations 6.35–6.37 and 6.39 hold over a very extended range of geometries and energies, sufficient to encompass most real S$_N$2 reactions. It is now necessary to determine whether these relationships are global in nature, or are more usefully grouped in families of ''related'' reactions. Throughout Chapters 5 and 6 we have argued that, regardless of the source of the data, the existence of *general* relationships is not accidental. This view is supported by a thermochemical analysis,[19] which shows that an increase in $\Delta E_f^{\ddagger}$ and $\Delta E_r^{\ddagger}$ will cause the transition structure to approach its dissociation limit more closely. Accordingly, the relationships of equations 6.35–6.37 and 6.39 appear to be physically sound, and it should be possible to devise experiments that permit them to be applied.

6.5.5   A Summary of Geometric Effects in Nonidentity Reactions

The geometry of a nonidentity S$_N$2 transition structure depends on the vertical electron transfer energy gaps ($I_{Y:} - A_{RX}$, $I_{X:} - A_{RY}$), the delocalization indices (f_R, f_P) of the charge transfer species ($Y \cdot /(R \overset{.}{\text{—}} X)^-$, $X \cdot /(R \overset{.}{\text{—}} Y)^-$), and on the reaction ergonicity, ΔE. The following trends have been found:

1. In general, large electron transfer energy gaps and delocalized odd electrons in the charge transfer species will give rise to high barriers and extensively stretched bonds in the transition structure. All of these effects reflect the movement of a single electron synchronized to bond coupling and interchange.

2. Specifically, for the reaction $Y^- + CH_3X \rightarrow YCH_3 + X^-$, $\%CX^{\ddagger}$ correlates linearly with the forward barrier, $\Delta E_f^{\ddagger}$, and $\%CY^{\ddagger}$ is linearly correlated with the reverse barrier, $\Delta E_r^{\ddagger}$. The looseness of the TS ($\%CX^{\ddagger} + \%CY^{\ddagger}$) correlates linearly with the sum of the barriers, $\Delta E_f^{\ddagger} + \Delta E_r^{\ddagger}$.

3. It can be shown that the sum of the barriers is approximated by $2\Delta E_0^{\ddagger}$. Thus, a higher intrinsic barrier reflects a looser transition structure.

4. The absolute asymmetry of the TS correlates linearly with the reaction ergonicity. Therefore, a highly exoergic reaction will be characterized by a highly asymmetric TS. This linear relationship remains valid despite breakdowns of rate–equilibrium relationships and, therefore, differs from the Bell–Evans–Polanyi–Leffler–Hammond hypothesis.

5. The bond order–asymmetry of the TS, $n_{CY}^{\ddagger} - n_{CX}^{\ddagger}$, is related to the position of the TS along a normalized reaction coordinate that varies between zero and one.

Absolute- and bond order-asymmetries are not the same, so that the linear relationship of 4 does not imply any specific "position of the TS along the reaction coordinate." The concept of bond order–asymmetry thus defines "early" and "late."

The linear correlations of 2–4 exist because C–X and C–Y stretching are the distinguished deformations that overcome the vertical electron transfer energy gaps required to reach the transition state.

6.6 CHARGE DISTRIBUTION IN THE TRANSITION STATE

The lowest four VB configurations of an identity S_N2 transition state are shown in **6.13**. The HL_1 and HL_2 configurations describe the Heitler–London covalent

$$X:^- \quad R\cdot\!\!-\!\!\cdot X \qquad X\cdot\!\!-\!\!\cdot R \quad X:^-$$
$$(HL_1) \qquad\qquad\qquad (HL_2)$$

$$X:^- \quad R^+ \quad X:^- \qquad X\cdot \quad R:^- \quad \cdot X$$
$$(R^+) \qquad\qquad\qquad (R^-)$$

6.13

bonds $R\cdot\!-\!\cdot X$ and $X\cdot\!-\!\cdot R$, respectively R^+ and R^- are intermediate configurations that do not involve RX bonding.

The $(X–R–X)^-$ transition state is reached at the crossing point of HL_1 and HL_2, as shown in **6.14**. For $R = CH_3$ the intermediate configurations are higher lying

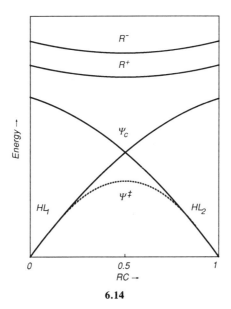

6.14

and the order of their energies for any X is $E(R^+) < E(R^-)$, as seen in **6.14**. At the crossing point, the wavefunction is a linear combination of HL_1 and HL_2, as defined in equation 6.40.

$$\Psi_C = 1/\sqrt{2}[HL_1 + HL_2] \tag{6.40}$$

Upon mixing of the secondary configurations, the wavefunction of the transition state becomes equation 6.41, where, if overlap is neglected, we have equation

$$\Psi^\ddagger = a(1/\sqrt{2}[HL_1 + HL_2]) + b[R^+] + c[R^-] \tag{6.41}$$

$$a^2 + b^2 + c^2 = 1 \tag{6.42}$$

6.42. The charge on X in the transition state (q_X) is given by equation 6.43, and the charge on CH_3 is given by equation 6.44.

$$q_X = -\tfrac{1}{2}(a^2 + 2b^2) < -\tfrac{1}{2} \tag{6.43}$$

$$q_{CH_3} = b^2 - c^2 \tag{6.44}$$

Since the R^- configuration has a higher energy than the R^+ configuration, $c^2 < b^2$, the charge on CH_3 will normally be positive, and the charge on X will be more negative than -0.5, as depicted in **6.15**. This charge distribution does

$$
\begin{array}{ccc}
-(0.5 + \delta) & +2\delta & -(0.5 + \delta) \\
X\!\!-\!\!\!-\!\!\!-\!\!\!-\!\!\!-\!\!\!&\!\!\!CH_3\!\!\!-\!\!\!-\!\!\!-\!\!\!-\!\!\!&\!\!\!X
\end{array}
$$

6.15

not depend upon the geometry of the transition state. Since the transition structure and the height of the barrier are related, it follows that the charge distribution need not depend upon the reaction barrier.

To illustrate this point, consider a transition state which has the charge distribution shown in **6.16**. This charge distribution could be obtained from several

$$
\begin{array}{ccc}
-0.55 & +0.10 & -0.55 \\
X\!\!-\!\!\!-\!\!\!-\!\!\!-\!\!\!&\!\!\!CH_3\!\!\!-\!\!\!-\!\!\!-\!\!\!&\!\!\!X
\end{array}
$$

6.16

combinations of the mixing weights a^2, b^2, and c^2, regardless of the energy or geometry of the transition state. We now allow **6.16** to dissociate into an equal mixture of $X:^- + CH_3\cdot + \cdot X$ and $X\cdot + \cdot CH_3 + :X^-$. This mixture has the charge distribution depicted in **6.17**. In the course of the dissociation, the transition

$$
\begin{array}{ccccc}
-0.5 & & 0.0 & & -0.5 \\
X & + & CH_3 & + & X
\end{array}
$$

6.17

state **6.16** undergoes the changes in energy and geometry shown in **6.18**. However,

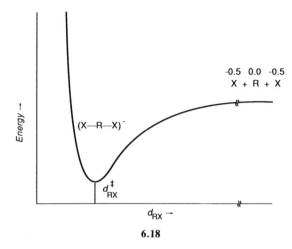

6.18

the charge distribution remains approximately constant. Clearly, there is no fundamental link between charge distribution and transition state geometry or energy.

The description **6.15** will extend to a nonidentity reaction,[11a, 20] and the crossing point will again be characterized by an equal mixture of the two Heitler–London configurations, equation 6.45. Since the crossing point is the point of highest en-

$$\Psi_C = 1/\sqrt{2}[(Y:^- \ R \cdot - \cdot X) \pm (Y \cdot - \cdot R \ :X^-)];$$

$$(HL_1 = Y:^- \ R \cdot - \cdot X; \ HL_2 = Y \cdot - \cdot R \ :X^-) \qquad (6.45)$$

ergy between reactants and products, it seems reasonable to assume that the geometry of the transition state, after the avoided crossing, will correspond closely to the geometry of the crossing point. In this case the wavefunction of the transition state will resemble that of an identity reaction, equation 6.46, and the charge distribution will be given by equation 6.47.

$$\Psi^{\ddagger} \approx a(1/\sqrt{2}[HL_1 + HL_2]) + b[R^+] + c[R^-];$$

$$a^2 + b^2 + c^2 = 1 \qquad (6.46)$$

$$q_Y \approx q_X = -\tfrac{1}{2}(a^2 + 2b^2); \qquad q_{CH_3} \approx b^2 - c^2 \qquad (6.47)$$

The important point is that, with two qualifications, the charge equality $q_Y \approx q_X$ holds, regardless of the position of the TS along the reaction coordinate, the bond-asymmetry, or the looseness of the transition structure:

1. Mixing of the very high-energy configurations $[Y^+R:^-:X^-]$ and $[Y:^-:R^-X^+]$ will alter the charge equality of the TS.[20a] This factor must be negligibly small.[21]

2. Curve skewing may lead to a breakdown of the assumption of charge equality.[11a, 20a] For example, as depicted in **6.19**, in very exoergic or endoergic

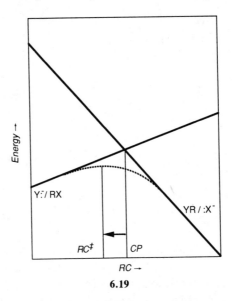

6.19

reactions, the position of the TS may not coincide with the position of the crossing point. In the case of **6.19**, the transition state would exhibit reactant-like charge distribution involving more of HL_1. Nevertheless, despite the caveat, studies of curve crossing suggest[22] that curve skewing is not a general effect.

Therefore, we do not agree with the widely held opinion that charge development in the transition state should correlate with the transition structure.

Although Boyd has reached a contrary conclusion from an integrated charge analysis of transition states optimized at the MP2/6-31++G* level,[23] his data actually support our views. Structures **6.20–6.22** summarize the calculated inte-

$$\overset{-0.64}{H}\text{———}R\text{———}\overset{-0.64}{OH}$$

$\Delta E = -26.8$ kcal/mol

6.20

$$\overset{-0.70}{H}\text{———}R\text{———}\overset{-0.68}{F}$$

$\Delta E = -46.6$ kcal/mol

6.21

$$\overset{-0.72}{H}\text{———}R\text{———}\overset{-0.51}{Cl}$$

$\Delta E = -89.4$ kcal/mol

6.22

grated charges of Y and X in $(Y-CH_3-X)^-$ transition states, and also the computed reaction ergonicities. As can be seen, two of these structures, for reactions having ΔE as large as -47 kcal/mol, are geometrically early, but exhibit charge equality $(q_Y \simeq q_X)$. The SCD model seems to be uniquely able to account for the existence of these cases. The third structure refers to a reaction in which $\Delta E = -89$ kcal/mol; although a reactant-like charge distribution is present here, this may well reflect case 2 above.

6.7   NONIDENTITY REACTIONS—CONCLUDING REMARKS

Much of S_N2 reactivity can be understood in terms of the deformations that are required to destabilize ground states and allow resonance to be achieved with charge transfer states. Barriers and transition structures are found to be dominated by the vertical electron transfer energy gaps, the delocalization indices, and the reaction ergonicity. The avoided crossing interaction B, which is a measure of the resonance energy in the transition state, appears to be large (ca. 14 kcal/mol), and reflects the strong bond coupling in this structure.

All of these findings are consistent with the description of the process as a single electron movement synchronized to the bond interchange and solvent reorganization. With its large B, the S_N2 reaction is a single-electron shift process[11,20,24] and should be distinguished from single-electron *transfer* processes, in which the single electron movement *precedes* the bond interchange. Such processes are characterized by $B \leq 1$ kcal/mol.[25]

REFERENCES

1. (a) R. P. Bell. *The Proton in Chemistry.* Cornell University Press, Ithaca, NY, 1973, Chapter 10; (b) M. G. Evans and M. Polanyi. *Trans. Faraday Soc.* **34,** 11 (1938).

2. (a) S. Wolfe, D. J. Mitchell, and H. B. Schlegel. *J. Am. Chem. Soc.* **103,** 7694 (1981); (b) D. J. Mitchell. *Theoretical Aspects of S_N2 Reactions.* PhD Thesis, Queen's University, 1981; (c) S. S. Shaik, H. B. Schlegel, and S. Wolfe. *J. Chem. Soc. Chem. Commun.* 1322 (1988).

3. (a) M. J. Pellerite and J. I. Brauman. *J. Am. Chem. Soc.* **102,** 5993 (1980); (b) M. J. Pellerite and J. I. Brauman. *J. Am. Chem. Soc.* **105,** 2672 (1983).

4. (a) W. J. Albery and M. M. Kreevoy. *Adv. Phys. Org. Chem.* **16,** 87 (1978); (b) W. J. Albery. *Ann. Rev. Phys. Chem.* **31,** 227 (1980).

5. (a) G. S. Hammond. *J. Am. Chem. Soc.* **77,** 334 (1955); (b) J. E. Leffler. *Science (Washington DC).* **117,** 340 (1953).

6. R. A. Marcus. *Ann. Rev. Phys. Chem.* **15,** 155 (1964).

7. J. A. Dodd and J. I. Brauman. *J. Am. Chem. Soc.* **106,** 5356 (1984).

8. (a) W. N. Olmstead and J. I. Brauman. *J. Am. Chem. Soc.* **99,** 4219 (1977); (b) G. Caldwell, T. F. Magnera, and P. Kebarle. *J. Am. Chem. Soc.* **106,** 959 (1984); (c) K. Tanaka, G. I. Mackay, J. D. Payzant, and D. K. Bohme. *Can. J. Chem.* **54,** 1643 (1976); (d) D. R. Anderson, V. M. Bierbaum, and C. H. DePuy. *J. Am. Chem. Soc.* **105,** 4244 (1983).

9. (a) R. C. Dougherty, J. Dalton, and J. D. Roberts. *Org. Mass. Spectrom.* **8,** 77 (1974); (b) R. C. Dougherty and J. D. Roberts. *Org. Mass. Spectrom.* **8,** 81 (1974); (c) R. C. Dougherty. *Org. Mass Spectrom.* **8,** 85 (1974).

10. (a) E. S. Lewis, *J. Phys. Chem.* **90,** 3757 (1986); (b) E. S. Lewis, T. A. Douglas, and M. L. McLaughlin. *Isr. J. Chem.* **26,** 331 (1985); (c) E. S. Lewis, T. A. Douglas, and M. L. McLaughlin. In *Nucleophilicity.* J. M. Harris and S. P. McManus, Editors, Advances in Chemistry Series, No. 215, American Chemical Society, Washington, DC, 1987.

11. (a) S. S. Shaik. *Progr. Phys. Org. Chem.* **15,** 197 (1985); (b) S. S. Shaik and A. Pross. *J. Am. Chem. Soc.* **104,** 2708 (1982).

12. A. J. Parker. *Chem. Revs.* **69,** 1 (1969).

13. D. J. McLennan. *Aust. J. Chem.* **31,** 1897 (1978).

14. (a) S. Winstein, L. G. Savedoff, S. Smith, I. D. R. Stevens, and J. S. Gall. *Tetrahedron Lett.* 24 (1960); (b) R. Fuchs and K. Mahendran. *J. Org. Chem.* **36,** 370 (1971); (c) R. F. Rodewald, K. Mahendran, J. L. Bear, and R. Fuchs. *J. Am. Chem. Soc.* **90,** 6698 (1968); (d) C. L. Liotta, E. E. Grisdale, and H. P. Hopkins. *Tetrahedron Lett.* 4205 (1975); (e) W. T. Ford, R. J. Hauri, and S. G. Smith. *J. Am. Chem. Soc.* **96,** 4316 (1974); (f) J. Hayami, N. Tanaka, and N. Hihara. *Bull. Inst. Chem. Res. Kyoto Univ.* **50,** 354 (1972); (g) A. J. Parker, *J. Chem. Soc. A.* 220 (1966); (h) A. J. Parker. *J. Chem. Soc. A.* 1328 (1961); (i) T. Thorstenson and J. Songstad. *Acta Chem. Scand.* **A32,** 133 (1978); (j) D. Cook and A. J. Parker. *J. Chem. Soc. B.* 142, (1968); (k) F. G. Bordwell and D. L. Hughes. *J. Org. Chem.* **46,** 3571 (1981); (l) C. Reichardt. *Solvent Effects in Organic Chemistry.* Verlag Chemie, Weinheim, W. Germany, 1979, pp. 148–150.

15. (a) G. B. Cox, G. R. Hedwig, A. J. Parker, and D. W. Watts. *Aust. J. Chem.* **27,** 477 (1974); (b) M. H. Abraham. *J. Chem. Soc. Perkin Trans. II* 1375 (1976).

16. E. M. Arnett and R. Reich. *J. Am. Chem. Soc.* **102,** 5892 (1980).

17. *Gas Phase Ion Chemistry.* M. T. Bowers, Editor, Academic Press, New York, 1979, Volume 2, p. 30.

18. I. H. Williams. *Bull. Soc. Chim.* 192 (1988).

19. S. S. Shaik, *J. Am. Chem. Soc.* **110,** 1127 (1988).

20. (a) A. Pross and S. S. Shaik. *Tetrahedron Lett.* **23,** 5467 (1982); (b) A. Pross and S. S. Shaik. *Acc. Chem. Res.* **16,** 363 (1983); (c) A. Pross. *Adv. Phys. Org. Chem.* **21,** 99 (1985).

21. For recent computational work on these configurations see J.-K. Hwang, S. King, S. Creighton, and A. Warshel. *J. Am. Chem. Soc.* **110,** 5297 (1988).

22. See, for example: (a) A. Warshel and S. Russell. *J. Am. Chem. Soc.* **108,** 6569 (1986); (b) A. Warshel. *Acc. Chem. Res.* **14,** 284 (1981); (c) A. Warshel. *Biochemistry* **20,** 3167 (1981).

23. Z. Shi and R. J. Boyd. *J. Am. Chem. Soc.* **111,** 1575 (1989); *ibid.,* **113,** 1072 (1991).

24. A. Pross. *Acc. Chem. Res.* **18,** 212 (1985).

25. L. Eberson. *Electron Transfer in Organic Chemistry.* Springer-Verlag, Heidelberg, 1987.

AUTHOR INDEX

Numbers in parentheses are reference numbers and indicate that the author's work is referred to although his (her) name is not mentioned in the text. Numbers in *italics* show the pages on which the complete references are cited.

SUBJECT INDEX